Stem Cell Therapy in Dermatological Disorders

Scrivener Publishing
100 Cummings Center, Suite 541J
Beverly, MA 01915-6106

Publishers at Scrivener
Martin Scrivener (martin@scrivenerpublishing.com)
Phillip Carmical (pcarmical@scrivenerpublishing.com)

Stem Cell Therapy in Dermatological Disorders

Edited by

Sumel Ashique

Bharat Technology, Howrah, West Bengal, India

Sanjay Dey

Department of Pharmaceutical Technology, School of Health and Medical Sciences, Adamas University

Biplab Debnath

Bharat Technology, Uluberia, Howrah, West Bengal

Mohhammad Ramzan

School of Pharmaceutical Sciences, Lovely Professional University, Phagwara, Punjab, India

and

Swarupananda Mukherjee

Department of Pharmaceutics, NSHM College of Pharmaceutical Technology, NSHM Knowledge Campus, Kolkata - Group of Institutions, Kolkata, West Bengal, India

Scrivener Publishing

WILEY

This edition first published 2025 by John Wiley & Sons, Inc., 111 River Street, Hoboken, NJ 07030, USA and Scrivener Publishing LLC, 100 Cummings Center, Suite 541J, Beverly, MA 01915, USA

For more information about Scrivener publications please visit www.scrivenerpublishing.com.

Wiley Global Headquarters
111 River Street, Hoboken, NJ 07030, USA

For details of our global editorial offices, customer services, and more information about Wiley products visit us at www.wiley.com.

The manufacturer's authorized representative according to the EU General Product Safety Regulation is Wiley-VCH GmbH, Boschstr. 12, 69469 Weinheim, Germany, e-mail: Product_Safety@wiley.com.

Library of Congress Cataloging-in-Publication Data

ISBN 978-1-394-39311-4

Cover image: Generated with AI using Adobe Firefly
Cover design by Russell Richardson

Set in size of 11pt and Minion Pro by Manila Typesetting Company, Makati, Philippines

Printed in the USA

10 9 8 7 6 5 4 3 2 1

Contents

Preface

This book offers an in-depth exploration of the evolving role of stem cell-based therapies in dermatology, highlighting their mechanisms, clinical applications, and future potential. Stem cells—characterized by their potency, self-renewal capacity, and clonality—have revolutionized regenerative medicine by introducing novel therapeutic approaches for a range of skin-related conditions.

Recent advancements in cell therapy have expanded its dermatological applications, offering promising solutions for disorders once considered difficult to treat. Stem cell-based treatments have shown remarkable potential in addressing autoimmune skin disorders such as psoriasis and vitiligo by modulating immune responses and promoting tissue regeneration. Their use in atopic dermatitis has similarly demonstrated anti-inflammatory effects and enhanced restoration of the skin barrier.

Beyond disease treatment, stem cell therapy is increasingly recognized in aesthetic and anti-aging dermatology. By stimulating collagen production, enhancing skin elasticity, and promoting cellular repair, stem cells have emerged as a powerful tool for reducing wrinkles and improving overall skin texture. Their regenerative capabilities are also evident in wound healing and scar management, where they help accelerate tissue repair, reduce inflammation, and minimize fibrosis.

This book delves into the molecular mechanisms underlying stem cell function, the various types of stem cells utilized in dermatology—including embryonic, mesenchymal, and induced pluripotent stem cells—and their therapeutic potential. It also examines recent clinical trials, emerging technologies, ethical considerations, and the challenges of translating these therapies into widespread clinical practice.

By offering a comprehensive analysis of the current landscape and future prospects of stem cell-based therapies in dermatology, this book serves as a valuable resource for researchers, clinicians, and medical professionals aiming to harness the power of regenerative medicine in skin health and disease management.

Special thanks to Martin Scrivener and the team at Scrivener Publishing for their support, guidance, and commitment to advancing scientific knowledge in this exciting and rapidly developing field.

The Editors
July 2025

1

Fundamentals of Stem Cell Therapy in Disease Prevention

Rajarshi Nath, Arijit Das, Shambo Panda, Souvik Biswas and Biplab Debnath*

Department of Pharmaceutical Technology, Bharat Technology, Uluberia, Howrah, WB, India

Abstract

Stem cell therapy is a possible method of treating most diseases and restoring damaged tissue. In this review, the history, mechanism, application, problems, and prospects of stem cell therapy are summarized. Self-renewal and differentiation into other cell types are key features of stem cells. There, they can be obtained from embryonic, adult, and induced pluripotent sources. Stem cells can replace damaged cells, modulate the immune system, and secrete factors that facilitate the healing of tissues. Stem cell therapy has been researched for the treatment of neurological diseases, cardiovascular diseases, autoimmune diseases, and other conditions. However, other problems have to be solved, such as tumorigenesis risk. Recent progress in technologies of induced pluripotent stem cells and gene-editing technologies, such as CRISPR-Cas9, has provided new opportunities for personalized and targeted stem cell therapies. In the future, we also need to develop safer, more effective, and more accessible stem cell–derived therapies. Consequently, laws should be created to guarantee the clinical implementation and safe evolution of these medications. Although stem cell treatment presents great possibilities for revolutionizing regenerative medicine and enhancing patient outcomes, its advantages can only be fully appreciated by coordinated efforts among scientists, doctors, and legislators.

Keywords: Stem cells, regenerative medicine, embryonic stem cells, adult stem cells, induced pluripotent stem cells, hematopoietic stem cells (HSCs), homology-directed repair

**Corresponding author*: biplab.d86@gmail.com; ORCID: https://orcid.org/0000-0002-3908-8121

Sumel Ashique, Sanjay Dey, Biplab Debnath, Mohhammad Ramzan and Swarupananda Mukherjee (eds.) Stem Cell Therapy in Dermatological Disorders, (1–18) © 2025 Scrivener Publishing LLC

1.1 Introduction to Stem Cell Therapy

Stem cells are a unique class of cells that are thought to have the ability to differentiate into a different cell type. They have a capacity of long-term renewal and proliferation, which could be effective in different regenerative medicine applications. Derived from a variety of sources, stem cell therapy is believed to be a promising treatment option for end-stage diseases that conventional therapies are unable to cure [1]. Since the 1950s and 1960s, interest in the field has been increasing; its history and major developments indicating therapeutic potential and anticipated diseases are discussed.

Conventional therapies for different diseases such as diabetes, myocardial infarction, bone disorders, etc., are generally focused on controlling the symptoms or curing them to a certain extent until the end-stage disease is reached. End-stage diseases have no further therapy, including surgery or transplantation, other than a fully viable organ or tissue transplantation. In the meantime, transplantation is limited due to a lack of supply, immune problems, and being an invasive procedure. Because of the efficacy in organogenesis and histogenesis, stem cells provide a promising treatment for these kinds of end-stage diseases. They can proliferate and differentiate into a variety of cell types. Moreover, they can self-renew and have a capacity for long-term renewal [2]. Adult or tissue-specific stem cells residing in various parts of the body have been successfully used for over 40 years in bone marrow transplantation. In summary, a variety of stem cells from multiple sources have been considered a potential treatment for many diseases, including differentiation into a wide range of cell types and appropriateness for different regenerative medicine applications [3].

1.1.1 Definition and Types of Stem Cells

Stem cells self-renew and differentiate into many different cell types, which means that they have potential applications in cell therapy, tissue repair, and disease modeling. Stem cells have two very special mechanisms that make them different from other cells; first, they can divide indefinitely while maintaining an undifferentiated state and are able to differentiate into multiple cell types. These properties are known as self-renewal potential and plasticity, respectively. There are several different types of stem cells that can be characterized by their origin, functioning activity, and applications [2]. The classification of stem cells is divided into embryonic stem cells, adult stem cells, and induced pluripotent stem cells (iPSCs).

The abundance of stem cells with a lot of potential application and the discovery of direct conversion methods provide a clear path for future research in stem cell biology and regenerative medicine. Stem cell regeneration is fundamentally regenerative during development and homeostasis due to the highly coordinated and tightly regulated nature of cellular signaling pathways, niche interactions, and epigenetic mechanisms that ensure precise cell fate decisions, tissue patterning, and organogenesis. During embryonic development, stem cells exhibit remarkable plasticity and proliferative capacity, enabling them to generate all tissue types in a highly orchestrated manner. In adult homeostasis, stem cells maintain and repair tissues by responding to routine cellular turnover and minor injuries, guided by cues from their microenvironment. However, this regenerative potential often diminishes or becomes dysregulated in response to severe injury, aging, or chronic disease. Since the first report, a lot of interest and numerous studies related to the biological impact and technical improvement of transdifferentiation already explain the conversions themselves, i.e., how the fibroblasts can be converted. Nevertheless, the vast untapped potential of stem cells in therapeutic applications is challenged by the limited availability of these primary sources. As an alternative, the conversion of mature cells into stem-like states has emerged as a promising strategy. However, this process often involves the use of heterokaryons and results in incomplete differentiation or imperfect cellular reprogramming. Consequently, achieving direct and functional replacement of damaged tissues or organs remains a complex and ongoing challenge, highlighting the need for further advancements in stem cell research and regenerative medicine [4].

1.1.2 Historical Background

The field of stem cell biology first gathered momentum with the groundbreaking research of A. Maximow in the late 19th century. However, the "classical era" of this science started with the isolation of hematopoietic stem cells (HSCs) by Till and McCulloch in 1962, which marked the start of an energetic and expanding investigation field in the biology of normal and cancer stem cells. In 1998, James Thomson and colleagues derived the first human embryonic stem (ES) cell line. Two years later, John Gearhart derived the first human embryonic germ cell line. These works opened new possibilities for cell-based therapies and led to an enormous scientific production [5]. A steering position of Chinese authorities was noticed as they gave substantial financial, media, and legal support for the use of HSCs. Regulations were drafted that blocked the expansion of large international clinical trials on adult stem cells in China. Indistinctively, Chinese HSC scientific community's voice has been raised against the utilization of the unsupported data. The academic debate between Chinese and Western HSC experts is a striking example of unbalanced competition in the global scientific context. Stem cell research reflects broader concerns such as health policy, the role of scientific expertise, and the relations between science and state. Ethical governance of stem cell research and therapy in Chinese regulatory provisions was analyzed, concerning not only adult but also ES and iPSCs [1].

Despite the considerable public interest in the field, there remain many scientific and technical hurdles still to be negotiated. Over the past decades, advances in human stem cell biology have come in fits and starts, some being the product of rigorous research, but many others of anecdotal use of poorly understood treatments. However, there is room for optimism given the powerful tools and techniques developed, as well as promising preliminary scientific results. Stem cell treatments should still be seen as investigational, and with the exceptions of bone marrow transplants and some skin transplants, any other treatments cannot be considered as originating from established medical practice. Nonetheless, much can be done to expedite the responsible progress of medical research using human stem cells. As a part of this effort, in the short term, more research is required to rigorously detect cancerous and precancerous formations in the products of stem cell differentiation. Longitudinal epidemiological studies of patients would also be a good practice, conducting an annual check-up and medical scan for at least a decade after cells are transplanted to detect solid tumor and leukemia formations [6].

1.2 Mechanisms of Stem Cell Therapy

A plenitude of human diseases is characterized by an effective attrition of cells and ultimately tissues. When this happens, the human organism is essentially unable to regenerate the damaged tissue, and the healing process is therefore doomed to heal *via* formation of scar tissue with impaired function. Many conditions such as neurodegenerative diseases, heart infarct and diabetes, to cite a few, are typified by this mechanism, and in such cases, adult stem cell therapy represents an appealing approach for curing these diseases. The concept behind stem cell–based therapy is that stem cells are culture expanded *in vitro* and subsequently implanted in the body. Once there, these cells would find favorable conditions for their differentiation into the required cell type to regenerate the damaged tissue. Another alternative hypothesis is that implanted stem cells might support the differentiation of the patient's resident stem cells into the tissue cell type in need of repair [7].

Stem cells are defined by their self-renewal capacity and their flexibility to generate different mature cells. In pluripotent stem cells, the master transcription factors induce the expression of a network of proteins that are responsible for the multiple routes that lead to self-renewal and pluripotency maintenance. Conversely, the same transcription factors induce the silencing of developmental and differentiation genes. Upon the onset of differentiation, this regulatory network is upset, and a new gene expression program is triggered. The molecular events that take place at the time of differentiation are intricate and only partly understood. The signals that provoke the silencing of self-renewal transcription factors and the activation of developmental ones are diversified, and they might differ depending on the cell type. When differentiation proceeds, the vinculation of gene promoters to activating or repressive transcription factors is firmly established. At the same time, the regulatory regions of genes that will be no longer expressed get increasingly methylated in lysine 27 of histone H3. There is fairly well agreement that most of these venues lead to a compaction of the chromatin, although the picture is complicated by the addition of up to three methyl groups, and it is controversial whether these "repressive" histone marks really block transcription. Stem cell therapy or stem cell–based therapy represents a new dimension of organ transplantation, a novel way that has been proven to repair defective tissues. The cells are stem/progenitor cells expanded from allogeneic or autologous sources used in transplantation and are expected to migrate to the sites of injury or inflammation where they undergo differentiation and promote tissue repair. But the realization is difficult because transplanted cells often encounter an unfamiliar environment where they have to compete with the host cells. The avascular phase and the rapid onset of hypoxia make the barrier even worse, as newly formed blood vessels and their permeability take a longer time to build. Under these circumstances, grafted cells more likely undergo apoptosis, have a decreased survival, and do not engraft (Figure 1.1) [6].

1.2.1 Differentiation and Pluripotency

Stem cells are capable of sustainable self-renewal and, at the same time, can differentiate into diverse specialized cell types. This dual property of stem cells has tantalized researchers for almost six decades. Discovering the mechanism of renewal of stem cells has laid the foundation for regenerative medicine, a promising field of research that has the potential to cure developmental diseases and regeneration of defective tissues. There are several signaling pathways and transcription factors involved in stem cell self-renewal. On the other hand, stem cells can also differentiate into lineage-specific lineages. One of the hallmarks of differentiation of stem cells is the silencing of the transcription factor OCT4. There are several signaling pathways and genes involved in controlling the differentiation of stem cells [8].

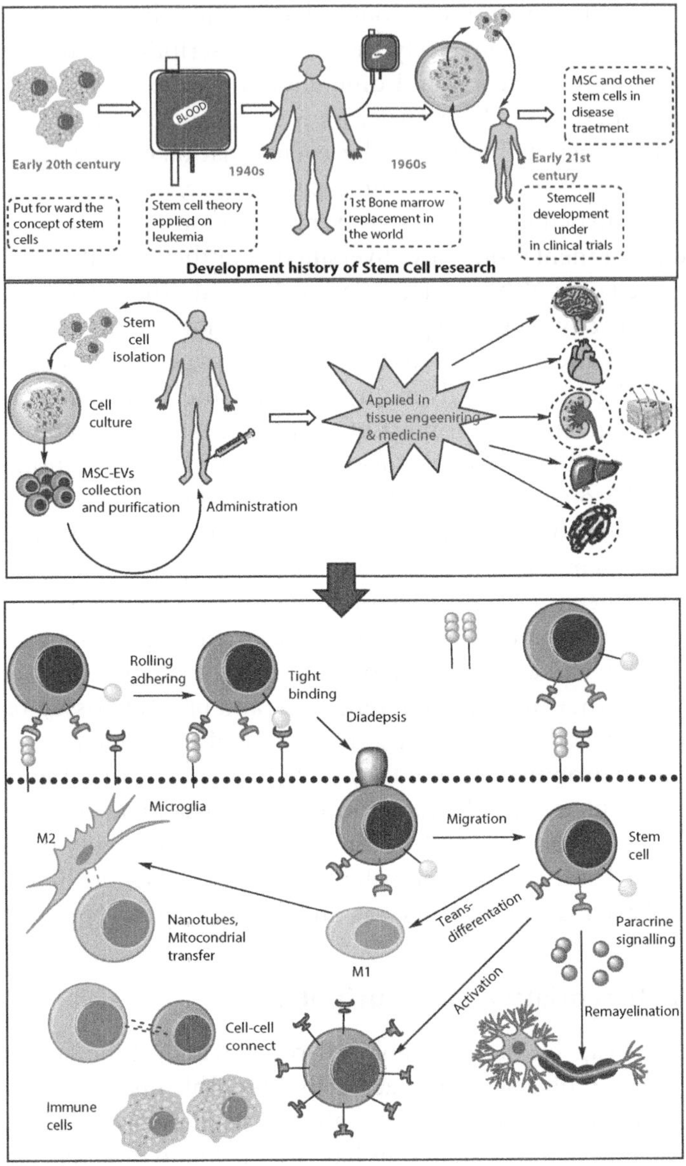

Figure 1.1 Development history of stem cell research as well as stem cell therapy and mechanism of stem cell therapy for neurodegenerative diseases.

In order to form the basis of enlightenment for regenerative medicine, this field discusses five aspects of stem cell biology, each of which is likely to revolutionize the development of new therapeutic approaches. The five aspects of stem cell study include the following: (1) the renewal of somatic cells,

(2) the progression and epigenetic empowerment of the stem cells, (3) uncovering the mechanism of reprogramming and transdifferentiation, (4) monitoring lineage tracing in the living organism, and (5) the possible side effects and means of reducing tumor formation from stem cell therapeutics and their derivatives. Ultimately, a complete understanding of the entire stem cell system constitutes the cornerstone of a new paradigm in interpreting the process of development and the formation of malignant cancers [9]. Additionally, having a better understanding of stem cell biology, whether in *Drosophila*, *Caenorhabditis elegans*, zebrafish, mouse, or human, enhances the likelihood of utilizing them for therapeutic tissue regenerative purposes.

1.2.2 Immunomodulation

Stem cells are among the most promising treatment options for diseases because they are able to prevent pathological changes in the body by forming desired cells. They have been intensely studied in recent years due to their different biological features and their capacity to repair damaged tissues. They are successfully used in various diseases due to their ability to differentiate into different cell types under the influence of appropriate stimulating factors. In addition, stem cells are also able to differentiate into the desired tissue macrophages and repair damaged tissues. These cells are capable of producing various growth factors and cytokines that prevent damage to tissues in degenerated tissues. In addition, certain molecular interactions between stem cells and immune system cells are important in the formation of this therapeutic effect. In particular, the effects of stem cells on macrophages and regulatory T cells (Tregs) of the immune mechanism associated with inflammation come to the fore, and it is emphasized that these interactions are mediated by certain cytokines. Therefore, stem cells have the potential to positively affect the treatment process of various diseases due to their effects that are required to be elucidated on immune response. Despite some difficulties, stem cell therapy has emerged as a special hope for treating various diseases, and important progress has been achieved in the development of this new and effective treatment method. Immunomodulatory properties are regarded due to the significant share of stem cell therapeutic benefits from the disease. At that point, it is necessary to explain how stem cells change the response of the immune system to reduce tissue damage. In addition, mesenchymal stem cells develop multiple treatments for the disease by studying the direct effects of the patient with immune cells and some molecules secreted by these cells. Nonetheless, most of these interactions have contributed to a therapeutic environment that is rapidly evolving. Stem cell therapy, in particular, has advanced our understanding and is paving the way for more specific and targeted therapies. As a result, research and development in this field are expected to accelerate significantly. This also aims to enhance properties associated with mesenchymal stem cells by utilizing an optimized induction method, which is becoming increasingly recognized in medicine for its immunomodulatory effects [6].

1.3 Applications of Stem Cell Therapy in Disease Prevention

Research on developmental biology has led to the discovery of different types of stem cells that can give rise to multiple tissue types. A stem cell is a cell that has the ability to develop into many different cell types during early life and growth. Stem cells grow up to be cells that make up the organs and tissues, such as muscles, red blood cells, and brain cells. Wrinkles of stem cells from the embryo implants into the uterus of the mother. There have been many different tissues that provide a new potential source for stem cells. For instance, is it possible to obtain stem cells from mature organs over the past 5 years? The research potential of embryonic stem cells has been long time known. Embryonic stem cells are isolated from the fertilized eggs in the blastocyst stage. This stage occurs at 3 to 5 days of the embryo development process [2]. The research in regenerative medicine appears to be focusing on the advances in molecular, developmental, and cellular biology. There is strong likelihood that in the future, a new perspective on the development of many tissues and organs will be provided.

Within the therapeutic landscape, stem cell therapy has shown considerable promise in averting the disease progression in many sectors of the domain. This article is aggregated with description, real-time instances, etc., of how stem cell therapy could have a broader role in prevention disease in and across various medical faculties. Clinicians, patients, and other health stakeholders are available to gain insight into how these approaches, either singularly or in combination, regulate diseases progress and host-related quality-of-life improvement. Ethical considerations and risks related to the south are also crushed. The piece ends up with a brief discussion about the safety practices of this approach [10].

1.3.1 Neurological Disorders

Neurological disorders represent a significant chronic health burden and present an ongoing challenge in medicine due to the limited regenerative capacity of the central nervous system [11]. Stem cell therapy offers hope for combatting many previously incurable diseases. Various types of stem cells have been intensively studied for regenerative therapy over the last decades. Therapeutically relevant cell types are of embryonic origin, including embryonic stem cells and more recently iPSCs. The properties and biology of these cells have been reviewed in depth. Other noteworthy types of stem cells found in adults are tissue-specific and so-called adult stem cells. Following injury, they differentiate to replace lost neurons or glial cells. Despite promising initial experiments, adult neural stem cells did not translate into successful clinical tools. In animal models, adult stem cells mostly migrate away from the lesion site and differentiate into scar tissue without extending projections or a native morphology. In contrast to nonimmunogenic iPSCs, the selection of donor-matched cells requires a source of healthy tissue, which is the main obstacle of fetal cells. Thus, an adult tissue-independent source of cells that can be obtained to provide long-lasting trophic support is advantageous. The main advantages of mesenchymal stem cells (MSCs) in the context of neurological disorders are their immunomodulatory and neuroprotective effects, as well as secretion of proregenerative factors. Furthermore, they are relatively easy to expand and modify *ex vivo*. This makes them an optimal vehicle for the delivery of therapeutic genes. Since the first bone marrow transplantation was performed in 1956, it has been established that bone marrow MSCs differentiate into specific tissues. On the other hand, following engraftment, MSCs very quickly disappear from the host tissue with half-lives of around 12 hours. This finding first suggested that the primary therapeutic mechanism may be secretion of bioactive factors rather than maturation into host tissue. Subsequently, it became clear that most stem cells act mainly through paracrine signaling. Indeed, the central focus of MSC research in recent years has shifted from transplantation to the educational effects of MSCs on the environment. Stimulated by maturing knowledge, there has been a significant increase in research activity and clinical translation with more than 120 clinical trials applying MSCs to treat neurological diseases. According to past data, these trials reach almost 1000 patients [12].

1.3.2 Cardiovascular Diseases

Cardiovascular diseases developed in the human body, such as diabetes, hypertension, and hypercholesterolemia, similar to cancer, are closely related to industrialization. These factors that increase the risk of developing a disease raise the national health burden by increasing the mortality rate of diseases. Cardiovascular diseases account for the large share of industrialization in global health due to an unbalanced lifestyle such as high-calorie diet, high stress, and cigarette. Cardiovascular dwarfs are the most common cause of death in industrialized countries, accounting for 44% of the total number of deaths. However, it is expected to become the leading cause of death overall in the world by 2020. Cardiovascular dwarfs include heart disease with damage of myocardium, arrhythmia inducing irregular heartbeat, congenital heart diseases, and vascular diseases.

In recent years, stem cell therapy has been applied in the fields of heart and vascular diseases in which the progress of traditional medicine is difficult. Recently, stem cells were injected into areas of myocardial infarction damaged by myocardial infarction, aspirating myocardium and regenerating cardiomyocytes. There are studies reporting that stem cells were transplanted successfully into a patient with myocardial infarction. Also, it was reported that bone marrow–derived stem cells were injected into patients with chronic heart failure and myocardial infarction, and the heart function was improved after 6 months. Several clinical trials are being performed to further validate the effect on chronic heart failure, and multiple institutions tested stem cells in severe chronic heart failure patients classified as New York Heart Association functional class III, IIV. Animal experiments with various types of stem cells continue to improve existing protocols. Some researchers genetically modified stem cells to promote angiogenesis and injected them into animal models, and as a result, they reduced scar formation while preventing predictable arrhythmias. Administered and drugs are added to maintain the cell function such as provitamin, erythropoietin, and α-cyclodextrin. However, most clinical treatments of cardiovascular diseases are still under development, and a number of problems such as low engraftment rate of stem cells, limited myocardial regeneration, and unsafe patient-specific factors are to be resolved. Evaluation of the optimal combination of stem cells with gene therapy, cell therapy, tissue engineering, and other experimental approaches is essential. Such integrated strategies may lead to more effective cardiomyocyte regeneration and help minimize the risk of unintended side effects associated with cardiac stem cell therapy [13].

1.3.3 Autoimmune Diseases

Some medical conditions are linked to an inappropriate response by the immune system, such as the case of rheumatoid arthritis, lupus, and type 1 diabetes, which are referred to as autoimmune diseases. Stem cell therapy has offered promising trials with the use of immune modulating cells that help prevent these unnecessary reactions. Different types of diseases and conditions can be treated with the application of stem cells. An increasingly wide number of trials and investigations are being carried out in relation to stem cell transplants, mainly aimed at developing curative options. In some conditions, stem cells boost the regeneration of impaired areas where tissue is damaged. Alternatively, stem cells can be utilized as carriers of substances that promote the self-repairing of tissues. The use of mesenchymal stem/stromal cells (MSCs) functions through the release of diffusible molecules. These MSCs also play essential roles in amplifying signal transduction for repair mechanisms, as well as recruiting specific cells to sites of inflammation. Thus, these cells can also modulate tissue homeostasis. In the case of considerably abundant inflammation, MSCs notice the presence of high levels of proinflammatory cytokines and redirect the cytokine yield toward a predominantly anti-inflammatory profile. This shift triggers the occurrence of macrophages that dampen inflammation. The inflammatory barrage is then disarmed in this manner, the impairment to adjacent tissues being kept to a bare minimum. MSCs encourage Treg generation by way of several mechanisms including prostaglandin E_2 and indoleamine 2,3-dioxygenase. Tregs play a pivotal role in immune homeostasis by enforcing quiescence on overanxious immune responses. Therefore, the transplantation of these cells will further intensify the resolution of inflammation [14].

1.4 Challenges and Limitations of Stem Cell Therapy

Cells within living organisms undergo some processes to recover from damage caused by diseases, injuries, or aging. The concept of tissue regeneration most likely first arose from the vivid observation of plants and animals being able to regrow parts. Many species of plants and lower animals show regeneration capabilities, varying from starfish to snails, newts, zebrafish, etc. In the 1950s, the first discovery of an adult murine tissue that exhibits regeneration capacity, the liver, became an inspiration to explore similar abilities of other tissues. This new area of developmental biology, called regenerative biology, focuses on understanding how damaged tissues are repaired or regenerated after injury and what is the mechanism and biology underlying this process. Stem cells have become a central focus in the investigation of tissue regeneration, because it is believed by many that this may hold the key to a never-ending supply of regenerative potential [15]. For many years, a significant amount of time and funding has been spent on expanding our knowledge and discovering new types of stem cells, to turn them into panaceas for a variety of health problems.

However, many challenges still hamper the successful application of stem cell therapy [16], ranging from the risks associated with tumor formation, immunological rejection, and ethical issues to problems relating to the translation from bench to bedside with many more sources but fewer high-quality stem cells. Shortly after the discovery of the first adult tissue-regenerating organ, a further observation of mouse liver regeneration also discovered that a dysregulated liver regenerative response could result in hepatocellular carcinoma. The first version of the somatic mutation hypothesis of cancer argued that the clonal expansion of cells carrying heritable point mutations generates tumor growth. This hypothesis was later expanded to include the development of transmissible cancers, which in essence is more complex genetically, invoking multiple mutations resulting in activation of oncogenes and inactivation of a number of opposing tumor suppressors. Regarding the emergence of tumor cells, genetics offer a compelling set of explanations. But the tumor cell concept has difficulties explaining the emerging stem cell view of cancers. Faced with the rapid degeneration of frequent adult cancer types in numerous species, the somatic mutation theory of cancer fails to explain the tumor cell population. Tumor cells must have extensive proliferative capacity just to be able to divide into large-enough clone sizes to harbor all necessary mutations [17].

1.4.1 Tumor Formation

Stem cell therapy is a promising new technology for various medical applications. However, significant risks associated with stem cell usage are still unresolved. One of the most substantial risks of stem cell therapy is the formation of tumors due to transplanted stem cells. The inhibitory mechanisms that prevent stem cells from forming tumors in their complicated natural cell environments are being lost or turned off during stem cell harvesting or *in vitro* manipulation. The loss of inhibitory functions results in unplanned spread and tumor formation of transplanted stem cells in their body environments. Investigation of the risk of tumor formation and development of new strategies to reduce that risk are required for the safe translation of stem cell therapy into clinical use [16]. More than four decades ago, human HSC transplantation was applied as a medical treatment. Today, it is a well-established method and is widely used in the treatment of leukemia and other cancers of the blood and lymphatic system. Despite numerous preclinical studies altering diseased populations' lives for the better, the translation of stem cell research into clinical therapy has been much slower.

Traditional drug development is based on protein target focus, small molecule, or antibody research, with biologics and regenerative medicine as recent additions. Over the last decade, however, public and private sectors have significantly invested in cutting-edge stem cell therapies based on their exciting preliminary phase results. At the same time, overall market pharmaceutical success rates have declined. Market success rates for AIDS, NYHA III, and cardiovascular disease therapies are 21%, 4%, and 12%, respectively, considerably lower than the 15% to 20% industry average. Clearly, continuing development of new approaches is required to maintain the financial viability of pharmaceutical research. The reception is roughly a study program of significant phenomena (treatment and prevented diseases, social intuition, native and folk theory). So, we can naturally contemplate observation of individuals with the capability to have empathy along with social skills of basic communication; psychosis such as various mental afflictions and abilities and Alzheimer disease and its debilitating symptoms prompt its reverence as a condition to be feared and avoided [18].

1.4.2 Immunological Rejection

After the discovery of embryonic stem cells and the subsequent generation of iPSCs, their potential application for treating a plethora of human ailments is currently being explored. A host of studies has recently been published reporting the therapeutic effectiveness of an array of "stem-like" cells in various disease settings; indeed, two of the most widely cited scientific papers in 2008 documented that human neural stem cells and adipose-derived stem cells could mitigate disease progression in rodent models of severe diseases. Stem cell technologies have now entered the consciousness of the general public and the arena of established pharmaceutical companies. The ultimate goal of stem cell research within the realm of cell transplantation would be the ability to use a patient's own cells to repair damaged or diseased tissue or organs. One method to achieve such personalized stem cell–based therapy would be to use a patient's somatic cells to generate iPSCs that could then be differentiated *in vitro* to a desired cell type. It was initially hoped that iPSCs could be used in an autologous setting, and therefore, immune rejection would not be a consideration. However, a recent study thoroughly contraindicated this early assumption by showing that iPSCs and their differentiated derivatives could be immunologically isolated. Moreover, concerns that human embryonic stem cells (hESCs) may induce a teratoma formation and an inflammatory response, while being vulnerable to immune rejection, have yet to be mollified [19]. The immune rejection barrier is perhaps the most significant as it imposes severe safety constraints, decreases the survival of the grafted cells, and thus reduces their therapeutic benefit. Overcoming immunorejection is one of the biggest challenges facing transplant biologists. Commonly used strategies to prevent immune rejection are generally based on suppressing or modulating the immune system. In the context of stem cell transplantation, this would require lifelong immunosuppression of the patient and could pose serious health issues. Drug therapy alone can be toxic for the patient, can increase the risk of infections and malignancies, and can have serious unknown long-term effects. Clearly, a compromise to minimize the patient's drug exposure while increasing the chance of successful engraftment is urgently required [20].

1.5 Future Perspectives and Innovations

The highly promising field of stem cell research and regenerative medicine has continued to flourish and revolutionize the current landscape of medicine. The emergence of iPSCs with properties equivalent to ESCs has facilitated vast and exciting opportunities for the development of personalized and regenerative therapeutics [21]. Their potential in overcoming various ethical issues associated with ESCs, as well as their successful applications in disease modeling, drug discovery, and cell-based regenerative therapies for complex disorders, has been extensively undertaken and highlighted over the last decade [22]. There is no doubt that the immanent development and application of gene-editing technologies may provide an unprecedented opportunity to advance the precision, efficacy, and safety of stem cell–based therapeutics. Apart from their direct potential as genetic correctives of single or multiple gene mutations, gene-editing approaches have also been successfully pursued to enhance the pluripotent or stemness properties of stem cells or to produce inducible caspase-9–based autodifferentiating logic gates in engineered ESC derivatives, thereby adding additional layers of safeguards to the cell transplantation strategy. Collectively, these methodologies provide a highly useful and combinatorial platform to address the onset and propagation of complex genetic diseases that would be unsuitable for current pharmacological or gene-transferring regimens. The emergence of gene-editing technology has revolutionized the manipulation of the genome by allowing specific changes in the genetic information quickly, easily, and cheaply, while also increasing the potential for genome editing in comparison to other gene-editing technologies. Therefore, gene-editing technologies could play a vital role in advancing the progress of iPSCs for a wide spectrum of their clinical applications in the mind run and in a regenerative therapeutic approach [23].

1.5.1 Induced Pluripotent Stem Cells

The discovery of human ESCs and, more recently, iPSCs poses a limitless source for regenerating any human tissue or organ cell type, offering unprecedented potentials for personalized medicine to treat aging-related diseases, congenital conditions, and traumatic injuries [22]. The prospect of enhancing preventive healthcare is tantalizing. Deriving iPSCs from patients, modeling the development or progression of complex diseases, and screening for the most effective drugs using basic CARDE catalogs, cellular or animal disease models together with clinical trial information or the electronic health records of individual patients are all within conceivable reach. The European Medicines Agency has already approved the first therapeutic drug for the treatment of cystic fibrosis on the basis of *in vitro* cell models. Reaping the full benefits of these transformative advances, however, depends on gaining an in-depth understanding of their mechanisms and surmounting a host of scientific, technical, regulatory, and ethical challenges [24].

The potential for widespread applications of iPSC technologies in drug discovery will also be delayed by the time, cost, and expertise required to build the necessary catalogs, as a similar fate awaits fulfillment of the widely anticipated therapeutic potential of human ESCs and iPSCs. Nonetheless, all of these issues have begun to receive attention, and initiatives to overcome them, led by researchers, clinicians, governments, and industry alike, have gained traction. Moreover, recent developments have begun to shed light on the unique properties of iPSCs obtained from different somatic cell types or reprogramming technologies and the relevance of these properties to their efficacy and safety. This review aims to provide an in-depth exploration of the field with the ultimate goal of spurring interest in further research and speeding the translation of progress into health benefits [25].

1.5.2 Gene-Editing Technologies

The intersection of human stem cell therapy and advanced gene-editing technologies holds great promise for the future advancement of attractive therapeutic approaches in regenerative medicine. Since iPSCs were first described in 2006, intense efforts have been devoted to the derivation of novel stem cell–based therapeutics that are able to replenish damaged tissues for disease prevention [26]. These could serve as an inexhaustible source of patient-specific therapeutic cells. In addition, gene-editing systems are powerful for precise manipulation of the genome. The gene-editing systems used for stem cell therapy were originally zinc finger nuclease

(ZNF) and transcription activator–like effector nuclease. However, the discovery of a more effective and convenient CRISPR-Cas9 system substituted for traditional gene-targeting technologies. The CRISPR-Cas9 system consists of a short guide RNA that directs the nuclease to a specific genome target and, in the presence of the Cas9 enzyme, generates a double-strand break. There is a sophisticated repair process for genomic damage known as homology-directed repair, which is utilized to generate purpose mutations or to correct mutations [27]. The recent development of novel gene-editing technologies that can avoid undesirable random gene disruption in gene-editing applications and result in new editing formats has driven significant efforts to improve the scope of gene-editing technologies as potential future therapeutics.

This review discusses these concepts and provides an insight into the future of stem cell–based gene-editing therapy. Moreover, the implications of combining gene-editing technologies and stem cell–based therapeutics for the prevention of biological issues including genetically inherited disorders and cancer are discussed. Furthermore, it is conceivable to envisage a new generation of personalized gene therapies composed of human induced pluripotent or embryonic stem cells (iPSCs or hESCs) gene-edited *ex vivo* and applied to patients. However, these models will require continued growth of technologies and the development of a safe and efficient method for precise, on-target gene modification. To this aim, the governance needs to prepare appropriate regulatory frameworks in response to evolving societal demands to ensure public confidence in the sociotechnical systems that support these technologies [28].

AI disclosure statement: The author confirms that no artificial intelligence (AI)-generated content or AI-assisted writing tools were used in the preparation of this chapter (including summaries and the abstract). All text, analysis, and conclusions are the result of the author's original work.

References

1. Sng, J. and Lufkin, T., Emerging Stem Cell Therapies: Treatment, Safety, and Biology, 1, 521343, 2012, ncbi.nlm.nih.gov.
2. Mirzapour, T., Azmi Tengku Ibrahim, T., Movahedin, M., Wahid Haron, A., *et al.*, Stem cells research and its application: a review, 1, 163–173, 2011, [PDF].

3. Nikokiraki, C., Psaraki, A., Roubelakis, M.G., The potential clinical use of stem/progenitor cells and organoids in liver diseases. *Cells*, 9, 1410, 2022, mdpi.com.
4. Menon, S., Shailendra, S., Renda, A., Longaker, M., *et al.*, An Overview of Direct Somatic Reprogramming: The Ins and Outs of iPSCs, 1, 141, 2016, ncbi.nlm.nih.gov.
5. Eguizabal, C., Aran, B., Chuva de Sousa Lopes, S.M., Geens, M., *et al.*, Two decades of embryonic stem cells: a historical overview, 1, hoy024, 2019, ncbi.nlm.nih.gov.
6. Mousaei Ghasroldasht, M., Seok, J., Park, H.S., Liakath Ali, F.B., Al-Hendy, A., Stem cell therapy: from idea to clinical practice. *Int. J. Mol. Sci.*, 23, 5, 2850, 2022 Mar 5, mdpi.com.
7. Rosner, M. and Hengstschläger, M., Stem Cell-Induced Cell Motility: A Removable Obstacle on the Way to Safe Therapies?, 2022, ncbi.nlm.nih.gov.
8. Poliwoda, S., Noor, N., Downs, E., Schaaf, A., Cantwell, A., Ganti, L., Kaye, A.D., Mosel, L.I., Carroll, C.B., Viswanath, O., Urits, I., Stem cells: a comprehensive review of origins and emerging clinical roles in medical practice. *Orthop. Rev.*, 14, 3, 37498, 2022 Aug 25, nih.gov.
9. Malecki, M., 'Above all, do no harm': safeguarding pluripotent stem cell therapy against iatrogenic tumorigenesis, 2, 1–16, 2014, ncbi.nlm.nih.gov.
10. Zayed, M.A., Sultan, S., Alsaab, H.O., Mohammad Yousof, S., *et al.*, Stem-Cell-Based Therapy: The Celestial Weapon against Neurological Disorders, 21, 3476, 2022, ncbi.nlm.nih.gov.
11. Andrzejewska, A., Dabrowska, S., Lukomska, B., Janowski, M., Mesenchymal Stem Cells for Neurological Disorders, 7, 2002944, 2021, ncbi.nlm.nih.gov.
12. Ling, A.P.K. and Wong, Y.P., Applying stem cell therapy in intractable diseases: a narrative review of decades of progress and challenges. *Stem Cell Invest.*, 4, 2022, nih.gov.
13. Zhao, M., Nakada, Y., Wei, Y., Bian, W., Chu, Y., Borovjagin, A.V., Xie, M., Zhu, W., Nguyen, T., Zhou, Y., Serpooshan, V., Cyclin D2 overexpression enhances the efficacy of human induced pluripotent stem cell–derived cardiomyocytes for myocardial repair in a swine model of myocardial infarction. *Circulation*, 144, 3, 210–28, 2021 Jul 20, ahajournals.org.
14. Ghobadinezhad, F., Ebrahimi, N., Mozaffari, F., Moradi, N., *et al.*, The emerging role of regulatory cell-based therapy in autoimmune disease, 13, 1075813, 2022, ncbi.nlm.nih.gov.
15. Poulos, J., The limited application of stem cells in medicine: a review, 9, 1–11, 2018, ncbi.nlm.nih.gov.
16. Mousavinejad, M., Andrews, P.W., Kargar Shoraki, E., Current Biosafety Considerations in Stem Cell Therapy, 2, 281, 2016, ncbi.nlm.nih.gov.
17. Hoang, D.M., Pham, P.T., Bach, T.Q., Ngo, A.T., Nguyen, Q.T., Phan, T.T., Nguyen, G.H., Le, P.T., Hoang, V.T., Forsyth, N.R., Heke, M., Stem cell-based therapy for human diseases. *Signal Transduction Targeted Ther.*, 7, 1, 272, 2022 Aug 6, nature.com.

18. Master, Z., Matthews, K.R.W., Abou-el-Enein, M., Unproven stem cell interventions: a global public health problem requiring global deliberation. *Stem Cell Rep.*, 6, 1435–45, 2021, cell.com.
19. Tan, Y., Ooi, S., Wang, L., Immunogenicity and Tumorigenicity of Pluripotent Stem Cells and their Derivatives: Genetic and Epigenetic Perspectives, 1, 63–72, 2014, ncbi.nlm.nih.gov.
20. Abdelkrim, H., Juan, D.B., Jane, W., Mohamed, A., *et al.*, The immune boundaries for stem cell based therapies: problems and prospective solutions, 8a, 1464–75, 2009, ncbi.nlm.nih.gov.
21. Martin, U., Therapeutic Application of Pluripotent Stem Cells: Challenges and Risks, 4, 229, 2017, ncbi.nlm.nih.gov.
22. Singh, V.K., Kalsan, M., Kumar, N., Saini, A., *et al.*, Induced pluripotent stem cells: applications in regenerative medicine, disease modeling, and drug discovery, 3, 2, 2015, ncbi.nlm.nih.gov.
23. Clavellina, D., Balkan, W., Hare, J.M., Stem cell therapy for acute myocardial infarction: Mesenchymal Stem Cells and induced Pluripotent Stem Cells. *Expert Opin. Biol. Ther.*, 23, 10, 951–67, 2023 Oct 3, nih.gov.
24. Ahmed, F.A., Advances in Regenerative Medicine: Stem Cell Therapy and Tissue Engineering. *Open Eur. J. Res. Med. Basic Sci. (OEJRMBS)*, 39–54, 2025 Jan 21, easdjournals.com.
25. Öztürk, S., Elçin, A.E., Koca, A., Elçin, Y.M., Therapeutic applications of stem cells and extracellular vesicles in emergency care: futuristic perspectives. *Stem Cell Rev. Rep.*, 390–410, 2021, springer.com.
26. Eun Yoon, D., Lee, H., Kim, K., Recent Research Trends in Stem Cells Using CRISPR/Cas-Based Genome Editing Methods, 1, 1–4, 2023, ncbi.nlm.nih.gov.
27. Schacker, M. and Seimetz, D., From fiction to science: clinical potentials and regulatory considerations of gene editing, 1, 27, 2019, ncbi.nlm.nih.gov.
28. Bolideei, M., Barzigar, R., Gahrouei, R.B., Mohebbi, E., Haider, K.H., Paul, S., Paul, M.K., Mehran, M.J., Applications of Gene Editing and Nanotechnology in Stem Cell-Based Therapies for Human Diseases. *Stem Cell Rev. Rep.*, 1–30, 2025 Feb 27, researchgate.net.

2

Stem Cell Therapy on Skin: Mechanisms and Recent Advances

Koushik Jana[1], Md. Kabirul Islam Mollah[1], Gouranga Dutta[1], Goutam Ghosh[2] and Goutam Rath[2*]

[1]Department of Pharmaceutical Technology, Bharat Technology, Uluberia, Howrah, West Bengal, India
[2]School of Pharmaceutical Sciences, Siksha 'O' Anusandhan (Deemed to be University), Kalinga Nagar, Bhubaneswar, Odisha, India

Abstract
Stem cell therapy constitutes an innovative treatment modality that harnesses the unique properties of stem cells, including self-renewal and differentiation, to facilitate the repair and regeneration of damaged tissues. This approach applies to regenerative medicine, neurology, dermatology, and orthopedics. The term "stem cell" was introduced in 1868 by German biologist Ernst Haeckel to explain how fertilized eggs develop into all cell types within an organism. Skin is the main organ for interacting with the environment, making up about 16% of body weight. Maintaining skin integrity is crucial for optimal function. Significant injuries or underlying conditions, such as diabetes, burns, trauma, and vascular insufficiency, may jeopardize the stability of the skin, thereby resulting in a myriad of diseases. Common disorders that pose treatment challenges include chronic wounds, ulcers, burns, skin injuries, scars, aging, autoimmune diseases, genetic disorders, and vitiligo. Skin stem cells, located within the epidermis, hair follicles, and dermis, play a crucial role in maintaining skin homeostasis, facilitating wound healing, and regenerating damaged tissue. Furthermore, stem cell therapy is being investigated for its efficacy in improving scars, including keloids and hypertrophic scars, as well as in addressing wrinkles and signs of skin aging. This therapeutic intervention is pivotal in enhancing skin texture, elasticity, and overall appearance. A brief overview of prevalent skin diseases and the corresponding stem cell therapies available for their management are presented.

**Corresponding author*: goutamrath123@gmail.com

Sumel Ashique, Sanjay Dey, Biplab Debnath, Mohhammad Ramzan and Swarupananda Mukherjee (eds.) Stem Cell Therapy in Dermatological Disorders, (19–42) © 2025 Scrivener Publishing LLC

Keywords: Stem cell therapy, skin disorders

2.1 Background and Historical Overview of Stem Cell Therapy

Stem cell therapy is an innovative treatment that harnesses the unique capabilities of stem cells, including self-renewal and differentiation, to repair and regenerate damaged tissues. This approach involves introducing exogenous stem cells into the body to replace impaired cells with healthy, fully functional ones [1]. It can be applied in various medical fields, including regenerative medicine, neurology, dermatology, and orthopedics, among others. Research studies demonstrate that stem cell therapy has the potential to treat spinal cord injuries, Parkinson disease, cardiac conditions, skin health disorders, and osteoarthritis by repairing damaged tissues and stimulating cellular renewal [2]. Additionally, it is used in hematology for bone marrow transplants to treat blood disorders, such as leukemia and lymphoma [3]. Therapeutic applications of stem cells include self-to-self therapy through the autologous use of patient tissues, as well as allogeneic implementations that utilize healthy donor cells [4]. The German biologist Ernst Haeckel introduced the term "stem cell" into the scientific literature in 1868 to explain how fertilized eggs develop into all the cells of an organism [5]. The timeline highlights significant milestones in stem cell research and therapy, beginning in 1888 when Theodor Heinrich Boveri and Valentin Haecker introduced the concept of stem cells [5]. In 1902, Franz Ernst Christian Neumann and Alexander A. Maximov described hematopoietic progenitor cells, laying the foundation for future research [6]. Scientists reported the first bone marrow transplantation for aplastic anemia in 1958, following George Mathe's successful use of stem cells to treat radiation exposure in 1958. In 1957, Dr. E. Donnall Thomas performed the first allogeneic hematopoietic stem cell transplantation. In 1969, he conducted the first bone marrow transplant in the United States. During the 1960s and 1970s, Friendenstein's team provided evidence for stromal stem cells, which A. I. Caplan later renamed mesenchymal stem cells (MSCs) in 1991. Researchers made significant strides in embryonic stem cells, with Evans, Kaufman, and G. R. Martin establishing murine embryonic stem cells in 1981 and James Thomson isolating the first human embryonic stem cells in 1998 [6]. In 2007, S. Yamanaka and K. Takahashi discovered induced pluripotent stem cells (iPSCs) revolutionizing regenerative medicine. Geron launched the first clinical trial using embryonic stem cell–derived OPC1 for spinal cord injury in 2010. In 2015, scientists reported the first case of human embryonic stem cell–derived cardiac progenitors treating severe heart disease, followed by the first use of iPSC-derived retinal cells for macular degeneration in 2017 [6]. In 2022, a report detailing the long-term effects of OPC1 in spinal cord injury treatment marked a decade of progress in stem cell therapy [6]. In 2023, scientists in China successfully reversed type 1 diabetes in a 25-year-old woman using a pioneering stem cell transplant, enabling her to regulate blood sugar levels without insulin injections [7]. In December 2024, the US Food and SDrug Administration approved Ryoncil (remestemcel-L), marking the first mesenchymal stromal cell therapy authorized for pediatric patients with steroid-refractory acute graft-versus-host disease [8]. These advancements underscore the transformative potential of stem cell therapies in treating previously intractable conditions. The historical overview of stem cell technology highlights its remarkable evolution from early discoveries in cell biology to groundbreaking advancements in regenerative medicine. From the identification of stem cells in the mid-20th century to the development of iPSCs and innovative therapeutic applications, this field has revolutionized medical research. Despite ethical and technical challenges, stem cell technology continues to offer immense potential for treating various diseases, regenerating damaged tissues, and shaping the future of personalized medicine.

2.2 Skin Diseases and Stem Cell Treatment

Humans rely on the skin as their fundamental organ for external communication with the environment [9]. It accounts for approximately 16% of the total body weight. The primary role of the skin is to protect against mechanical, chemical, and physical damage while serving as a barrier against bacterial invasion [10]. The epidermis, which includes hair follicles, sebaceous glands, and sweat glands, plays a crucial role in maintaining the skin's structural integrity, immunity, and nourishment. The dermis is rich in extracellular matrix (ECM), blood vessels, and mechanoreceptors, contributing to skin support and sensory functions. Meanwhile, the subcutaneous adipose tissue serves as an energy reservoir, providing a steady supply of growth factors essential for skin regeneration [11]. Maintaining skin integrity is critical for its proper function. However, when the body sustains severe injuries or experiences internal issues such as diabetes, burns, trauma, or vascular insufficiency, the skin's structural and functional stability can be compromised [12]. These lead to various skin diseases, particularly those involving tissue damage, chronic wounds, and immune system dysfunction. Chronic wounds and ulcers, burns and skin injuries, scars and skin aging, autoimmune skin diseases, genetic skin disorders, and vitiligo are some of the common skin disorders that face challenges in their treatment [13, 14]. Chronic wounds and ulcers are typically managed with wound dressings, debridement, antibiotics, negative-pressure wound therapy, and skin grafts [15]. Burns and skin injuries often require wound care, pain management, skin grafting, and advanced therapies such as bioengineered skin substitutes [16, 17]. Scars and skin aging can be treated with laser therapy, microneedling, chemical peels, dermal fillers, and topical treatments such as retinoids [18]. Autoimmune skin diseases, such as psoriasis and systemic sclerosis, are managed with immunosuppressants, corticosteroids, biological therapies, and phototherapy [19, 20]. Genetic skin disorders such as epidermolysis bullosa may require symptomatic management with specialized wound care, and protein replacement therapy [21]. Vitiligo treatment focuses on repigmentation through corticosteroids, topical calcineurin inhibitors, phototherapy, and melanocyte transplantation in severe cases [22]. Despite advancements in dermatology, many skin diseases lack effective long-term treatments, with conventional therapies often proving inadequate or causing side effects. In this context, stem cell therapy offers a promising approach to treating the abovementioned diseases (Figure 2.1). Its regenerative potential makes it a valuable option where traditional treatments fall short.

Stem cells possess the distinct capacity to develop into specialized cell types while renewing themselves over a long period and remaining undifferentiated [23]. They have different origins, starting from embryonic stem cells taken from the early embryo together with adult stem cells extracted from tissue, bone marrow, and skin and additionally including iPSCs, which are reprogrammed adult cells [24]. Stem cell formation begins with early embryonic development, where they retain pluripotency, allowing them to give rise to all body tissues. Their key characteristics include self-renewal, differentiation potential, and the ability to repair damaged tissues [25]. Special features of stem cells include their plasticity, ability to modulate the immune system, and role in tissue regeneration. Skin stem cells, found in the epidermis, hair follicles, and dermis, play a crucial role in maintaining skin homeostasis, repairing wounds, and regenerating damaged tissue [26]. Stem cell therapy has shown promise in promoting wound healing by enhancing tissue regeneration, reducing inflammation, and stimulating the formation of new blood vessels, making it beneficial for chronic

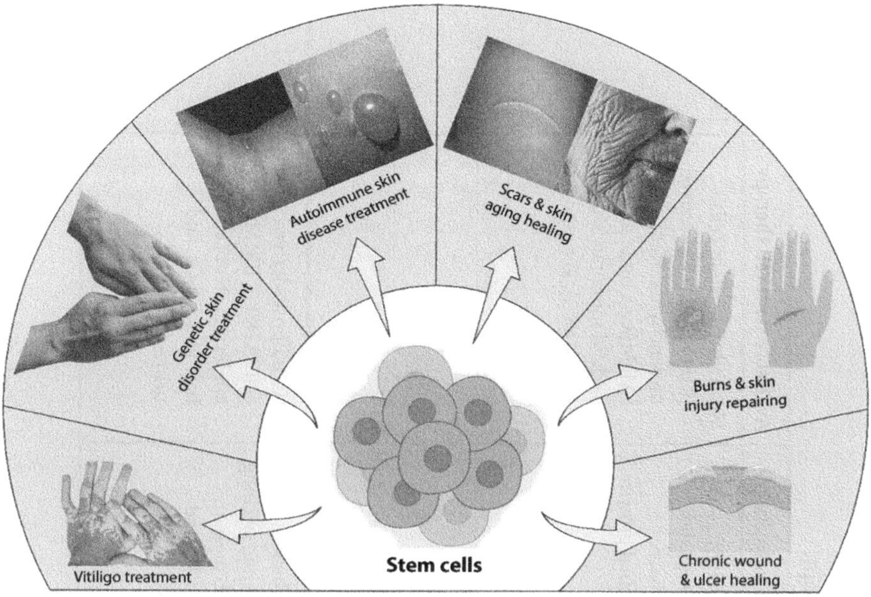

Figure 2.1 Different uses of stem in the treatment of skin diseases.

wounds such as diabetic ulcers, pressure sores, and venous ulcers [27]. In cases of burns and skin injuries, including severe burns and radiation-induced skin damage, stem cells play a crucial role in regenerating skin layers, minimizing scarring, and enhancing elasticity [28]. Autoimmune skin diseases such as systemic sclerosis, lupus-related skin lesions, and psoriasis (still under research) may benefit from stem cells, which help modulate the immune system and reduce inflammation [29]. Genetic skin disorders, such as epidermolysis bullosa—a rare condition characterized by fragile, blistering skin—have shown promise in clinical trials using gene-corrected stem cell therapy [30]. Additionally, conditions such as vitiligo may be treated with skin or bone marrow stem cells to restore pigmentation by regenerating melanocytes [31]. Stem cell therapy is also being explored for improving scars, including keloids and hypertrophic scars, as well as for addressing wrinkles and skin aging; stem cell therapy plays a key role in enhancing skin texture, elasticity, and overall appearance. A brief overview of overmaintained skin diseases and available stem cell therapies for their treatment is presented in Table 2.1.

Table 2.1 Characteristics of various skin diseases and available stem cell therapy.

Sl. no	Disease type	Causes	Pathway involved	Clinical indication	Available stem cell therapy	References
1	Chronic wounds and ulcers	Poor blood circulation, diabetes, pressure injuries, infections, immune disorders, etc.	Overproduction of proinflammatory cytokines (tumor necrosis factor α [TNF-α], interleukin [IL] 1β, IL-6), reduced vascular endothelial growth factor (VEGF), reduced transforming growth factor β (TGF-β), excessive MMP activity, etc.	Tissue injury, bleeding, and permissive microbial growth	Mesenchymal stem cells (MSCs), adipose-derived mesenchymal stem cell (ADMSCs), bone marrow–derived mononuclear cells (BMMNCs), keratinocytes	[32, 33]
2	Burns and skin injuries	Thermal, chemical, electrical, etc. burns, cuts and lacerations, abrasions, puncture wounds, etc.	High levels of ROS, clotting cascade activation, VEGF, collagen overproduction, etc.	Pain, redness, blistering, swelling, and changes in skin color and texture	MSC, adipose-derived stem cell (ADSC), BMMNCs, epidermal stem cells (SC), hair follicle SCs, and urine-derived SCs	[34, 35]

(Continued)

Table 2.1 Characteristics of various skin diseases and available stem cell therapy. (*Continued*)

Sl. no	Disease type	Causes	Pathway involved	Clinical indication	Available stem cell therapy	References
3	Scars and skin aging	Injury and trauma, inflammation, excessive fibrosis, genetic factors, UV-radiation, loss of collagen and elastin, etc.	TGF-β, CTGF, nuclear factor κB (NF-κB) signaling, Wnt/β-catenin, Nrf2/Keap1, mTOR, etc.	Keloids, wrinkles, sagging, loss of elasticity, and changes in pigmentation, texture, and thickness	ADSCs, MSCs, induced pluripotent stem cells (iPSCs), and umbilical cord-derived mesenchymal stem cell (UCMSCs)	[36, 37]
4	Autoimmune skin diseases	Genetic factors, immune system dysfunction, environmental triggers, infections, medications, chronic stress, etc.	NF-κB, JAK-STAT, mitogen-activated protein kinase (MAPK), TLR signaling, IL-23/helper T 17 (T_H17), TGF-β signaling, PI3K-AKT, Notch signaling, TNF-α signaling, interferon signaling pathway, etc.	Systemic sclerosis and sclerodermatous graft-versus-host disease	MSCs and iPSCs	[38]

(*Continued*)

Table 2.1 Characteristics of various skin diseases and available stem cell therapy. (*Continued*)

Sl. no	Disease type	Causes	Pathway involved	Clinical indication	Available stem cell therapy	References
5	Genetic skin disorders	Inherited mutations, spontaneous mutations, chromosomal abnormalities, defective protein production, autosomal recessive inheritance, metabolic disorders, etc.	Hedgehog signaling, Wnt/β-catenin, Notch signaling, TGF-β signaling, MAPK/ERK, PI3K-AKT, mTOR signaling, NF-κB, JAK-STAT, Integrin signaling pathway, etc.	Skin blistering and inflammation	MSCs, keratinocytes, unrestricted somatic SCs, iPSCs, and hematopoietic SCs	[39]
6	Vitiligo	Autoimmune response, genetic factors, emotional and physical stress, hormonal imbalances, vitamin deficiencies, etc.	NF-κB, JAK-STAT, MAPK/ERK, PI3K-AKT, Wnt/β-catenin, TGF-β signaling, p53 signaling, oxidative stress, autophagy, IL-23/T_H17 pathway, etc.	Innate immune response, melanocyte damage, adaptive immune response, and immune-inflammatory microenvironment disorders	MSCs, noncultured epidermal SCs, and follicular SCs	[40]

2.3 Types of Stem Cells and Their Mechanism in Skin Diseases

Stem cells are mainly classified into two types: embryonic stem cells and somatic stem cells. Embryonic stem cells come from early-stage embryos (blastocysts) and are pluripotent, meaning they can develop into any cell type from the three germ layers ectoderm, endoderm, and mesoderm [41]. Somatic stem cells, on the other hand, are found in mature organs and tissues. Some are multipotent, but most are limited in what they can become. For example, hematopoietic stem cells produce only blood cells, whereas neural stem cells develop into brain-related cells [42]. A breakthrough by Professor Yamanaka's laboratory in Kyoto led to the creation of iPSCs. These are reprogrammed adult cells that behave like embryonic stem cells. The process uses four key transcription factors (Oct-3/4, Sox2, c-Myc, and Klf4) and allows iPSCs to generate many different cell types [43]. Skin stem cells belong to the somatic stem cell category. Due to the complex nature of skin, scientists have discovered various skin stem cell types in recent years, aided by advanced molecular tools. They are listed in Table 2.2.

Stem cells exert their therapeutic effects on skin diseases through multiple mechanisms, engaging diverse pathways and receptors. They promote healing by modulating inflammation, enhancing regeneration, and restoring skin integrity. MSCs support the proliferative phase of regeneration by releasing growth factors, such as vascular endothelial growth factor, basic fibroblast growth factor, and keratinocyte growth factor, which enhance granulation tissue formation and epithelialization. Additionally, they help regulate wound remodeling by facilitating the structured deposition of the ECM during tissue replacement [51]. Adipose-derived stem cells (ASCs) have an immunosuppressive effect, which regulates the immune response. This occurs through direct cell–cell interactions with immune cells, including T cells, macrophages, and B cells, as well as indirect modulation *via* the secretion of cytokines, interleukins, growth factors, and extracellular vesicles [52]. The hair follicle stem cells in the bulge region regulate skin homeostasis, wound healing, and regeneration by controlling pathways such as Wnt/β-catenin, transforming growth factor β/bone morphogenetic protein (BMP), notch, and hedgehog, which are crucial for skin repair [53]. Additionally, iPSCs enable skin regeneration and disease modeling by differentiating into various skin cell types, including keratinocytes and melanocytes. Bone marrow–derived mononuclear cells aid skin disease treatment through direct differentiation and paracrine signaling, enhancing regeneration by stimulating angiogenesis, reducing

Table 2.2 An overview of different types of stem cells.

Sl. no.	Types of stem cell	Location	Specific cell marker	Special characters	Ref.
1	Epidermal stem cells (ESC)	The basal layer of the interfollicular epidermis	p63, b1 high/melanoma chondroitin sulfate proteoglycan þ (MCSPþ), a6high/ CD71dim	Strong self-renewal ability, slow cycling, and adhesion to the basement membrane of the epidermis	[44]
2	Follicular stem cells (FSC)	Bulge region	K15, CD34, Lgr5, Sox9, Lhx2, NFATC1, NFIB, K15, PHLDA1, CD200, K19, bromodeoxyuridine dye retention	Self-renewal ability and differentiate into follicle cells and escort cells, with their fate regulated by signaling pathways and spatial location	[26]
3	Melanocyte stem cells	Hair follicle bulge region and hair germ	Dct, Sox, Pax3	Small, oval-shaped cells with a low proliferative rate and lack melanin granules	[45]
4	Mesenchymal stem cell	Dermis	CD70, CD90, CD105	Multipotent differentiation potential, immunomodulatory properties, and ability to promote tissue repair and regeneration	[46]

(Continued)

Table 2.2 An overview of different types of stem cells. (*Continued*)

Sl. no.	Types of stem cell	Location	Specific cell marker	Special characters	Ref.
5	Neural progenitor cells	follicle dermal papillae	Nestin	Ability to proliferate and differentiate into neurons and glial cells	[47]
6	Hematopoietic stem cells	Follicle dermal papillae	CD34, CD49f, and integrin subunit α6 (CD49f)	Unique ability to self-renew and differentiate into all blood cell types	[48]
7	Sebaceous gland stem cells	Mid dermis	GPR39	Ability to proliferate and differentiate	[49]
8	Adipose-derived stem cells	Stromal vascular fraction	CD44, CD90, CD73, CD105, CD166, CD29, CD13, CD10, and CD200	Ability to self-renew, proliferate rapidly, and differentiate into various cell types, including adipocytes, chondrocytes, and osteoblasts	[50]

inflammation, and modulating immunity [54]. Umbilical cord stem cells (USCs) promote skin regeneration and repair in skin diseases through paracrine mechanisms. On the other hand, neural progenitor cells exert their action by inhibiting the Shh, fibroblast growth factor, Wnt, and BMP signaling pathways, with the help of epigenetic modifications. Hematopoietic stem cells are a promising target for cancer and leukemia immunotherapy by introducing genes encoding T-cell receptors or chimeric antigen receptors against tumor-associated antigens [55]. Sebaceous stem cells regulate sebum production, which is influenced by peroxisome proliferator-activated receptor γ and androgen signaling, with imbalances linked to various skin disorders. Adipose-derived stem cells aid skin regeneration by differentiating into skin cells and promoting healing through autocrine and paracrine signaling. These are the key mechanisms of stem cells in skin diseases. A detailed overview is provided in Table 2.3.

2.4 Recent Advancements in Skin Disease Stem Cell Therapy

Recent advancements in stem cell therapy have revolutionized multiple fields, including regenerative medicine, neurology, dermatology, oncology, and others. Stem cell treatments have successfully cured or significantly improved conditions of various disorders. Unlike traditional treatments, stem cell therapy provides long-term regeneration rather than just symptom relief, making it a groundbreaking approach. Accessibility is improving through ongoing research and advancements in induced iPSCs, MSCs, and other cell types, thereby reducing ethical concerns and donor dependence. Advances in bioreactors and 3D bioprinting have improved stem cell production, making treatments more scalable [64]. Scientists have conducted several clinical trials to explore it as an alternative treatment for skin diseases. A phase I clinical trial is advancing with autologous epidermal cellular sheets containing type VII collagen for the treatment of patients with recessive dystrophic epidermolysis bullosa [65, 66]. Researchers genetically modified keratinocytes to potentially behave like stem cells derived from junctional epidermolysis bullosa patients in severe combined immune deficiency mouse models, examining skin proliferation and cellular regeneration [52]. The possibility of somatic stem cells switching their lineage under specific environmental conditions is attracting increasing interest in regenerative medicine applications [67]. As a result, multiple types of somatic stem cells are receiving clinical investigation alongside

Table 2.3 Overview of skin stem cell types, their niches, signaling pathways, and roles.

Sl. no.	Skin stem cell type	Location/niche	Key signaling pathways	Key transcription factors	Key regulatory molecules/factors	Major functions/roles	Regulatory mechanisms	Impact of dysregulation	Ref.
1	Epidermal basal stem cells	Basal epidermis; adjacent to basement membrane	Notch, Wnt/β-catenin, TGF-β	p63, AP-1, c-*Myc*	EGF, TGF-β, laminin-332	Epidermal homeostasis, wound healing	Differentiation toward keratinocytes, cell proliferation	Impaired wound healing, SCC	[56]
2	Follicle stem cells (bulge)	Hair follicle bulge; surrounded by dermal sheath cells	Wnt/β-catenin, Shh, bone morphogenetic protein (BMP)	Lef1, Gli1, NFATc1	Wnt ligands, BMPs, keratinocyte growth factor	Hair follicle cycling, epidermal regeneration	Activation of hair growth, quiescence maintenance	Alopecia, skin tumors	[57]
3	Melanocyte stem cells	Hair follicle bulge (outer root sheath)	Wnt, endothelin-1 (EDN1), c-Kit	MITF, SOX10, PAX3	Stem cell factor (SCF), EDN1, α-MSH, β-catenin	Repopulation of melanocytes in hair follicles, skin pigmentation	Wnt-β-catenin signaling regulates proliferation and differentiation, EDN1 modulates differentiation. c-Kit survival	Premature graying (loss of melanocyte stem cells), melanoma development (uncontrolled proliferation	[58]
4	Mesenchymal stem cell–like cells	Bone marrow, adipose tissue, umbilical cord, periosteum	Wnt, TGF-β, BMP, fibroblast growth factor (FGF), Notch	RUNX2, Peroxisome proliferator-activated receptor γ (PPARγ), SOX9, OCT4	Cytokines (e.g., IL-6, TNF-α), growth factors (e.g., TGF-β1, basic fibroblast growth factor), microRNAs (miRNAs)	Tissue repair and regeneration, Immunomodulation, differentiation into osteoblasts, chondrocytes, adipocytes	Paracrine signaling, cell–cell contact, extracellular matrix interactions, epigenetic modifications	Fibrosis, impaired tissue regeneration, tumor microenvironment modulation	[59]
5	Neural progenitor cells	Neural tube (embryonic), subventricular zone, and subgranular zone	Notch, Shh, FGF, Wnt, BMP	Pax6, Sox2, Olig2, Nestin	Growth factors (EGF, FGF), neurotransmitters, miRNAs	Generation of neurons, astrocytes, and oligodendrocytes; brain development and repair	Epigenetic modifications	Neurodevelopmental disorders	[60]

(Continued)

Table 2.3 Overview of skin stem cell types, their niches, signaling pathways, and roles. (*Continued*)

Sl. no.	Skin stem cell type	Location/niche	Key signaling pathways	Key transcription factors	Key regulatory molecules/factors	Major functions/roles	Regulatory mechanisms	Impact of dysregulation	Ref.
6	Hematopoietic stem cell	Bone marrow	Notch, Wnt, TGF-β, cytokine receptors (e.g., IL-3, SCF)	GATA1, PU.1, RUNX1, SCL	Cytokines (e.g., SCF, TPO, Flt3L), chemokines, adhesion molecules like VCAM-1	Production of all blood cell types (erythrocytes, leukocytes, platelets)	Quiescence, self-renewal, differentiation through lineage commitment	Leukemia (uncontrolled proliferation), anemia (insufficient red blood cell production), Immunodeficiency (impaired immune cell production), bone marrow failure	[61]
7	Sebaceous gland stem cells	Sebaceous gland, infundibulum of hair follicle	Wnt/β-catenin, sonic hedgehog, PI3K/Akt/mTOR	c-Myc, Lhx2, blimp	miRNAs, long noncoding RNAs	Sebaceous gland maintenance (sebum production), Skin repair, Potential roles in basal cell carcinoma (BCC) and sebaceous carcinoma	Epigenetic modifications (e.g., DNA methylation, histone acetylation)	Sebaceous gland hyperplasia	[62]
8	Adipose-derived stem cells	Dermal adipose tissue	Wnt signaling, TGF-β signaling, vascular endothelial growth factor signaling, PDGF signaling	PPARγ, Runx2, C/EBPα	Exosomes, cytokines (e.g., IL-6, IL-8), growth factors (e.g., FGF2, HGF)	Promote wound healing, contribute to tumor angiogenesis	Sebaceous gland maintenance (sebum production), skin repair, Potential roles in BCC and sebaceous carcinoma.	Promotion or inhibition of tumor growth (context-dependent), enhanced metastasis	[63]

epidermal stem cells for skin and soft tissue restoration. Furthermore, whereas early clinical trials primarily focused on autologous engraftment, recent trials have begun to explore allogeneic applications [68]. As research progresses, stem cell therapy continues to hold immense promise for previously incurable diseases, offering a safer, more effective, and personalized treatment option. An incidence in 2015, following the mass-burn incident at Formosa Fun Coast Water Park in Taiwan, the advanced autologous transplantation technique JACE, developed by J-TEC (now merged with Fuji), was introduced to aid victims. Initially pioneered by Prof. Howard Green and later transferred to J-TEC by Prof. Minoru Ueda, this technology enables the isolation of keratinocytes from a 1-cm^2 skin sample, which are then cultured on a fibroblast feeder to produce a 1000-cm^2 epidermal sheet within 2 weeks. During the incident, five patients benefited from JACE®, highlighting its effectiveness. However, in large-scale burn emergencies, both autologous and allogeneic skin products are essential to meet high-demand treatment needs [69]. Later, other products gained market approvals. Epicel® (Vericel Corporation, USA) is a cultured epidermal autograft for severe burns, whereas Holoderm® (Molteni Therapeutics, Italy) provides autologous keratinocyte sheets for burn and wound healing [70]. Kaloderm™ (Tego Science, South Korea) is an allogeneic keratinocyte-based therapy for burns and chronic wounds, and ReCell® (Avita Medical, USA) is an autologous cell harvesting device for skin regeneration [71]. StrataGraft® (Mallinckrodt Pharmaceuticals, USA) is an allogeneic bioengineered skin substitute for burns, whereas Dermagraft® (Organogenesis, USA) is a human fibroblast-derived dermal substitute for chronic wounds [72]. In recent years, several patents have been granted for advancements in stem cell therapy. In 2022, James R. Musick, Tiana States, and Raymond Bunch received a patent for using stem cell transplants and activation to treat various medical conditions [73]. That same year, Shinya Yamanaka and Keisuke Okita patented a method to create safer iPSCs [74]. More recently, in 2024, Therapeutic Solutions International patented gene-edited pluripotent stem cells designed to help treat degenerative and autoimmune diseases. All these patents were approved by the United States Patent and Trademark Office, reflecting the growing interest in stem cell–based therapies [75]. In 2022, ExoCoBio Inc. received a patent from the Japanese Patent Office for their work on adipose stem cell–derived exosomes aimed at alleviating dermatitis [76]. This innovation utilizes stem cell–based therapy to enhance skin conditions and promote healing. Furthermore, the global stem cell therapy market, valued at $259 million in 2022, is projected to grow at a compound annual growth rate (CAGR) of 16.5%, reaching approximately $615 million by 2028. Meanwhile, the

Indian stem cell therapy market, valued at $395.15 million in 2024, is expected to expand at a CAGR of 12.15% through 2030, driven by increasing awareness of its potential. Notably, the global market for stem cell therapy, including applications in skin diseases, was estimated at $14.15 billion in 2023 and is projected to surge to $48.89 billion by 2033, with a CAGR of 13.2%.

2.5 Challenges and Regulatory Aspects

Before a new medical product is tested on humans, it must be proven safe. Stem cell–based therapies, one of the latest advancements in medicine, are complicated to bring to market because many factors need to be considered. The major challenges in stem cell therapy include ensuring controlled differentiation of stem/progenitor cells into desired phenotypes while preventing karyotypic abnormalities during long-term culture [77]. Identifying reliable biomarkers for cellular purity before finalizing cell products remains critical [78]. The use of adjuvants and biomaterials to create a suitable microenvironment raises concerns about unintended immunological responses [79]. Selecting the optimal grafting method and delivery approach for patients is crucial for therapeutic success. The long-term survival of transplanted cells and the development of noninvasive tracking methods *in vivo* remain unresolved [80]. Lastly, comparing benefits and risks against standard treatments and defining clinical endpoints to evaluate graft functionality poses significant challenges in assessing therapeutic efficacy [81]. Key regulatory aspects include clinical trial approvals, which necessitate preclinical data and phased trials; Good Manufacturing Practices for stem cell processing; considerations for patient safety and informed consent; ethical issues such as the use of embryonic stem cells and donor consent; product classification as drugs, biologics, or advanced therapy medicinal products; licensing and oversight by regulatory agencies such as the US FDA, EMA, ANVISA, CDSCO, and PMDA; long-term follow-up requirements; and marketing restrictions to prevent unproven claims. Compliance with these regulations is essential for the development of legal and safe therapies [82].

2.6 Conclusion and Future Outlooks

Stem cells can develop into various types of adult cells and play a crucial role in maintaining healthy skin. When they interact with damaged skin, they release growth factors, reduce inflammation, support the immune system, prevent oxidative stress, and promote the growth of new blood vessels and cells. Research has demonstrated efficient treatment opportunities for autoimmune skin diseases, as well as skin aging, vitiligo, scars, and wound healing, with minimal adverse effects. The therapeutic results have shown limited success for challenging conditions, including scleroderma alongside pemphigus and epidermolysis bullosa, because they produce significant adverse effects along with treatment failure. Future developments in stem cell therapy show promise through ongoing studies that aim to enhance safety and effectiveness. The development of gene editing combined with 3D bioprinting methods will create opportunities for creating personalized medical solutions. The research direction focuses on developing strategies to minimize treatment side effects and boost cell survival rates for improved therapeutic results. Stem cell therapy appears promising as a dependable medical intervention for various diseases, but it requires comprehensive regulatory control and extensive studies to be used safely. Research needs significant expansion to establish the absolute safety profile and effectiveness levels of these treatments for each condition. The therapeutic potential of stem cells offers an effective medical intervention for skin-related diseases, as current treatments often provide minimal or no therapeutic benefits.

AI disclosure statement: The author confirms that no artificial intelligence (AI)-generated content or AI-assisted writing tools were used in the preparation of this chapter (including summaries and the abstract). All text, analysis, and conclusions are the result of the author's original work.

References

1. Mandacı Şanlı, N. and Timuroğlu, A., Current Advances in Stem Cell-Based Therapies: Adult Stem Cell Applications, in: *Stem Cell Transplantation* IntechOpen, 2024.
2. Zeng, C.-W., Advancing Spinal Cord Injury Treatment through Stem Cell Therapy: A Comprehensive Review of Cell Types, Challenges, and Emerging Technologies in Regenerative Medicine. *IJMS*, *24*, 14349, 2023.
3. Schubert, M.-L., Hückelhoven, A., Hoffmann, J.-M., Schmitt, A., Wuchter, P., Sellner, L., Hofmann, S., Ho, A.D., Dreger, P., Schmitt, M., Chimeric Antigen Receptor T Cell Therapy Targeting CD19-Positive Leukemia and Lymphoma in the Context of Stem Cell Transplantation. *Hum. Gene Ther.*, *27*, 758–771, 2016.

4. Srijaya, T.C., Ramasamy, T.S., Kasim, N.H.A., Advancing Stem Cell Therapy from Bench to Bedside: Lessons from Drug Therapies. *J. Transl. Med., 12,* 243, 2014.
5. Ramalho-Santos, M. and Willenbring, H., On the Origin of the Term "Stem Cell." *Cell Stem Cell, 1,* 35–38, 2007.
6. Hoang, D.M., Pham, P.T., Bach, T.Q., Ngo, A.T.L., Nguyen, Q.T., Phan, T.T.K., Nguyen, G.H., Le, P.T.T., Hoang, V.T., Forsyth, N.R., Heke, M., Nguyen, L.T., Stem Cell-Based Therapy for Human Diseases. *Signal Transduction Targeted Ther., 7,* 272, 2022.
7. Kumar, D. and Tanwar, R., World's First: Stem Cell Therapy Reverses Diabetes. *Stem Cell Res. Ther., 15,* 487, 2024.
8. FDA Approves First Mesenchymal Stromal Cell Therapy to Treat Steroid-refractory Acute Graft-versus-host Disease, FDA, https://www.fda.gov/news-events/press-announcements/fda-approves-first-mesenchymal-stromal-cell-therapy-treat-steroid-refractory-acute-graft-versus-host (Accessed Mar 17, 2025).
9. Hanson, S.E., Bentz, M.L., Hematti, P., Mesenchymal Stem Cell Therapy for Nonhealing Cutaneous Wounds. *Plast. Reconstr. Surg., 125,* 510–516, 2010.
10. Smith, R., Russo, J., Fiegel, J., Brogden, N., Antibiotic Delivery Strategies to Treat Skin Infections When Innate Antimicrobial Defense Fails. *Antibiotics, 9,* 56, 2020.
11. Leclère, F.M., The Use of Integra® Dermal Regeneration Template Versus Flaps for Reconstruction of Full-Thickness Scalp Defects Involving the Calvaria: A Cost–Benefit Analysis. *Aesthetic Plast. Surg., 41,* 472–473, 2017.
12. Zhao, Y., Wang, M., Liang, F., Li, J., Recent Strategies for Enhancing the Therapeutic Efficacy of Stem Cells in Wound Healing. *Stem Cell Res. Ther., 12,* 588, 2021.
13. Mahajan, V.K. and Handa, S., Diagnosis of Skin Diseases, in: *Atlas of Dermatology, Dermatopathology and Venereology,* B. Smoller, and N. Bagherani, (Eds.), pp. 45–107, Springer International Publishing, Cham, 2022.
14. Peate, I., *The Skin,* John Wiley & Sons, 2025.
15. Eriksson, E., Liu, P.Y., Schultz, G.S., Martins-Green, M.M., Tanaka, R., Weir, D., Gould, L.J., Armstrong, D.G., Gibbons, G.W., Wolcott, R., Olutoye, O.O., Kirsner, R.S., Gurtner, G.C., Chronic Wounds: Treatment Consensus. *Wound Repair Regener., 30,* 156–171, 2022.
16. Augustine, R., Kalarikkal, N., Thomas, S., Advancement of Wound Care from Grafts to Bioengineered Smart Skin Substitutes. *Prog. Biomater., 3,* 103–113, 2014.
17. Ghosh, S., Bishal, A., Ghosh, S.K., Jana, K., Gayen, B., Sahu, S., Debnath, B., Herbal Medicines: A Potent Approach to Human Diseases, Their Chief Compounds, Formulations, Present Status, and Future Aspects. *IJMST, 10,* 442–464, 2023.

18. Sethupathi, S.K., Poole, M., Darji, K., Fehlman, J., Cosmetic Dermatology Concerns in Older Adults. *Clin. Geriatr. Med.*, *40*, 197–210, 2024.
19. Jana, K., Ghosh, S., Debnath, B., Kumar Rout, S., Kumar Sarangi, M., Halder, J., Kanti Rajwar, T., Pradhan, D., Kumar Rai, V., Dash, P., Das, C., Kar, B., Ghosh, G., Rath, G., Recent Advancements of Hyaluronic Acid Nanoscaffolds in Arthritis Management. *ChemistrySelect*, *9*, e202402035, 2024.
20. Kanatoula, D.D., Bodner, E., Ghoreschi, K., Meier, K., Solimani, F., Non-biologic Immunosuppressive Drugs for Inflammatory and Autoimmune Skin Diseases. *J. Dtsch. Dermatol. Ges.*, *22*, 400–421, 2024.
21. Welponer, T., Prodinger, C., Pinon-Hofbauer, J., Hintersteininger, A., Breitenbach-Koller, H., Bauer, J.W., Laimer, M., Clinical Perspectives of Gene-Targeted Therapies for Epidermolysis Bullosa. *Dermatol. Ther. (Heidelb)*, *11*, 1175–1197, 2021.
22. Srivastav, Y., Singh, J., Hameed, A., Srivastav, A., Ahmad, M.I., A Comprehensive Description of Vitiligo Disease and Its Medication Based Management. *IJMS*, *2*, 8–47, 2024.
23. Stem Cells, Definition, Classification and Sources, in: *Frontiers in Pluripotent Stem Cells Research and Therapeutic Potentials*, Sidhu, K.S. (Ed.), pp. 3–15, Bentham Science Publishers, 2012.
24. Abarrategi, A., Tornin, J., Martinez-Cruzado, L., Hamilton, A., Martinez-Campos, E., Rodrigo, J.P., González, M.V., Baldini, N., Garcia-Castro, J., Rodriguez, R., Osteosarcoma: Cells-of-Origin, Cancer Stem Cells, and Targeted Therapies. *Stem Cells Int.*, *2016*, 3631764, 2016.
25. Arias, A.M., Drosophila Melanogaster and the Development of Biology in the 20th Century, in: *Drosophila*, vol. 420, C. Dahmann, (Ed.), Methods in Molecular Biology, pp. 1–25, Totowa, NJ, Humana Press, 2008.
26. Ojeh, N., Pastar, I., Tomic-Canic, M., Stojadinovic, O., Stem Cells in Skin Regeneration, Wound Healing, and Their Clinical Applications. *IJMS*, *16*, 25476–25501, 2015.
27. Yu, Q., Qiao, G.-H., Wang, M., Yu, L., Sun, Y., Shi, H., Ma, T.-L., Stem Cell-Based Therapy for Diabetic Foot Ulcers. *Front. Cell Dev. Biol.*, *10*, 812262, 2022.
28. Lukomskyj, A.O., Rao, N., Yan, L., Pye, J.S., Li, H., Wang, B., Li, J.J., Stem Cell-Based Tissue Engineering for the Treatment of Burn Wounds: A Systematic Review of Preclinical Studies. *Stem Cell Rev. Rep.*, *18*, 1926–1955, 2022.
29. Zaripova, L.N., Midgley, A., Christmas, S.E., Beresford, M.W., Pain, C., Baildam, E.M., Oldershaw, R.A., Mesenchymal Stem Cells in the Pathogenesis and Therapy of Autoimmune and Autoinflammatory Diseases. *Int. J. Mol. Sci.*, *24*, 16040, 2023.
30. Søgaard, P.P., Lind, M., Christiansen, C.R., Petersson, K., Clauss, A., Caffarel-Salvador, E., Future Perspectives of Oral Delivery of Next Generation Therapies for Treatment of Skin Diseases. *Pharmaceutics*, *13*, 1722, 2021.

31. Esquivel, D., Mishra, R., Srivastava, A., Stem Cell Therapy Offers a Possible Safe and Promising Alternative Approach for Treating Vitiligo: A Review. *CPD*, *26*, 4815–4821, 2020.
32. Raghuram, A.C., Yu, R.P., Lo, A.Y., Sung, C.J., Bircan, M., Thompson, H.J., Wong, A.K., Role of Stem Cell Therapies in Treating Chronic Wounds: A Systematic Review. *WJSC*, *12*, 659–675, 2020.
33. Saleh, M., Sohrabpour, A.A., Mehrabi, M.R., Seyhoun, I., Vaezi, A.A., Therapeutic Approach of Adipose-Derived Mesenchymal Stem Cells in Refractory Peptic Ulcer. *Stem Cell Res. Ther.*, *12*, 515, 2021.
34. Francis, E., Kearney, L., Clover, J., The Effects of Stem Cells on Burn Wounds: A Review. *Int. J. Burns Trauma*, *9*, 1–12, 2019.
35. Jo, H., Brito, S., Kwak, B.M., Park, S., Lee, M.-G., Bin, B.-H., Applications of Mesenchymal Stem Cells in Skin Regeneration and Rejuvenation. *Int. J. Mol. Sci.*, *22*, 2410, 2021.
36. Chang, L., Fan, W., Pan, X., Zhu, X., Stem Cells to Reverse Aging. *Chin. Med. J.*, *135*, 901–910, 2022.
37. Chen, H., Hou, K., Wu, Y., Liu, Z., Use of Adipose Stem Cells Against Hypertrophic Scarring or Keloid. *Front. Cell Dev. Biol.*, *9*, 823694, 2022.
38. Yang, H., Cheong, S., He, Y., Lu, F., Mesenchymal Stem Cell-Based Therapy for Autoimmune-Related Fibrotic Skin Diseases—Systemic Sclerosis and Sclerodermatous Graft-versus-Host Disease. *Stem Cell Res. Ther.*, *14*, 372, 2023.
39. Larsimont, J. and Blanpain, C., Single Stem Cell Gene Therapy for Genetic Skin Disease. *EMBO Mol. Med.*, *7*, 366–367, 2015.
40. Zhang, M., Xia, T., Lin, F., Yu, J., Yang, Y., Lei, W., Zhang, T., Vitiligo: An Immune Disease and Its Emerging Mesenchymal Stem Cell Therapy Paradigm. *Transplant Immunol.*, *76*, 101766, 2023.
41. Thomson, J.A., Itskovitz-Eldor, J., Shapiro, S.S., Waknitz, M.A., Swiergiel, J.J., Marshall, V.S., Jones, J.M., Embryonic Stem Cell Lines Derived from Human Blastocysts. *Science*, *282*, 1145–1147, 1998.
42. Clarke, D.L., Johansson, C.B., Wilbertz, J., Veress, B., Nilsson, E., Karlström, H., Lendahl, U., Frisén, J., Generalized Potential of Adult Neural Stem Cells. *Science*, *288*, 1660–1663, 2000.
43. Takahashi, K. and Yamanaka, S., Induction of Pluripotent Stem Cells from Mouse Embryonic and Adult Fibroblast Cultures by Defined Factors. *Cell*, *126*, 663–676, 2006.
44. Abbas, O. and Mahalingam, M., Epidermal Stem Cells: Practical Perspectives and Potential Uses. *Br. J. Dermatol.*, *161*, 228–236, 2009.
45. Lee, J.H. and Fisher, D.E., Melanocyte Stem Cells as Potential Therapeutics in Skin Disorders. *Expert Opin. Biol. Ther.*, *14*, 1569–1579, 2014.
46. Lin, C.-S., Xin, Z.-C., Dai, J., Lue, T.F., Commonly Used Mesenchymal Stem Cell Markers and Tracking Labels: Limitations and Challenges. *Histol. Histopathol.*, *28*, 1109–1116, 2013.

47. Liu, F., Zhang, C., Hoffman, R.M., Nestin-Expressing Stem Cells from the Hair Follicle Can Differentiate Into Motor Neurons and Reduce Muscle Atrophy after Transplantation to Injured Nerves. *Tissue Eng. Part A*, 20, 3–4, 656–62, 2013, 131106060201007.
48. Krebsbach, P.H. and Villa-Diaz, L.G., The Role of Integrin A6 (CD49f) in Stem Cells: More than a Conserved Biomarker. *Stem Cells Dev.*, *26*, 1090–1099, 2017.
49. Guo, C., Liu, H., Zhang, B.-H., Cadaneanu, R.M., Mayle, A.M., Garraway, I.P., Epcam, CD44, and CD49f Distinguish Sphere-Forming Human Prostate Basal Cells from a Subpopulation with Predominant Tubule Initiation Capability. *PLoS One*, *7*, e34219, 2012.
50. Ong, W.K., Chakraborty, S., Sugii, S., Adipose Tissue: Understanding the Heterogeneity of Stem Cells for Regenerative Medicine. *Biomolecules*, *11*, 918, 2021.
51. Ogliari, K.S., Marinowic, D., Brum, D.E., Loth, F., Stem Cells in Dermatology. *An. Bras. Dermatol.*, *89*, 286–291, 2014.
52. Mavilio, F., Pellegrini, G., Ferrari, S., Di Nunzio, F., Di Iorio, E., Recchia, A., Maruggi, G., Ferrari, G., Provasi, E., Bonini, C., Capurro, S., Conti, A., Magnoni, C., Giannetti, A., De Luca, M., Correction of Junctional Epidermolysis Bullosa by Transplantation of Genetically Modified Epidermal Stem Cells. *Nat. Med.*, *12*, 1397–1402, 2006.
53. Lee, J.H. and Choi, S., Deciphering the Molecular Mechanisms of Stem Cell Dynamics in Hair Follicle Regeneration. *Exp. Mol. Med.*, *56*, 110–117, 2024.
54. Bilousova, G. and Roop, D.R., Induced Pluripotent Stem Cells in Dermatology: Potentials, Advances, and Limitations. *Cold Spring Harbor Perspect. Med.*, *4*, a015164–a015164, 2014.
55. Gschweng, E., De Oliveira, S., Kohn, D.B., Hematopoietic Stem Cells for Cancer Immunotherapy. *Immunol. Rev.*, *257*, 237–249, 2014.
56. Piccione, M. and Di Liddo, R., Skin Stem Cells in Cancer, in: *Cancer Stem Cells: New Horizons in Cancer Therapies*, S. Pathak, and A. Banerjee, (Eds.), pp. 111–124, Springer Singapore, Singapore, 2020.
57. Perez-Losada, J. and Balmain, A., Stem-Cell Hierarchy in Skin Cancer. *Nat. Rev. Cancer*, 3, 434–443, 2003.
58. Qiu, W., Chuong, C., Lei, M., Regulation of Melanocyte Stem Cells in the Pigmentation of Skin and Its Appendages: Biological Patterning and Therapeutic Potentials. *Exp. Dermatol.*, *28*, 395–405, 2019.
59. Wu, M., Mi, J., Qu, G., Zhang, S., Jian, Y., Gao, C., Cai, Q., Liu, J., Jiang, J., Huang, H., Role of Hedgehog Signaling Pathways in Multipotent Mesenchymal Stem Cells Differentiation. *Cell Transplant.*, *33*, 09636897241244943, 2024.
60. Faigle, R. and Song, H., Signaling Mechanisms Regulating Adult Neural Stem Cells and Neurogenesis. *Biochim. Biophys. Acta, (BBA) - Gen. Subj.*, *1830*, 2435–2448, 2013.

61. Mann, Z., Sengar, M., Verma, Y.K., Rajalingam, R., Raghav, P.K., Hematopoietic Stem Cell Factors: Their Functional Role in Self-Renewal and Clinical Aspects. *Front. Cell Dev. Biol.*, *10*, 664261, 2022.
62. Niemann, C. and Horsley, V., Development and Homeostasis of the Sebaceous Gland. *Semin. Cell Dev. Biol.*, *23*, 928–936, 2012.
63. Mazini, L., Rochette, L., Admou, B., Amal, S., Malka, G., Hopes and Limits of Adipose-Derived Stem Cells (ADSCs) and Mesenchymal Stem Cells (MSCs) in Wound Healing. *IJMS*, *21*, 1306, 2020.
64. Ong, C.S., Yesantharao, P., Huang, C.Y., Mattson, G., Boktor, J., Fukunishi, T., Zhang, H., Hibino, N., 3D Bioprinting Using Stem Cells. *Pediatr. Res.*, *83*, 223–231, 2018.
65. Siprashvili, Z., Nguyen, N.T., Bezchinsky, M.Y., Marinkovich, M.P., Lane, A.T., Khavari, P.A., Long-Term Type VII Collagen Restoration to Human Epidermolysis Bullosa Skin Tissue. *Hum. Gene Ther.*, *21*, 1299–1310, 2010.
66. Uitto, J., McGrath, J.A., Rodeck, U., Bruckner-Tuderman, L., Clare Robinson, E., Progress in Epidermolysis Bullosa Research: Toward Treatment and Cure. *J. Invest. Dermatol.*, *130*, 1778–1784, 2010.
67. Wagers, A.J. and Weissman, I.L., Plasticity of Adult Stem Cells. *Cell*, *116*, 639–648, 2004.
68. Aly, R.M., Current State of Stem Cell-Based Therapies: An Overview. *Stem Cell Investig.*, *7*, 8–8, 2020.
69. Chu, G.-Y., Chen, Y.-F., Chen, H.-Y., Chan, M.-H., Gau, C.-S., Weng, S.-M., Stem Cell Therapy on Skin: Mechanisms, Recent Advances and Drug Reviewing Issues. *J. Food Drug Anal.*, *26*, 14–20, 2018.
70. Ayala, R.D., Chowdhry, T., Hodge, J., Outcomes of Using Cultured Epidermal Autograft (CEA) and Autologous Spray Cell Suspension (ASCS) in Addition to Split Thickness Skin Graft (STSG) on the Same Burn Wound: Grady Hospital Experience. *Burns Open*, *6*, 32–36, 2022.
71. Kim, E.H. and Lee, S.H., Efficacy of Cultured Allogenic Keratinocytes in Treatment of Deep Second-Degree Burn. *J. Burn Care Res.*, *42*, 533–537, 2021.
72. Centanni, J.M., Straseski, J.A., Wicks, A., Hank, J.A., Rasmussen, C.A., Lokuta, M.A., Schurr, M.J., Foster, K.N., Faucher, L.D., Caruso, D.M., Comer, A.R., Allen-Hoffmann, B.L., StrataGraft Skin Substitute Is Well-Tolerated and Is Not Acutely Immunogenic in Patients With Traumatic Wounds: Results From a Prospective, Randomized, Controlled Dose Escalation Trial. *Ann. Surg.*, *253*, 672–683, 2011.
73. Musick, J.R., States, T., Bunch, R., Treatment of Medical Conditions by Stem Cell Transplants and Stem Cell Activation, US20220389377A1, December 8, 2022.
74. Yamanaka, S. and Okita, K., Method of Making Induced Pluripotent Stem Cells, US9499797B2, November 22, 2016.

75. Wei, L., Yan, W., Shah, W., Zhang, Z., Wang, M., Liu, B., Xue, Z., Cao, Y., Hou, X., Zhang, K., Yan, B., Wang, X., Advancements and Challenges in Stem Cell Transplantation for Regenerative Medicine. *Heliyon, 10*, 1–21, e35836, 2024.
76. Rodriguez, C., Porcello, A., Chemali, M., Raffoul, W., Marques, C., Scaletta, C., Lourenço, K., Abdel-Sayed, P., Applegate, L.A., Pelissier Vatter, F., Laurent, A., Medicalized Aesthetic Uses of Exosomes and Cell Culture-Conditioned Media: Opening an Advanced Care Era for Biologically Inspired Cutaneous Prejuvenation and Rejuvenation. *Cosmetics, 11*, 154, 2024.
77. Zakrzewski, W., Dobrzyński, M., Szymonowicz, M., Rybak, Z., Stem Cells: Past, Present, and Future. *Stem Cell Res. Ther., 10*, 68, 2019.
78. Saito, S., Ku, C.-C., Wuputra, K., Pan, J.-B., Lin, C.-S., Lin, Y.-C., Wu, D.-C., Yokoyama, K.K., Biomarkers of Cancer Stem Cells for Experimental Research and Clinical Application. *JPM, 12*, 715, 2022.
79. Kelly, S.H., Shores, L.S., Votaw, N.L., Collier, J.H., Biomaterial Strategies for Generating Therapeutic Immune Responses. *Adv. Drug Deliv. Rev., 114*, 3–18, 2017.
80. Mathiasen, A.B. and Kastrup, J., Non-Invasive *in-Vivo* Imaging of Stem Cells after Transplantation in Cardiovascular Tissue. *Theranostics, 3*, 561–572, 2013.
81. Conte, M.S., Geraghty, P.J., Bradbury, A.W., Hevelone, N.D., Lipsitz, S.R., Moneta, G.L., Nehler, M.R., Powell, R.J., Sidawy, A.N., Suggested Objective Performance Goals and Clinical Trial Design for Evaluating Catheter-Based Treatment of Critical Limb Ischemia. *J. Vasc. Surg., 50*, 1462–1473.e3, 2009.
82. George, B., Regulations and Guidelines Governing Stem Cell Based Products: Clinical Considerations. *Perspect. Clin. Res., 2*, 94–99, 2011.

3

Stem Cell Therapy: Significant Option for Psoriasis Management

Joy Das[1], Utpal Bhui[2], Saistha Anjum[3], Sumel Ashique[4], Mohammad Muztaba[5] and Sanjay Dey[6*]

[1]School of Pharmaceutical Sciences, Lovely Professional University, Phagwara, Punjab, India
[2]NSHM Institute of Pharmaceutical Technology, NSHM Knowledge Campus, Durgapur, West Bengal, India
[3]Department of Pharmacology, Al-Ameen College of Pharmacy, Karnataka, India
[4]Department of Pharmaceutical Technology, Bharat Technology, Uluberia, West Bengal, India
[5]Department of Pharmacology, Praduman Singh SPS Pharmacy College, Phutahiya Sansarpur, Basti, Uttar Pradesh, India
[6]Department of Pharmaceutical Technology, School of Health & Medical Sciences, Adamas University, Kolkata, West Bengal, India

Abstract
Psoriasis is a chronic inflammatory skin condition characterized by the presence of white scales as well as erythematous plaques and distinct areas of redness on the skin. It affects approximately 125 million people around the world, or about 2% to 3% of the population. Particularly, its onset is influenced by genetic, environmental, and ethnic factors. Biological therapies against interleukin 17 (IL-17) and IL-23p19 have revolutionized treatment of moderate to severe psoriasis, but these therapies are not curative. Moreover, the traditional psoriasis treatments have not worked for all patients with severe symptoms. They need more effective treatment options such as stem cell therapy. Hence, stem cell therapy now represents an attractive treatment choice for psoriasis patients who need new options to manage their severe symptoms. In this chapter, we examine several stem cell–based treatments, including hematopoietic stem cell transplantation, mesenchymal stem cells, and regulatory T cells. In addition, this chapter also examines current

**Corresponding author*: sanju1980dey@gmail.com; ORCID: 0000-0003-3412-0417

Sumel Ashique, Sanjay Dey, Biplab Debnath, Mohhammad Ramzan and Swarupananda Mukherjee (eds.) Stem Cell Therapy in Dermatological Disorders, (43–78) research while discussing ongoing clinical trials and regulatory requirements to show how stem cell therapy can change severe psoriasis treatment.

***Keywords*:** Psoriasis, erythematous, interleukins, stem cell therapy, mesenchymal stem cells, T cells

3.1 Introduction

3.1.1 Psoriasis

Psoriasis is a lifelong inflammatory condition that primarily affects the skin, but it has systemic implications [1, 2]. It is characterized by plaque, guttate, pustular, erythrodermic, and flexural psoriasis [1, 3, 4]. Among all of those classification, plaque is the most commonly diagnosed form of psoriasis [5, 6]. The conditions affect an estimated 125 million people worldwide in which around 1.52% of the UK population is affected by this [7, 8]. It is an underlying cause that involves genetic predisposition and immune system dysfunction, which primarily targets the interleukin 23 (IL-23)/T helper 17 cell (T_H17) pathway. It is primarily responsible for chronic inflammation and overproduction of skin cells [9–12]. There are several environmental factors or triggers that can further exacerbate the condition, and those factors are stress, obesity, and certain medications [13]. In addition to being a skin disease, psoriasis can have a major effect on general health. In fact, it is associated with a number of other conditions such as psoriatic arthritis, and about 30% of people with psoriasis will be affected by this condition [1, 14, 15]. But if not treated, it may cause extreme joint pain and disability [16]. In addition to that, people with psoriasis are more prone to suffering from cardiovascular diseases, diabetes, and mental disorders such as anxiety, depression, and so on. It is particularly stigmatizing to have visible lesions because otherwise emotional pain that causes stigma and social isolation often results [17–20]. Holistic care is very important for this. There are a variety of treatments for psoriasis including topical creams (i.e., corticosteroids or vitamin D analogs) or phototherapy or systemic medications [21–23]. The biological therapies that target the specific inflammatory pathways such as TNF (tumor necrosis factor), IL-17, and IL-23 are proven to help in case of moderate to severe disease. Beneficial outcomes have been achieved in these patient groups using these biological therapies [24–27]. For those who require systemic treatment, several newer drugs such as apremilast and dimethyl fumarate provide an additional options [28–30]. The advancements represent a current understanding of how complex and systemic psoriasis actually is.

It is necessary to view psoriasis as more than a skin disorder in order to provide effective care. Advanced imaging techniques such as fluorodeoxyglucose positron emission tomography/computed tomography show that inflammation occurs in psoriasis and also affects the other organs, which include the liver and heart. It has also put us at risk for other diseases such coronary artery disease and stroke [31–33]. Several studies have shown that effective treatments, especially biologics, are effective at reducing systemic inflammation and improve skin symptoms side by side [24, 34]. It highlights the importance of targeted therapies. Psoriasis not only makes skin look weird, but also it has a detrimental effect on mental health as well as the quality of life (QoL) [35]. Patients with psoriasis have physical discomforts such as itching, pain, and bleeding associated with the emotional burden of stigma, which contributes to the higher rates of depression and anxiety [36]. The proper treatment greatly improves both the physical and psychological symptoms. So, the need of such treatments involves the dermatologists, rheumatologists, and mental health professionals. Psoriasis is a result of excessive growth of keratinocytes in response to an overactive immune response at a cell level [37, 38]. Among these important participants are dendritic cells, T cells, and cytokine IL-17 that do play a key role in this scenario [38, 39]. Also, genetic cause is involved. Some other hereditary traits affect immune function and skin cell regulation, which may also be genetic [40, 41]. This shows the need for personalized treatment strategies because it involves this complicated interplay of genetics, immune dysregulation, and environmental factors.

3.1.2 Stem Cell Therapy

Stem cell therapy is a new and rapidly evolving form of medical treatment in which the cells' special ability to repair damaged and replace or regenerate destroyed tissues and organs is used. The reason that stem cells are so unique is because they can differentiate into many different kinds of cell types, and they can also self-renew [42, 103]. This explains why they serve as a good treatment for many diseases and other conditions. According to the classes, the stem cells are generally classified into two: pluripotent stem cells and multipotent stem cells [42, 103]. For instance, pluripotent stem cells can become any cell type throughout the body, whereas multipotent stem cells are a little bit more restricted and can turn into certain cell types. Cells of such type can be obtained from different sources such as adult tissues and umbilical blood or embryos, etc. Treatment of psoriasis using stem cell therapy has been proved successful. In addition, it is important in the treatment of the above condition [42, 43, 60, 103]. In addition, human mesenchymal stem cells (MSCs) are originally made from bone marrow, skin tissues, or umbilical cord blood. They are commonly used because of their specialized ability to reduce inflammation and regulate the immune system. These properties will help maintain the immune imbalance that causes psoriasis. MSCs are also offering hope for people affected with severe or treatment-resistant psoriasis, which promotes the healing process of damaged skin [42–44, 60, 103].

Researchers are also exploring the potential of MSC-derived exosomes in their relevant studies. Importantly, MSC-derived exosomes are tiny particles that help to reduce inflammation and support the skin repair of damaged skin in a less invasive way [44–47, 107]. However, stem cell therapy holds great promise for improving the treatment of psoriasis and many other diseases while it is still being researched to find out their more possibilities to contribute to treating different kinds of incurable diseases. Moreover, stem cells can potentially regenerate damaged tissues across multiple systems including the cardiovascular, musculoskeletal, and nervous systems beyond dermatology. So far, they prove promising results in treating severe vulnerable conditions such as myocardial infarction, spinal cord injuries, and neurodegenerative diseases [42, 45, 48, 103]. Despite these advancements, there are still challenges to overcome such as immune rejection, cost constraints, and the requirements for robust clinical validation [49]. All of these challenges together could not overshadow the future of stem cell therapy. Therefore, further research is now required to fully understand its long-term effects and its optimal applications. As the more advancements of stem cell therapy occur, soon it will become a better regenerative medicine as well as a better personalized medicine treatment.

3.2 Advantages of Stem Cell Therapy for Psoriasis

Stem cell therapy has been considered as a treatment for psoriasis due to the advantages when compared with conventional treatment [42, 103]. Psoriasis is a type of skin condition that causes too many skin cells to form and become inflamed [1, 2]. Corticosteroids, biologics, and phototherapy control symptoms but may not be a long-term cure and have side effects [21–24, 34]. Regenerative and immunomodulatory stem cell therapy in the form of a combined treatment has the potential to be a viable, altered course of treatment for the underlying pathology in psoriasis. The most important advantage of stem cell therapy for psoriasis is that it regulates the immune system [42, 103]. It has been shown that MSCs derived from bone marrow, adipose tissue, or umbilical cord blood are also able to inhibit the autoimmune sensitization triggered by psoriasis [43, 60, 103]. Anti-inflammatory cytokines produced by these cells decrease inflammation and help modulate immune cell activity to have an immunomodulatory effect of restoring immune system balance [10–12, 33]. Unlike conventional immunosuppressant drugs that impair immune function with concomitant risk of infection, MSCs are not immunosuppressive and possess immune regulation on the target but with little side effects. Skin regeneration is also done with stem cell therapy [45, 46, 107]. Psoriasis breaks the skin barrier and produces scales, redness, and irritation [50]. Induced pluripotent stem cells (iPSCs) and MSCs both have the ability to differentiate into skin-related cells and help stimulate tissue repair. It has been shown that stem cells stimulate collagen and elastin synthesis and maintain skin integrity and reduce lesion formation [43, 44, 50, 51]. Regeneration is potentially from stem cell therapy and is valuable in severe or psoriasis-resistant disease [52]. Second, exosomes transporting bioactive molecules such as proteins, RNA, and lipids have been used little in relation to treat patients with psoriasis, and finally, there has been little discussion about using stem cell–derived exosomes in its treatment. These exosomes help to bring together cells and offer a means of reducing inflammation and encouraging healing of the tissue. Exosome therapy is cell-free, therefore less likely to cause immune rejection, and is a safer, less invasive option for patients [53–55]. Stem cell therapy provides a long-term therapeutic effect as well. Some conventional psoriasis treatments have to be used constantly; it is difficult to stop treatment, and the symptoms will return. On the other hand, stem cell therapy has shown prolonged benefits by reprogramming the immune responses and boosting skin repair mechanisms. Stem cell therapy could offer extended relief instead of only short-term symptom management, according to some studies that have reported sustained improvements in patients some months after treatment [42, 103]. Another advantage is the potential for personalized treatment. Stem cell therapy can be tailored to individual patients, with autologous stem cell transplantation (using the patient's own cells) reducing the risk of immune rejection [56]. This approach enhances treatment compatibility and effectiveness. Although research continues, personalized stem cell therapies may become more refined, enabling better psoriasis outcomes for those who do not respond well to standard treatments [57]. Other therapies for severe psoriasis, such as radiation, are also considerably more invasive than a stem cell therapy procedure [58]. Although systemic medications come with very serious side effects, and biologic injections have to be given continuously, stem cell therapy can be simple cell infusion or injection, which makes it much more comfortable for patients [59, 60]. Delivery methods are still being optimized, for example, topical applications and targeted injections to try and give us the most effectiveness with the least amount of invasiveness. These advantages exist, but on behalf of making stem cell therapy available for psoriasis, patient treatment to the wider population has been an important challenge [42, 103]. There are hurdles of high cost, regulatory concern, and need for validation in the clinic [49]. Although scientific progress and the results of clinical trials on the efficacy and safety of stem cells are increasing, stem cell therapy can revolutionize how psoriasis is treated if it truly can be used in the long term [50, 61].

3.3 Types of Stem Cell Therapy in Psoriasis

Different stem cell therapy approaches serve as treatments for psoriasis. Through regulatory T cells (Tregs), patients can gain control over their immune responses along with reduced inflammation to better manage their psoriasis symptoms. The anti-inflammatory biologic capabilities and tissue repair properties of MSCs make these cells effective for treatment of damaged skin. When treating blood disorders, hematopoietic stem cells (HSCs) function to reset the immune system and minimize autoimmune responses. Stem cell treatments seem to offer exciting benefits for treating psoriasis symptoms combined with skin regeneration potential.

3.3.1 Regulatory T Cells

Tregs maintain immune system balance by preventing the overly inflammatory immune responses. Because Tregs with psoriasis appear to have a reduced number or function, their characteristic ability to limit the effects of effector T cells, such as T cells of the T_H17 lineage that promote inflammation, is diminished [61, 62]. It may be possible to reduce inflammation and thus psoriasis symptoms by restoring Treg function [62, 63]. Several research studies have tried to see how Treg activity can be improved to optimize psoriasis treatment. The first is to try to have an increased number of Tregs simply using adoptive cell therapy, where Tregs are isolated and those from outside the body are expanded and then reintroduced them back into the patient to help modulate the response of the immune system [64–66]. There is another type of method that involves the use of drugs or biologics to expand or activate Tregs; for example, low-dose IL-2 treatment has been shown to increase Treg numbers at the expense of reducing inflammation [67, 68]. Lastly, MSCs have been examined for their capacity to differentially promote Treg differentiation and anti-inflammatory pathways in psoriasis

[69, 70]. Cells that release important immune system–regulating molecules that indirectly support Treg function are dendritic cells, which produce cytokines such as IL-10 and TGF-β, creating an immunosuppressive microenvironment that promotes regulatory T cell stability and activity [70]. More research and clinical trials are required to understand how well Treg-based therapies might work in the long term, but targeting these cells has promise in improving psoriasis patient outcomes and in combating the disease itself.

3.3.2 Mesenchymal Stem Cells

As therapy for psoriasis, MSCs have emerged as a promising therapy because they can be used to regulate the immune responses and promote tissue repair [49, 50]. Psoriasis is a type of chronic autoimmune condition where skin cell production and inflammation are excessive [1]. Skin regeneration is supported by strong immunomodulatory properties that MSCs have of being derived from bone marrow, adipose tissue, or umbilical cord blood [42, 44, 71]. MSC therapy for psoriasis turns out to be one of the key benefits, as it suppresses overactive immune responses. MSCs suppress proinflammatory cells (T_H17 and T_H1) and support the function of suppressive T cells (Treg) that control inflammation [72, 73]. Moreover, they release the anti-inflammatory cytokines IL-10 and transforming growth factor β (TGF-β) and thereby diminish the severity of psoriatic lesions

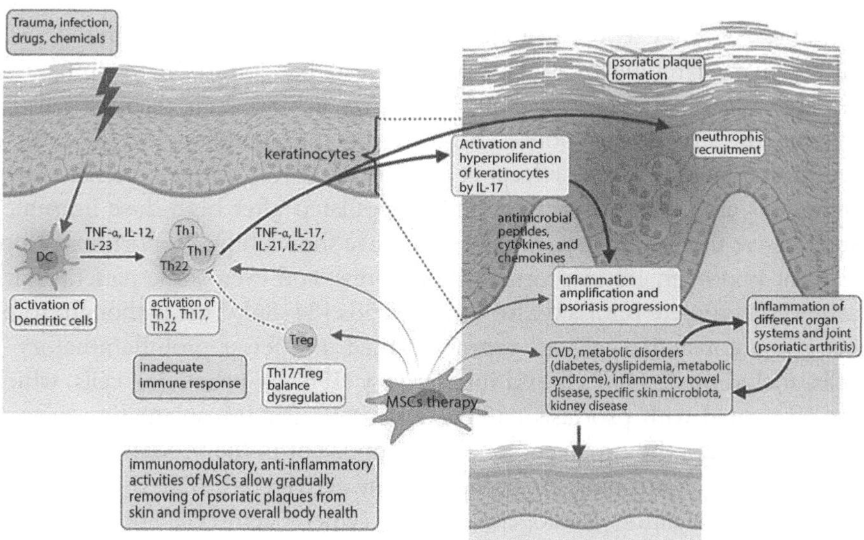

Figure 3.1 Application of MSCs in psoriasis management [78].

Figure 3.1 [56, 60]. Moreover, exosomes derived from MSCs have been further investigated and are shown to suppress T-cell proliferation *via* the p27kip1/Cdk2 signaling pathway and thus could be an attractive cell-free therapy for psoriasis [53, 54, 74]. Encouraging results of clinical trials have been shown. An adipose-derived MSC pilot study proved that these cells have therapeutic potential to improve psoriatic lesions [75]. A phase I clinical trial of subcutaneous injections of allogeneic adipose-derived MSCs was also conducted, with the treatment found to be efficacious in reducing lesion severity and a safe treatment [76]. Additionally, an animal model study showed efficient alleviation of skin inflammation by MSCs or their exosomes [77]. MSC therapy promises great benefit, but further research establishing its long-term safety and efficacy is needed. However, in cases where MSCs are highly effective at regulating immune function and directing skin repair, they also could change the management of psoriasis and thus improve the QoL for psoriasis patients.

3.3.3 Hematopoietic Stem Cells

HSCs have been considered as a therapeutic option for psoriasis given their ability to reset the immune system and reestablish the homeostatic immune balance [79]. With HSC transplantation (HSCT), various studies have shown amazingly good results with targeting the underlying immune dysfunction in psoriasis, an immune-mediated inflammatory disease [80]. As HSCs are present in the bone marrow and have the potential to generate all blood and immune cells, they represent an ideal immune system manipulation tool [81]. Although researching treatments for psoriasis has long been documented, research has also suggested that HSCT may achieve long-term benefit by altering the immune response and lowering inflammation [82, 83]. There are several studies on the effects of autologous HSCT where a patient's own stem cells are collected, treated, and repopulated after high-dose immunosuppressive therapy. In fact, clinical trials have shown high proportions of significant improvement in psoriatic skin lesions, and even some patients have achieved complete remission for long periods [84–86]. This is thought to be because autoreactive cells are reprogrammed into fewer proinflammatory T cells, including T_H17, T_H1, and interferon γ (IFN-γ)–producing cells, which are associated with psoriasis pathogenesis [87]. The transplantation process also encourages the expansion of Tregs, important in the process of immune tolerance and controlling excessive immune activity [61, 62]. There has also been an investigation of allogeneic HSCT (where stem cells are taken from a healthy donor) to treat severe and refractory psoriasis. Reports of studies have indicated that patients who undergo allogeneic HSCT have decreased inflammatory cytokines, such as TNF-α, IL-17, and IFN-γ, which are all heavily implicated in the disease process [88, 89]. These cytokines were suppressed, and there was decreased keratinocyte hyperproliferation and reduced inflammation in psoriatic plaques [90]. Also, several cases with long-term disease control after transplantation showed that donor-derived immune cells can help maintain immune stability in affected individuals [91, 92].

Further studies with animal models have also confirmed the role of HSCs in treating the condition of psoriasis [93, 94]. Bone marrow–derived HSCT produced a very significant effect on disease severity in murine models of psoriasis [95, 96]. And the transplanted cells helped with immune reconstitution and reduced the activation of immune cells that lead to excessive inflammation in humans [97]. HSCT also stimulated the production of critical immune regulatory cytokines such as IL-10 and TGF-β [98]. However, despite promising results, HSCT for the treatment of psoriasis is not widely used. As such, the procedure has potential risks, including graft versus host disease, infection, and prolonged immune suppression. Nevertheless, improvements in transplantation protocols, selection of better patients, and conditioning regimens have limited some of these risks [99, 100]. In addition, newer approaches, including the use of exosomes from HSC-derived products, are being explored as less invasive alternatives to traditional transplantation [81]. Bioactive molecules that modulate immune function and decrease inflammation are present in these exosomes without the demand for full cell transplantation [101]. Although HSCT is not yet a standard of care for psoriasis, ongoing work is continued to determine the safety and efficacy of HSCT. In clinical trials, the optimal dosing strategy, clinical trials patient eligibility criteria, and long-term outcomes of this approach are being investigated. If further studies confirm its benefits and safety, HSCT could become a valuable treatment option to relieve psoriasis in severely affected or treatment-resistant patients for many years [82].

3.4 Therapeutic Value of Stem Cell

Stem cell therapy presents itself as a promising therapeutic approach for psoriasis treatment because it controls immune responses and maintains tissue repair functions while restoring skin equilibrium [102]. Studies indicate that stem cells have robust immunomodulatory effects that help control the extreme inflammation that characterizes psoriasis. Stem cells achieve effective reduction of psoriatic lesions by inhibiting proinflammatory cytokines including TNF-α, IFN-γ, and IL-17. These therapeutic agents promote the rise in anti-inflammatory cytokines IL-10 and TGF-β to help control the immune system as well as sustain the disease outcome over time [103]. Stem cell therapy stands out because it repairs harmed skin tissues as a primary therapeutic benefit. The healing process occurs when stem cells develop into keratinocytes and fibroblasts to regenerate skin tissue according to experimental findings [104]. Research indicates that exosomes derived from stem cells reveal potential as psoriasis treatments because they contain bioactive molecules, including growth factors, proteins, and microRNAs that decrease keratinocyte hyperproliferation [105]. Exosomes act as an effective therapeutic approach, which provides substantial medical benefits by using a procedure that excludes cell transplantation and its associated safety hazards [106]. Multiple extensive research studies have established the credibility of stem cell therapy as a treatment approach for psoriasis. The impact of MSC treatment was examined on patients having moderate to severe psoriasis in a clinical study. The clinical infusion of MSC cells resulted in important decreases of Psoriasis Area and Severity Index scores, and these benefits were sustained for months [107]. The research shows MSCs downgrade immune system activity through cytokine suppression of TNF-α along with IL-17 and IL-23, which function as key agents in psoriasis development [108]. Laboratory experiments have reinforced the effectiveness of exosomes produced by MSCs. Research using imiquimod-induced psoriasis mice showed that exosomes from MSCs produced substantial reductions in both inflammation and cell proliferation symptoms. The treatment functioned through blocking the activation process of T_H17 and T_H1 cells alongside activating Tregs [109]. The scientific evidence demonstrates that exosomes derived from MSC cells present a safe cellular-free replacement for traditional stem cell transplant procedures. For treating severe psoriasis, doctors investigate the method of using HSCT as a therapeutic approach. Peripheral stem cell transplantation with autologous cells has led to sustained symptom absence in certain patients according to clinical study findings with remission periods reaching several years [110]. The procedure of HSCT enables a resetting of immune system function by deleting autoreactive T cells and promoting immune tolerance, which minimizes psoriasis recurrence. Research support shows that adipose-derived stem cells (ADSCs) may effectively treat psoriasis. Studies showed that ADSC subcutaneous delivery into psoriatic lesions resulted in major skin inflammation reduction besides minimizing epidermal thickness [111]. Strong anti-inflammatory properties present in ADSCs were responsible for repairing immune equilibrium and aiding tissue recovery. Table 3.1 describes several *in vivo* and *in vitro* model for stem cell therapy [115–127], and Table 3.2 depicts ongoing clinical trial investigation of stem cells against psoriasis [128–136].

Table 3.1 *In vivo* and *in vitro* model for stem cell therapy.

Model type	Model name	Significance	References
In vivo	Imiquimod (IMQ)-induced psoriasis mouse model	Mimics human psoriasis pathology, including epidermal hyperplasia and immune activation	[112]
In vivo	Xenograft human psoriatic skin transplant on immunodeficient mice	Allows direct testing of stem cell therapy on human psoriatic tissue	[113]
In vivo	K14-VEGF transgenic mouse model	Stimulates angiogenesis and inflammation in psoriasis, useful for testing mesenchymal stem cells (MSCs)	[114]
In vitro	Keratinocytes–fibroblast coculture system	Useful for studying epidermal stem cell therapy effects	[115]
In vitro	3D skin organotypic model	Mimic skin architecture and psoriasis pathophysiology for evaluating stem cell–based treatment	[116]
In vitro	Induced pluripotent stem cell–derived keratinocytes	Provides a patient-specific model to study genetic and molecular mechanism	[117]
In vitro	MSC-conditioned media model	Uses MSC-secreted factors to study their immunomodulatory effects on psoriatic keratinocytes	[118]

(*Continued*)

Table 3.1 *In vivo* and *in vitro* model for stem cell therapy. (*Continued*)

Model type	Model name	Significance	References
In vitro	T-cell and keratinocyte coculture model	Investigates the immunological mechanisms of psoriasis and how stem cell therapy modulates inflammatory pathway	[119]
In vitro	Organ-on-chip psoriasis model	Integrates stem cell therapy testing in a microfluidic system, closely mimicking *in vivo* conditions	[120]
In vivo	CD18 hypomorphic mouse model	Displays chronic psoriatic-like skin inflammation due to immune dysregulation, useful for testing stem cell immunomodulation	[121]
In vivo	SCID mouse engrafted with human immune cells	Humanized mouse model that replicates psoriasis pathogenesis, enabling stem cell therapy evaluation in a human immune system context	[122]
In vivo	HLA-B27 transgenic rat model	Exhibits psoriasis-like skin lesions, providing an alternative species for stem cell therapy evaluation	[123]
In vivo	Agr2-deficient mouse model	Mimics IL-17/IL-23 axis dysregulation, a key pathway in psoriasis, making it useful for testing stem cell–based immunotherapies	[124]

Table 3.2 Recent studies on stem cell therapy in psoriasis management.

Study number	Type of study	Sample size	Route of administration	Type of stem cells	Therapeutic advantages	Therapeutic disadvantages	References
1	Single-center, open-label pilot study	7 Patients	Intravenous	Mesenchymal stem cells (MSCs)	The results indicate that intravenous injection of AD-MSCs is safe and holds potential as a therapeutic option for treating psoriasis	No serious adverse events (SAEs) were found except transient fever, headache and pharyngitis	[125]
2	Phase I open-label study	10 healthy adults	Topical	MSCs	The result indicates that the MSC exosome ointment was well tolerated for topical use	NA	[126]
3	Phase I clinical study	5 patients (3 male/2 female)	Subcutaneous	Mesenchymal stromal cells (MSCs)	Adipose-derived stem cell injection could be considered as a safe and effective therapeutic approach for psoriatic plaques	No SAEs were found except burning, pain, itching	[76]

(Continued)

Table 3.2 Recent studies on stem cell therapy in psoriasis management. *(Continued)*

Study number	Type of study	Sample size	Route of administration	Type of stem cells	Therapeutic advantages	Therapeutic disadvantages	References
4	Phase I/II clinical trial	12 patients	Intradermal (psoriasis plaque)	MSCs	The study concludes that exosomes are safe for early-phase psoriasis treatment, with optimal efficacy at lower doses, as they help reduce immune cell infiltration and alleviate symptoms	NA	[127]
5	Case report	2 patients (1 male/1 female)	Intravenous	MSCs from subcutaneous abdominal fat	No SAEs were noted for either patient as a result of MSC infusions	NA	[128]

(Continued)

Table 3.2 Recent studies on stem cell therapy in psoriasis management. (*Continued*)

Study number	Type of study	Sample size	Route of administration	Type of stem cells	Therapeutic advantages	Therapeutic disadvantages	References
6	Case report	47-year old man	Intravenous	Minimally manipulated umbilical cord–derived MSCs	Psoriasis, unresponsive to treatments for years, completely resolved after three rounds of minimally manipulated umbilical cord–derived MSCs, with no recurrence. The patient reported having experienced no adverse reactions or side effects, including dermatitis, fever, chills, or nausea.	NA	[129]

(*Continued*)

Table 3.2 Recent studies on stem cell therapy in psoriasis management. (*Continued*)

Study number	Type of study	Sample size	Route of administration	Type of stem cells	Therapeutic advantages	Therapeutic disadvantages	References
7	Case report	19-year-old male college student	Normal saline prior to bolus injection	Mesenchymal stromal cells (MSCs)	Allogeneic human gingival MSCs demonstrated a safe and effective treatment for severe, treatment-resistant plaque psoriasis, leading to complete regression without adverse effects and sustained remission over three years	NA	[130]
8	Case report	38-year-old male patient	Topical	MSC conditioned media (MSC-CM)	MSC-CM shows promise as an effective alternative therapy for psoriasis, potentially overcoming the limitations of cell-based treatment	NA	[131]

(*Continued*)

Table 3.2 Recent studies on stem cell therapy in psoriasis management. (*Continued*)

Study number	Type of study	Sample size	Route of administration	Type of stem cells	Therapeutic advantages	Therapeutic disadvantages	References
9	Case report	43-year-old male patient	Intravenous	Stromal vascular fraction (a mixture of adipose-derived stem cells/MSCs, endothelial/ progenitors, pericytes, fibroblasts, and other cells obtained from fat tissue)	This case suggests that MSC treatment may be a feasible and effective option for improving psoriasis and patient quality of life	NA	[132]

3.5 Challenges in Cost, Availability, and Clinical Application

Blood system diseases have traditionally used stem cell transplants, benefiting from the self-renewal, multipoint differentiation, and proliferation abilities of stem cells. Transplanting stem cells is more feasible for blood system diseases than for other diseases, such as solid tumors. Blood system diseases treated with transplantable stem cells include hematological diseases such as acute myeloid leukemia, non-Hodgkin lymphoma, and chronic lymphocytic leukemia. There is a risk of transplant-associated complications in any transplant, including risks of donor rejection or complications of immunosuppressive drug therapies [133]. Stem cell therapy constitutes a multifaceted service that is delivered by clinics, companies that supplies the equipment and materials for stem cell culture, insurers, and other related services. Furthermore, regulatory bodies are actively involved in every stage of the process, from overseeing initial research to ensuring the safe and effective translation of findings into clinical applications. The application of stem cell knowledge in the clinical space is happening at a very quick pace in order to treat life-threatening diseases. The number of businesses providing services in regenerative medicine (including stem cell therapy) worldwide increased from 772 in 2016 to more than 1550 by 2024. Over the past 50 years, more than 1,000,000 HSC transplants have been performed on patients suffering from hematologic malignancies. These therapies are routinely used to cure the likes of situations such as leukemias, aplastic anemia, thalassemia, sickle cell anemia, and hereditary metabolic disorders. CAGR stands for Compound Annual Growth Rate. It is a measure used to describe the mean annual growth rate of an investment, revenue, or other financial metric over a specified time period longer than one year, assuming the value grows at a steady rate each year. Stem cell treatment per patient is estimated to cost between US $10,000 and $60,000. The treatment of autoimmune disorders and some other segments such as wound treatment and cosmetics are expected to be the fastest-growing segment in the stem cell therapy markets, achieving a CAGR of 22.1% from 2021 to 2026 [134]. With the increasing number of clinical trials that use stem cells as therapeutic approaches, guidelines and standards for their regulatory implementation have become a matter of utmost urgency to ensure the safety of patients. Nonetheless, being a relatively new field, stem cell therapy has recently encountered the controversies of science, ethics, and law that all await regulation [135].

Although stem cell therapies have made significant progress in their availability and clinical application, challenges remain to realizing widespread access and effective treatment outcomes. Legislative obstacles continue to impede this area, with many countries putting in place stringent approval mechanisms for stem cell–based treatments that slow the translation of these therapies into the clinical [136]. In an even greater problem, owing to this variability in treatment response across different medical centers, an absence of standardized protocols for medication administration on the one hand and the spectrum of conditions that stem cell therapies can potentially address on the other hand have continued to hinder clinical usage of human stem cells further [137]. Safety issues in the use of stem cells in clinical trials continue to raise concerns regarding immune rejection, tumor formation, and long-term efficacy of therapy [138]. In addition to that, the availability of quality-ensured stem cells is quite limited, especially in low-resource settings where specialized facilities for cell culture and differentiation may not be available [139]. The ethical concerns in the use of embryonic stem cells continue to exist, and these issues affect their global accessibility and the prevailing tendency of patients to accept this treatment modality [140]. The economic viability of stem cell therapies is another impediment, as the cost of personalized treatment procedures, especially autologous therapies, may be extremely high [141]. Insurance coverage for stem cell therapies in high-resource settings is often limited, and reimbursement is rarely available, reducing accessibility for patients [142]. Another factor aggravating the situation is the absence of a harmonious global regulatory framework, which results in uneven access to stem cell therapies in different regions [143]. With adequate regulation and standardized treatment protocols, advances in stem cell technology offer promising prospects to resolve many of these aforementioned challenges [144]. Nonetheless, some of these hurdles are already receiving significant attention, and with that, research and innovation in stem-cell biology and clinical applications continually expand the horizons for regenerative medicine [145].

Stem cell therapy shows great potential to treat various diseases, but big hurdles still stand in the way of its widespread use in clinics. The ethical debate about using embryonic stem cells tops the list of challenges pushing scientists to look at other options such as iPSCs [146]. It is also crucial to keep stem cells stable during growth and expansion to avoid mutations that might cause tumors [147]. Making stem cell products is another worry, as differences can change how well the therapy works [148]. The chance of immune rejection remains a roadblock especially when using stem cells from others, so we need ways to make the immune system accept them or use the patient's own cells [149]. On top of that, getting enough stem cells for treatments costs a lot and takes time, which raises questions about scaling up and making them available to everyone [150].

Table 3.3 Ongoing clinical trials of cell therapy for psoriasis.

NCT number	Study title	Study URL	Study status	Conventions	Interventions	Sponsor	Collaborators	Study type
NCT03765957	Clinical Research on Treatment of Psoriasis by Human Umbilical Cord-Derived Mesenchymal Stem Cells	https://clinicaltrials.gov/study/NCT03765957	Unknown	Psoriasis	Biological: mesenchymal stem cells (MSCs)	Central South University	Central South University	Interventional
NCT02918123	Safety of FURESTEM-CD Inj. in Patients With Moderate to Severe Plaque-Type Psoriasis	https://clinicaltrials.gov/study/NCT02918123	Unknown	Psoriasis	Biological: FURESTEM-CD Inj.	Kang Stem Biotech Co., Ltd.	Kang Stem Biotech Co., Ltd.	Interventional
NCT03745417	Efficacy and Safety of Expanded UCMSCs on Patients With Moderate to Severe Psoriasis	https://clinicaltrials.gov/study/NCT03745417	Unknown	Mesenchymal stromal cells/ psoriasis/ drug effect/ drug toxicity	Biological: Umbilical cord mesenchymal stem cells	Guangdong Provincial Hospital of Traditional Chinese Medicine	Peking Union Medical College Hospital	Interventional [158]
NCT05858645	Correction of Psoriatic T Cell Signatures by Deucravacitinib	https://clinicaltrials.ucsf.edu/trial/NCT05858645	RECRUITING	Psoriasis		University of California, San Francisco	University of California, San Francisco	Interventional
NCT04645355	Guselkumab Immunogenetics	https://clinicaltrials.ucsf.edu/trial/NCT04645355	Unknown	Psoriasis and guttate psoriasis		University of California, San Francisco	University of California, San Francisco	Interventional
NCT05858632	Immune Spatial Features of Guselkumab Cutaneous Response	https://clinicaltrials.ucsf.edu/trial/NCT05858632	Unknown	Psoriasis		University of California, San Francisco	University of California, San Francisco	Interventional
NCT05858632	Immune Spatial Features of Guselkumab Cutaneous Response	https://clinicaltrials.ucsf.edu/trial/NCT05858632	Unknown	Psoriasis		University of California, San Francisco	University of California, San Francisco	Interventional

(Continued)

Table 3.3 Ongoing clinical trials of cell therapy for psoriasis. (*Continued*)

NCT number	Study title	Study URL	Study status	Conventions	Interventions	Sponsor	Collaborators	Study type
NCT05390515	Psoriatic Immune Response to Tildrakizumab	https://clinicaltrials.ucsf.edu/trial/NCT05390515	Unknown	Psoriasis		University of California, San Francisco	University of California, San Francisco	Interventional
NCT05102474	Dermatologic Research Evaluation and Monitoring of Sleep in Moderate-to-Severe Psoriasis	https://clinicaltrials.ucsf.edu/trial/NCT05102474	Unknown	Psoriasis		University of California, San Francisco	University of California, San Francisco	OBSERVATIONAL
NCT06220604	A Study of JNJ-77242113 for the Treatment of Participants With Moderate to Severe Plaque Psoriasis (ICONIC-ADVANCE 2)	https://clinicaltrials.ucsf.edu/trial/NCT06220604	Unknown	Psoriasis		University of California, San Francisco	University of California, San Francisco	Interventional
NCT03662685	Immunogenetic Profiling of Goeckerman Therapy in the Treatment of Psoriasis Vulgaris	https://clinicaltrials.ucsf.edu/trial/NCT03662685	Unknown	Psoriasis		University of California, San Francisco	University of California, San Francisco	Interventional
NCT05272150	Study of Guselkumab in Skin of Color Participants With Moderate-to-Severe Plaque and/or Scalp Psoriasis	https://clinicaltrials.ucsf.edu/trial/NCT05272150	Unknown	Psoriasis		University of California, San Francisco	University of California, San Francisco	Interventional [159]

Stem cell–based treatments also face tough rules, as different global standards and guidelines slow down the creation and approval of new therapies Table 3.3 [151]. Besides technical and regulatory hurdles, money problems often hold back investment in stem cell research in places with few resources [152]. Not enough clinical proof about the long-term safety and how well stem cell treatments work also makes people slow to use these therapies [153]. Also, we need standard methods to make sure we can repeat and turn stem cells into the tissue types we want [154]. The chance of tumors forming stays a big worry, so we need careful screening and long-term check-ups for patients treated with stem cells [155]. In poorer countries, problems with logistics and infrastructure, including not having enough trained people, also slow down progress in stem cell therapies [156]. How much people accept stem cell therapy differs a lot in different places often because of cultural and religious beliefs, which can slow down the widespread use of these treatments [157].

3.6 Regulatory Aspects of Stem Cell Therapy

Stem cell therapy shows great potential to treat various diseases, but rules and regulations have a big impact on its widespread use in clinics. These rules make sure stem cell–based products are safe, work well, and are made, while also stopping possible misuse in the market. Rules differ a lot between regions, which makes it hard to keep things consistent and work together across countries in stem cell research. In the United States, the Food and Drug Administration (FDA) has a key part to play in controlling stem cell therapies. The FDA says stem cell–based products must go through tough laboratory tests and trials with people to prove they are safe and effective before they can be sold [160, 161]. In Europe, it is similar. The European Medicines Agency (EMA) controls stem cell–based products as advanced therapy medicinal products. The EMA has strict standards for quality, safety, and how well they work before they can be sold [162]. Nevertheless, even with these established frameworks, regulatory issues persist, particularly in nations with underdeveloped regulatory systems.

In India, the Central Drugs Standard Control Organization is responsible for overseeing stem cell therapies. Yet, the absence of clear, enforceable guidelines and the inconsistent enforcement of regulations have resulted in the rise of unregulated stem cell clinics offering treatments that lack scientific validation [163, 164]. The necessity for strong and enforceable regulations to shield patients from unverified or hazardous treatments is underscored by these issues. Conversely, China's regulatory environment is known for its quicker approval processes. Although this speed facilitates faster access to new treatments, it has also sparked concerns about patient safety due to a lack of adequate regulatory oversight in certain areas [165]. Ethical issues related to stem cell research present another challenge that intersects with regulatory systems. Different countries have varying stances on the ethical aspects of using embryonic stem cells [166]. Although some regions permit research and therapies involving these cells, others, such as the United States, emphasize alternatives such as adult stem cells or iPSCs. The integration of gene-editing technologies, such as CRISPR, with stem cell therapy adds layers of complexity to international regulatory efforts.

Although gene editing holds considerable promise for therapeutic applications, its regulation is inconsistent worldwide, with differing standards regarding safety, effectiveness, and ethical issues [167, 168]. Harmonizing international regulations is crucial for promoting global cooperation in stem cell research and guaranteeing the safe development and administration of therapies. The International Society for Stem Cell Research has championed the establishment of standardized guidelines to ensure that stem cell therapies are subjected to thorough clinical trials and adhere to ethical standards internationally [169]. Despite ongoing efforts to synchronize regulatory practices, a considerable disparity remains between nations with robust regulatory frameworks and those lacking such systems, highlighting the need for continued dialogue and collaboration among global stakeholders.

In summary, although there have been advancements in regulatory frameworks for stem cell therapies, challenges remain, including inconsistent implementation, ethical concerns, and differing standards across various regions. To safeguard patient safety and foster scientific progress, it is crucial to maintain international collaboration and establish unified guidelines for the future of stem cell therapy [170].

3.7 Conclusion and Future Directions

There is a myriad of reasons why psoriasis stands as one of the most troublesome diseases to manage; it not only greatly affects the QoL, but also comes with a range of other inflammatory skin diseases as well. The use of stem cells, on the other hand, has enabled the potential for developing advanced safe and efficient treatment for psoriatic patients. Currently, cell therapy with adipose-derived MSCs may provide a safer approach for patients suffering from psoriasis. There is still a lot more to learn, but at this foray, MSCs stand as a hopeful psoriasis treatment option. With an increase in the number of patients suffering from psoriasis who are refractory to modern biological and nonbiological systemic therapies and with a decrease in new pharmacological approaches emerging to the disease, there is a need to explore more novel holistic and advanced therapeutic options.

Among all the interest in advanced therapeutics, it is worth noting that cell therapy is not necessarily the only available option, although perhaps the most effective approach, for the management, albeit not the cure, of severe psoriasis. Psoriasis can also be managed with medically assisted lifestyle changes and other forms of time-modulated intervention to halt or reverse the disease's progression, offering the possibility of long-term remission for addressing the psychosocial impact of the disease. Of the various treatment options available to immune modulator, advanced cell therapy is the only one that has the capability of completely curing the disease. On the contrary, psoriasis presents some significant considerations regarding the risk–benefit ratio of psoriasis treatment cost and treatment complexity, which includes important regulatory barriers wherein "cells" in this case are treated as drugs. A safer and more appealing approach to psoriasis treatment is the use of MSCs as a form of cell therapy.

Other cell-based therapies using Tregs, fibroblasts, Muse cells, and iPSCs need consideration too. There is a bright future for psoriatic treatment with the development of new biologic agents targeting specific pathways, including IL-23 inhibitor mirikizumab. Other biologic agents are being developed as RORγt inhibitors—reflecting a whole new way of thinking about the treatment efficacy and control of disease. Advances in topical treatments, especially microneedles and nanoparticle drug delivery, can improve drug accessibility and ultimately increase the efficacy for psoriatic plaques. In particular, these advances will lead to new therapies that may be more targeted and effective, resulting in fewer adverse effects and ultimately improving patient compliance.

Identifying and validating biomarkers with multiomics will help facilitate the steps to personalized treatment. Biomarkers can aid in predicting treatment response, measure disease activity, and make therapeutic decisions to help lead toward more personalized treatment strategies. Although these advances all have potential and promise, collaboration across disciplines, including research, clinical, and industry, is critical to bridging the gap between discovery and clinical practice.

AI disclosure statement: The author confirms that no artificial intelligence (AI)-generated content or AI-assisted writing tools were used in the preparation of this chapter (including summaries and the abstract). All text, analysis, and conclusions are the result of the author's original work.

References

1. Korman, N.J., Management of psoriasis as a systemic disease: what is the evidence? *Br. J. Dermatol.*, 182, 4, 840–8, 2020.
2. Yan, B.X., Chen, X.Y., Ye, L.R., Chen, J.Q., Zheng, M., Man, X.Y., Cutaneous and Systemic Psoriasis: Classifications and Classification for the Distinction. *Front. Med.*, 8, October, 1–9, 2021.
3. Orzan, O.A., Tutunaru, C.V., Ianoşi, S.L., Understanding the Intricate Pathophysiology of Psoriasis and Related Skin Disorders. *Int. J. Mol. Sci.*, 26, 2, 1–26, 2025.
4. Tashiro, T. and Sawada, Y., Psoriasis and Systemic Inflammatory Disorders. *Int. J. Mol. Sci.*, 23, 8, 4457, 2022.
5. Dairov, A., Issabekova, A., Sekenova, A., Shakhatbayev, M., Ogay, V., Prevalence, incidence, gender and age distribution, and economic burden of psoriasis worldwide and in Kazakhstan. *J. Clin. Med. Kaz.*, 21, 2, 18–30, 2024.
6. Frysz, M., Patel, S., Li, M.O.Y., Griffiths, C.E.M., Warren, R.B., Ashcroft, D.M., Prevalence, incidence, mortality and healthcare resource use for generalized pustular psoriasis, palmoplantar pustulosis and plaque psoriasis in England: a population-based cohort study. *Br. J. Dermatol.*, 4, 529–38, May, 2024.
7. Zhang, Y., Dong, S., Ma, Y., Mou, Y., Burden of psoriasis in young adults worldwide from the global burden of disease study 2019. *Front. Endocrinol. (Lausanne)*, 15, February, 1–11, 2024.
8. Damiani, G., Bragazzi, N.L., Karimkhani Aksut, C., Wu, D., Alicandro, G., McGonagle, D., *et al.*, The Global, Regional, and National Burden of Psoriasis: Results and Insights From the Global Burden of Disease 2019 Study. *Front. Med.*, 8, 743180, December, 2021.
9. Harden, J.L., Krueger, J.G., Bowcock, A.M., The immunogenetics of Psoriasis: A comprehensive review. *J. Autoimmun.*, 64, 66–73, 2015.
10. Dhabale, A. and Nagpure, S., Types of Psoriasis and Their Effects on the Immune System. *Cureus*, 14, 9, 9, 2022.
11. Bugaut, H. and Aractingi, S., Major Role of the IL17/23 Axis in Psoriasis Supports the Development of New Targeted Therapies. *Front. Immunol.*, 12, February, 1–8, 2021.
12. Vecellio, M., Hake, V.X., Davidson, C., Carena, M.C., Wordsworth, B.P., Selmi, C., The IL-17/IL-23 Axis and Its Genetic Contribution to Psoriatic Arthritis. *Front. Immunol.*, 11, January, 1–10, 2021.

13. Kamiya, K., Kishimoto, M., Sugai, J., Komine, M., Ohtsuki, M., Risk factors for the development of psoriasis. *Int. J. Mol. Sci.*, 20, 18, 1–14, 2019.
14. Winchester, R. and FitzGerald, O., The many faces of psoriatic arthritis: Their genetic determinism. *Rheumatol. (United Kingdom)*, 59, I4–9, 2020.
15. Mease, P.J., Husni, M.E., Siegel, E., Furst, D.E., Mcilraith, M., Strand, V., *et al.*, What Matters in Psoriatic Arthritis: A Comparison of Patient and Clinician Perspectives. *ACR Open Rheumatol.*, 7, 1, 1–10, 2025.
16. Sepehri, N.Z., Raeisi, T., Razi, B., Janmohammadi, P., Darand, M., Alizadeh, S., The association between psoriasis and psoriatic arthritis with the risk of osteoporosis, osteopenia and bone fractures: A systematic review and meta-analysis. *Int. J. Clin. Pract.*, 75, 11, 1–15, 2021.
17. Song, Y., Yao, L., Li, S., Zhou, J., Psoriasis comorbidity management in the COVID era: a pressing challenge. *Front. Microbiol.*, 14, November, 1–8, 2023.
18. Daugaard, C., Iversen, L., Hjuler, K.F., Comorbidity in Adult Psoriasis: Considerations for the Clinician. *Psoriasis Targets Ther.*, 12, March, 139–50, 2022.
19. Almenara-Blasco, M., Gracia-Cazaña, T., Poblador-Plou, B., Laguna-Berna, C., Carmona-Pírez, J., Navarro-Bielsa, A., *et al.*, Multimorbidity of Psoriasis: A Large-Scale Population Study of Its Associated Comorbidities. *J. Clin. Med.*, 13, 2, 492, 2024.
20. Yamazaki, F., Psoriasis: Comorbidities. *J. Dermatol.*, 48, 6, 732–40, 2021.
21. Lee, H.J. and Kim, M., Challenges and Future Trends in the Treatment of Psoriasis. *Int. J. Mol. Sci.*, 24, 17, 13313, 2023.
22. Gao, Y., Xu, T., Wang, Y., Hu, Y., Yin, S., Qin, Z., *et al.*, Pathophysiology and Treatment of Psoriasis: From Clinical Practice to Basic Research. *Pharmaceutics*, 17, 1, 56, 2025.
23. An, P., Zhao, Q., Hao, S., Wang, X., Tian, J., Ma, Z., Recent Advancements and Trends of Topical Drug Delivery Systems in Psoriasis: A Review and Bibliometric Analysis. *Int. J. Nanomed.*, 19, July, 7631–71, 2024.
24. Brownstone, N.D., Hong, J., Mosca, M., Hadeler, E., Liao, W., Bhutani, T., *et al.*, Biologic treatments of psoriasis: An update for the clinician. *Biol. Targets Ther.*, 15, 39–51, 2021.
25. Boczar, K.E., Beanlands, R.S., Glassman, S.J., Wang, J., Zeng, W., deKemp, R.A., *et al.*, Anti-inflammatory effect of biologic therapy in patients with psoriatic disease: A prospective cohort FDG PET study. *J. Nucl. Cardiol.*, 30, 4, 1642–52, 2023.
26. Guo, J., Zhang, H., Lin, W., Lu, L., Su, J., Chen, X., Signaling pathways and targeted therapies for psoriasis. *Signal Transduct. Target. Ther.*, 8, 1, 437, 2023.
27. Wu, B., Chen, Q., Cao, R., Zhu, L., Zhu, H., Comparative effectiveness of combined biologic agents versus standard therapies in the treatment of plaque psoriasis: a retrospective analysis. *Front. Med.*, 11, September, 1–7, 2024.

28. Rajagopalan, M., Dogra, S., Saraswat, A., Varma, S., Banodkar, P., The Use of Apremilast in Psoriasis: An Indian Perspective on Real-World Scenarios. *Psoriasis Targets Ther.*, 11, 109–22, 2021.
29. Pellacani, G., Bigi, L., Parodi, A., Burlando, M., Lanna, C., Campione, E., *et al.*, Efficacy and Safety of Dimethyl Fumarate in Patients with Moderate-to-Severe Plaque Psoriasis: DIMESKIN-2, a Multicentre Single-Arm Phase IIIb Study. *J. Clin. Med.*, 11, 16, 1–15, 2022.
30. Campione, E., Mazzilli, S., Di Prete, M., Dattola, A., Cosio, T., Lettieri Barbato, D., *et al.*, The Role of Glutathione-S Transferase in Psoriasis and Associated Comorbidities and the Effect of Dimethyl Fumarate in This Pathway. *Front. Med.*, 9, February, 1–7, 2022.
31. Kleinrensink, N.J., Spierings, J., Vonkeman, H.E., Seddiqi, N., Herman, A., Suijkerbuijk, K.P.M., *et al.*, Increased vascular inflammation on PET/CT in psoriatic arthritis patients in comparison with controls. *RMD Open*, 10, 1, 1–8, 2024.
32. Man, A.M., Orăsan, M.S., Hoteiuc, O.A., Olănescu-Vaida-Voevod, M.C., Mocan, T., Inflammation and Psoriasis: A Comprehensive Review. *Int. J. Mol. Sci.*, 24, 22, 16095, 2023.
33. Kaiser, H., Kvist-Hansen, A., Krakauer, M., Gørtz, P.M., Henningsen, K.M.A., Wang, X., *et al.*, Association between vascular inflammation and inflammation in adipose tissue, spleen, and bone marrow in patients with psoriasis. *Life*, 11, 4, 305, 2021.
34. Campanati, A., Marani, A., Martina, E., Diotallevi, F., Radi, G., Offidani, A., Psoriasis as an immune-mediated and inflammatory systemic disease: From pathophysiology to novel therapeutic approaches. *Biomedicines*, 9, 11, 1511, 2021.
35. Zhong, H., Yang, H., Mao, Z., Chai, X., Li, S., Impact of moderate-to-severe psoriasis on quality of life in China: a qualitative study. *Health Qual. Life Outcomes*, 19, 1, 1–9, 2021.
36. Lee, L.L., Huo, A.P., Chen, S.L., Embodied suffering: Uncovering the illness experiences of patients with severe psoriasis. *Scand. J. Caring Sci.*, 38, 1041–9, September, 2024.
37. Afonina, I.S., Van Nuffel, E., Beyaert, R., Immune responses and therapeutic options in psoriasis. *Cell. Mol. Life Sci.*, 78, 6, 2709–27, 2021.
38. Vičić, M., Kaštelan, M., Brajac, I., Sotošek, V., Massari, L.P., Current concepts of psoriasis immunopathogenesis. *Int. J. Mol. Sci.*, 22, 21, 11574, 2021.
39. Kamata, M. and Tada, Y., Dendritic Cells and Macrophages in the Pathogenesis of Psoriasis. *Front. Immunol.*, 13, June, 1–10, 2022.
40. Babaie, F., Omraninava, M., Gorabi, A.M., Khosrojerdi, A., Aslani, S., Yazdchi, A., *et al.*, Etiopathogenesis of Psoriasis from Genetic Perspective: An updated Review. *Curr. Genomics*, 23, 3, 163–74, 2022.

41. Uppala, R., Tsoi, L.C., Harms, P.W., Wang, B., Billi, A.C., Maverakis, E., *et al.*, "Autoinflammatory psoriasis"—genetics and biology of pustular psoriasis. *Cell. Mol. Immunol.*, 18, 2, 307–17, 2021.
42. Lwin, S.M., Snowden, J.A., Griffiths, C.E.M., The promise and challenges of cell therapy for psoriasis*. *Br. J. Dermatol.*, 185, 5, 887–98, 2021.
43. Dairov, A., Sekenova, A., Alimbek, S., Nurkina, A., Shakhatbayev, M., Kumasheva, V., *et al.*, Psoriasis: The Versatility of Mesenchymal Stem Cell and Exosome Therapies. *Biomolecules*, 14, 11, 1–25, 2024.
44. Cheng, L., Wang, S., Peng, C., Zou, X., Yang, C., Mei, H., *et al.*, Human umbilical cord mesenchymal stem cells for psoriasis: a phase 1/2a, single-arm study. *Signal Transduct. Target. Ther.*, 7, 1, 263, 2022.
45. Zhang, B., Lai, R.C., Sim, W.K., Choo, A.B.H., Lane, E.B., Lim, S.K., Topical application of mesenchymal stem cell exosomes alleviates the imiquimod induced psoriasis-like inflammation. *Int. J. Mol. Sci.*, 22, 2, 1–13, 2021.
46. Chen, Y., Liu, H., He, Y., Yang, B., Lu, W., Dai, Z., Roles for Exosomes in the Pathogenesis, Drug Delivery and Therapy of Psoriasis. *Pharmaceutics*, 17, 1, 1–20, 2025.
47. Quiñones-Vico, M.I., Sanabria-de la Torre, R., Sánchez-Díaz, M., Sierra-Sánchez, A., Montero-Vílchez, T., Fernández-González, A., *et al.*, The Role of Exosomes Derived From Mesenchymal Stromal Cells in Dermatology. *Front. Cell Dev. Biol.*, 9, April, 1–18, 2021.
48. Hoang, D.M., Pham, P.T., Bach, T.Q., Ngo, A.T.L., Nguyen, Q.T., Phan, T.T.K., *et al.*, Stem cell-based therapy for human diseases. *Signal Transduct. Target. Ther.*, 7, 1, 272, 2022.
49. Zeng, C.W., Advancing Spinal Cord Injury Treatment through Stem Cell Therapy: A Comprehensive Review of Cell Types, Challenges, and Emerging Technologies in Regenerative Medicine. *Int. J. Mol. Sci.*, 24, 18, 14349, 2023.
50. Diotallevi, F., Di Vincenzo, M., Martina, E., Radi, G., Lariccia, V., Offidani, A., *et al.*, Mesenchymal Stem Cells and Psoriasis: Systematic Review. *Int. J. Mol. Sci.*, 23, 23, 15080, 2022.
51. Yu, H., Feng, H., Zeng, H., Wu, Y., Zhang, Q., Yu, J., *et al.*, Exosomes: The emerging mechanisms and potential clinical applications in dermatology. *Int. J. Biol. Sci.*, 20, 5, 1778–95, 2024.
52. Khandpur, S., Gupta, S., Gunaabalaji, D.R., Stem cell therapy in dermatology. *Indian J. Dermatol. Venereol. Leprol.*, 87, 6, 753–67, 2021.
53. Pinto, D., Giuliani, G., Mondadori, G., Rinaldi, F., Exosomes in Dermatological Research: Unveiling Their Multifaceted Role in Cellular Communication, Healing, and Disease Modulation, 12, 1, 16, 2025.
54. Tang, B., Bi, Y., Zheng, X., Yang, Y., Huang, X., Yang, K., *et al.*, The Role of Extracellular Vesicles in the Development and Treatment of Psoriasis: Narrative Review. *Pharmaceutics*, 16, 12, 1586, 2024.
55. Shi, H., Wang, M., Sun, Y., Yang, D., Xu, W., Qian, H., Exosomes: Emerging Cell-Free Based Therapeutics in Dermatologic Diseases. *Front. Cell Dev. Biol.*, 9, October, 1–17, 2021.

56. Műzes, G. and Sipos, F., Mesenchymal Stem Cell-Derived Secretome: A Potential Therapeutic Option for Autoimmune and Immune-Mediated Inflammatory Diseases. *Cells*, 11, 15, 2300, 2022.
57. Fu, Y., Feng, C., Qin, S., Xing, Z., Liu, C., Liu, Z., *et al.*, Breaking barriers: advancing cellular therapies in autoimmune disease management. *Front. Immunol.*, 15, November, 1–15, 2024.
58. Thatiparthi, A., Martin, A., Liu, J., Wu, J.J., Risk of Skin Cancer with Phototherapy in Moderate-to-Severe Psoriasis. *J. Clin. Aesthet. Dermatol.*, 15, 6, 68–75, 2022.
59. Piragine, E., Petri, D., Martelli, A., Janowska, A., Dini, V., Romanelli, M., *et al.*, Adherence and Persistence to Biological Drugs for Psoriasis: Systematic Review with Meta-Analysis. *J. Clin. Med.*, 11, 6, 1506, 2022.
60. Kuang, Y.H., Zhu, W., Lin, G., Cheng, L.M., Qin, Q., Huang, Z.J., *et al.*, Expert Consensus on the Application of Stem Cells in Psoriasis Research and Clinical Trials. *Aging Dis.*, 16, 1363, June, 2024.
61. Nussbaum, L., Chen, Y.L., Ogg, G.S., Role of regulatory T cells in psoriasis pathogenesis and treatment. *Br. J. Dermatol.*, 184, 1, 14–24, 2021.
62. Pietraforte, I. and Frasca, L., Autoreactive T-Cells in Psoriasis: Are They Spoiled Tregs and Can Therapies Restore Their Functions? *Int. J. Mol. Sci.*, 24, 5, 4348, 2023.
63. Zhong, L., Luo, N., Zhong, X., Xu, T., Hao, P., The immunoregulatory effects of natural products on psoriasis *via* its action on Th17 cells versus regulatory T cells balance. *Int. Immunopharmacol.*, 110, May, 109032, 2022.
64. Sulaimani, J., Cluxton, D., Clowry, J., Petrasca, A., Molloy, O.E., Moran, B., *et al.*, Dimethyl fumarate modulates the Treg–Th17 cell axis in patients with psoriasis*. *Br. J. Dermatol.*, 184, 3, 495–503, 2021.
65. Goswami, T.K., Singh, M., Dhawan, M., Mitra, S., Emran, T.B., AA, R., *et al.*, Regulatory T cells (Tregs) and their therapeutic potential against autoimmune disorders–Advances and challenges. *Hum. Vaccines Immunother.*, 18, 1, 2035117, 2022.
66. Younis, N., Puigmal, N., El, K.A., Badaoui, A., Zhang, D., Morales-Garay, C., *et al.*, Microneedle-Mediated Delivery of Immunomodulators Restores Immune Privilege in Hair Follicles and Reverses Immune-Mediated Alopecia. *Adv. Mater.*, 36, 31, 1–19, 2024.
67. Nadeem, A., Ahmad, S.F., Al-Harbi, N.O., Ibrahim, K.E., Alqahtani, F., As Sobeai, H.M., *et al.*, Inhibition of interleukin-2-inducible T-cell kinase causes reduction in imiquimod-induced psoriasiform inflammation through reduction of Th17 cells and enhancement of Treg cells in mice. *Biochimie*, 179, 146–56, 2020.
68. Qiao, Z., Zhao, W., Liu, Y., Feng, W., Ma, Y., Jin, H., Low-dose Interleukin-2 For Psoriasis Therapy Based on the Regulation of Th17/Treg Cell Balance in Peripheral Blood. *Inflammation*, 46, 6, 2359–73, 2023.
69. Cuesta-Gomez, N., Medina-Ruiz, L., Graham, G.J., Campbell, J.D.M., IL-6 and TGF-β-Secreting Adoptively-Transferred Murine Mesenchymal Stromal

Cells Accelerate Healing of Psoriasis-like Skin Inflammation and Upregulate IL-17A and TGF-β. *Int. J. Mol. Sci.*, 24, 12, 10132, 2023.
70. Halliday, N., Dyson, J.K., Thorburn, D., Lohse, A.W., Heneghan, M.A., Review article: experimental therapies in autoimmune hepatitis. *Aliment. Pharmacol. Ther.*, 52, 7, 1134–49, 2020.
71. Ha, D.H., Kim, H.-K., Lee, J., Kwon, H.H., Park, G.-H., Yang, S.H., *et al.*, Mesenchymal Stem/Stromal Cell-Derived Exosomes for Immunomodulatory Therapeutics and Skin Regeneration. *Cells*, 9, 1157, 1–45, 2020.
72. Zhou, X., Tang, B., Huang, Q., Yang, S., Jiang, Y., Xu, L., *et al.*, Engineered Mesenchymal Stem Cell-Derived Extracellular Vesicles Scavenge Self-Antigens for Psoriasis Therapy *via* Modulating Metabolic and Immunological Disorders. *Adv. Sci.*, 2410067, 1–18, 2024.
73. Terraza-Aguirre, C., Campos-Mora, M., Elizondo-Vega, R., Contreras-López, R.A., Luz-Crawford, P., Jorgensen, C., *et al.*, Mechanisms behind the immunoregulatory dialogue between mesenchymal stem cells and th17 cells. *Cells*, 9, 7, 1–21, 2020.
74. Müller, L., Tunger, A., Wobus, M., von Bonin, M., Towers, R., Bornhäuser, M., *et al.*, Immunomodulatory Properties of Mesenchymal Stromal Cells: An Update. *Front. Cell Dev. Biol.*, 9, February, 1–9, 2021.
75. Bellei, B., Migliano, E., Picardo, M., Therapeutic potential of adipose tissue-derivatives in modern dermatology. *Exp. Dermatol.*, 31, 12, 1837–52, 2022.
76. Bajouri, A., Dayani, D., Taj Sharghi, A., Karimi, S., Niknejadi, M., Bidgoli, K.M., *et al.*, Subcutaneous Injection of Allogeneic Adipose-Derived Mesenchymal Stromal Cells in Psoriasis Plaques: Clinical Trial Phase I. *Cell J.*, 25, 6, 363–71, 2023.
77. https://kintarocells.com/blog/a-curative-treatment-for-psoriasis.
78. Levy, O., Kuai, R., Siren, E.M.J., Bhere, D., Milton, Y., Nissar, N., *et al.*, Shattering barriers toward clinically meaningful MSC therapies. *Sci. Adv.*, 6, 30, 1–18, 2020.
79. Liu, J., Liu, Z., Pang, Y., Zhou, H., The interaction between nanoparticles and immune system: application in the treatment of inflammatory diseases. *J. Nanobiotechnol.*, 20, 1, 1–25, 2022.
80. Mukhatayev, Z., Ostapchuk, Y.O., Fang, D., Le Poole, I.C., Engineered antigen-specific regulatory T cells for autoimmune skin conditions. *Autoimmun. Rev.*, 20, 3, 102761, 2021.
81. Schirrmacher, V., Bone Marrow: The Central Immune System. *Immuno.*, 3, 3, 289–329, 2023.
82. Jensen, N.C., Robins, J., Snyder, A.M., Harris, R., Ferris, L.K., Johnson, L., Psoriasis in the transplant population. *Curr. Bladder Dysfunct. Rep.*, 315, 5, 1109–28, 2023.
83. Bulliard, Y., Freeborn, R., Uyeda, M.J., Humes, D., Bjordahl, R., de Vries, D., *et al.*, From promise to practice: CAR T and Treg cell therapies in autoimmunity and other immune-mediated diseases. *Front. Immunol.*, 15, December, 1–20, 2024.

84. Von Niederhäusern, V., Ruder, J., Ghraichy, M., Jelcic, I., Müller, A.M., Schanz, U., *et al.*, B-Cell Reconstitution After Autologous Hematopoietic Stem Cell Transplantation in Multiple Sclerosis. *Neurol. Neuroimmunol. Neuroinflamm.*, 9, 6, e200027, 2022.
85. Sen, T. and Thummer, R.P., The Impact of Human Microbiotas in Hematopoietic Stem Cell and Organ Transplantation. *Front. Immunol.*, 13, July, 1–25, 2022.
86. Konen, F.F., Schwenkenbecher, P., Jendretzky, K.F., Gingele, S., Grote-Levi, L., Möhn, N., *et al.*, Stem Cell Therapy in Neuroimmunological Diseases and Its Potential Neuroimmunological Complications. *Cells*, 11, 14, 2165, 2022.
87. Kuczyńska, M., Moskot, M., Gabig-Cimińska, M., Insights into Autophagic Machinery and Lysosomal Function in Cells Involved in the Psoriatic Immune-Mediated Inflammatory Cascade. *Arch. Immunol. Ther. Exp. (Warsz)*, 72, 1, 1–24, 2024.
88. Buxbaum, N.P. and Pavletic, S.Z., Autoimmunity Following Allogeneic Hematopoietic Stem Cell Transplantation. *Front. Immunol.*, 11, 2017, August, 2020.
89. Vale, N., Pereira, M., Mendes, R.A., Systemic Inflammatory Disorders, Immunosuppressive Treatment and Increase Risk of Head and Neck Cancers—A Narrative Review of Potential Physiopathological and Biological Mechanisms. *Cells*, 12, 17, 2192, 2023.
90. Thatikonda, S., Pooladanda, V., Sigalapalli, D.K., Godugu, C., Piperlongumine regulates epigenetic modulation and alleviates psoriasis-like skin inflammation *via* inhibition of hyperproliferation and inflammation. *Cell Death Dis.*, 11, 1, 21, 2020.
91. Borna, S., Meffre, E., Bacchetta, R., FOXP3 deficiency, from the mechanisms of the disease to curative strategies. *Immunol. Rev.*, 322, 1, 244–58, 2024.
92. Strobl, J., Gail, L.M., Krecu, L., Madad, S., Kleissl, L., Unterluggauer, L., *et al.*, Diverse macrophage populations contribute to distinct manifestations of human cutaneous graft-versus-host disease. *Br. J. Dermatol.*, 190, 3, 402–14, 2024.
93. Xu, F., Fei, Z., Dai, H., Xu, J., Fan, Q., Shen, S., *et al.*, Mesenchymal Stem Cell-Derived Extracellular Vesicles with High PD-L1 Expression for Autoimmune Diseases Treatment. *Adv. Mater.*, 34, 1, 1–34, 2022.
94. Mizoguchi, E., Low, D., Ezaki, Y., Okada, T., Recent updates on the basic mechanisms and pathogenesis of inflammatory bowel diseases in experimental animal models. *Intest. Res.*, 18, 2, 151–67, 2020.
95. Hsu, L.Y., Rosenbaum, J.T., Lindner, S., Allanore, Y., Cauvet, A., Pawar, R.D., *et al.*, Soquelitinib, A Selective Inhibitor of Interleukin-2-Inducible T Cell Kinase (ITK), is Active in Several Murine Models of T Cell-Mediated Inflammatory Disease, bioRxiv, 1–23, 2023.
96. Zhang, J., Hodges, A., Chen, S.H., Pan, P.Y., Myeloid-Derived Suppressor Cells as Cellular Immunotherapy in Transplantation and Autoimmune Diseases. *Cell. Immunol.*, 362, 104300, 2021.

97. Xu, Y., Murphy, A.J., Fleetwood, A.J., Hematopoietic Progenitors and the Bone Marrow Niche Shape the Inflammatory Response and Contribute to Chronic Disease. *Int. J. Mol. Sci.*, 23, 4, 2234, 2022.
98. Nedelkopoulou, N., Dhawan, A., Xinias, I., Gidaris, D., Farmaki, E., Interleukin 10: the critical role of a pleiotropic cytokine in food allergy. *Allergol. Immunopathol. (Madr)*, 48, 4, 401–8, 2020.
99. Annaloro, C., Serpenti, F., Saporiti, G., Galassi, G., Cavallaro, F., Grifoni, F., *et al.*, Viral Infections in HSCT: Detection, Monitoring, Clinical Management, and Immunologic Implications. *Front. Immunol.*, 11, January, 1–23, 2021.
100. Müskens, K.F., Lindemans, C.A., Belderbos, M.E., Hematopoietic dysfunction during graft-versus-host disease: A self-destructive process? *Cells*, 10, 8, 1–17, 2021.
101. Gangadaran, P., Madhyastha, H., Madhyastha, R., Rajendran, R.L., Nakajima, Y., Watanabe, N., *et al.*, The emerging role of exosomes in innate immunity, diagnosis and therapy. *Front. Immunol.*, 13, January, 1–20, 2023.
102. Kaisar, S. and Badreldin, Y., Stem cell therapy as a potential treatment option for psoriasis. *Cytotherapy*, 26, 6, e2–3, 2024 Jun 1.
103. Naik, P.P., Stem cell therapy as a potential treatment option for psoriasis. *An. Bras. Dermatol.*, 97, 4, 471–7, 2022 Jul 29.
104. Wu, S., Sun, S., Fu, W., Yang, Z., Yao, H., Zhang, Z., The role and prospects of mesenchymal stem cells in skin repair and regeneration. *Biomedicines*, 12, 4, 743, 2024 Mar 27.
105. Wang, S., Han, Q., Wang, J., Peng, X., Zhou, L., Niu, X., Proteomic Analysis of Mesenchymal Stem Cell-derived Exosomes in Psoriasis. *Curr. Mol. Med.* 25, e15665240304551, 2025
106. Kim, H.R., Lee, S.Y., You, G.E., Kim, H.O., Park, C.W., Chung, B.Y., Adipose-Derived Stem Cell Exosomes Alleviate Psoriasis Serum Exosomes-Induced Inflammation by Regulating Autophagy and Redox Status in Keratinocytes. *Clin. Cosmet. Invest. Dermatol.*, 16, 3699–711, 2023 Dec 31.
107. Zhang, Y., Yan, J., Li, Z., Zheng, J., Sun, Q., Exosomes derived from human umbilical cord mesenchymal stem cells alleviate psoriasis-like skin inflammation. *J. Interferon Cytokine Res*, 42, 1, 8–18, 2022 Jan 1.
108. Jiang, M., Fang, H., Shao, S., Dang, E., Zhang, J., Qiao, P., Yang, A., Wang, G., Keratinocyte exosomes activate neutrophils and enhance skin inflammation in psoriasis. *FASEB J.*, 33, 12, 13241–53, 2019 Dec.
109. Hazrati, A., Malekpour, K., Soudi, S., Hashemi, S.M., Mesenchymal stromal/stem cells spheroid culture effect on the therapeutic efficacy of these cells and their exosomes: A new strategy to overcome cell therapy limitations. *Biomed. Pharmacother.*, 152, 113211, 2022 Aug 1.
110. Yin, K., Wang, S., Zhao, R.C., Exosomes from mesenchymal stem/stromal cells: a new therapeutic paradigm. *Biomarker Res.*, 7, 1–8, 2019 Dec.
111. Janockova, J., Slovinska, L., Harvanova, D., Spakova, T., Rosocha, J., New therapeutic approaches of mesenchymal stem cells-derived exosomes. *J. Biomed. Sci.*, 28, 1–26, 2021 Dec.

112. Pischon, H., Radbruch, M., Ostrowski, A., Schumacher, F., Hönzke, S., Kleuser, B., Hedtrich, S., Fluhr, J.W., Gruber, A.D., Mundhenk, L., How effective is tacrolimus in the imiquimod-induced mouse model of psoriasis? *J. Invest. Dermatol.*, 138, 2, 455–8, 2018 Feb 1.
113. Di Domizio, J., Conrad, C., Gilliet, M., Xenotransplantation model of psoriasis, in: *Inflammation: Methods and Protocols*, pp. 83–90, 2017.
114. Hvid, H., Teige, I., Kvist, P.H., Svensson, L., Kemp, K., TPA induction leads to a Th17-like response in transgenic K14/VEGF mice: a novel *in vivo* screening model of psoriasis. *Int. Immunol.*, 20, 8, 1097–106, 2008 Aug 1.
115. Ojeh, N.O. and Navsaria, H.A., An *in vitro* skin model to study the effect of mesenchymal stem cells in wound healing and epidermal regeneration. *J. Biomed. Mater. Res. Part A*, 102, 8, 2785–92, 2014 Aug.
116. Niehues, H. and Van Den Bogaard, E.H., Past, present and future of *in vitro* 3D reconstructed inflammatory skin models to study psoriasis. *Exp. Dermatol.*, 27, 5, 512–9, 2018 May.
117. Coutier, J., Bonnette, M., Martineau, S., Mercadier, A., Domingues, S., Saidani, M., Jarrige, M., Polveche, H., Darle, A., Holic, N., Hadj-Rabia, S., Human-induced pluripotent stem cell–derived keratinocytes, a useful model to identify and explore the pathological phenotype of epidermolysis bullosa simplex. *J. Invest. Dermatol.*, 142, 10, 2695–705, 2022 Oct 1.
118. Walter, M.N., Wright, K.T., Fuller, H.R., MacNeil, S., Johnson, W.E., Mesenchymal stem cell-conditioned medium accelerates skin wound healing: an *in vitro* study of fibroblast and keratinocyte scratch assays. *Exp. Cell. Res.*, 316, 7, 1271–81, 2010 Apr 15.
119. Rosa Fortin, M.M., Poubelle, P.E., Soucy, J., Pouliot, R., Cellular interactions *in vitro*: psoriatic keratinocytes enhance T lymphocyte survival, in: *Psoriasis Forum*, 2010 Mar, vol. 16, no. 1, SAGE Publications, Sage CA, Los Angeles, CA, pp. 12–15.
120. Ubago-Rodríguez, A., Quiñones-Vico, M.I., Sánchez-Díaz, M., Sanabria-de la Torre, R., Sierra-Sánchez, A., Montero-Vílchez, T., Fernández-González, A., Arias-Santiago, S., Challenges in psoriasis research: a systematic review of preclinical models. *Dermatology (Basel Switzerland)*, 240, 620–652, 2024.
121. Wang, H., Peters, T., Kess, D., Sindrilaru, A., Oreshkova, T., Van Rooijen, N., Stratis, A., Renkl, A.C., Sunderkötter, C., Wlaschek, M., Haase, I., Activated macrophages are essential in a murine model for T cell–mediated chronic psoriasiform skin inflammation. *J. Clin. Investig.*, 116, 8, 2105–14, 2006 Aug 1.
122. Kundu-Raychaudhuri, S., Datta-Mitra, A., Abria, C.J., Peters, J., Raychaudhuri, S.P., Severe combined immunodeficiency mouse-psoriatic human skin xenograft model: a modern tool connecting bench to bedside. *Indian J. Dermatol. Venereol. Leprol.*, 80, 204, 2014 May 1.
123. Attia, S.S., Rafla, M., El-Nefiawy, N.E., Hamid, H.A., Amin, M.A., Fetouh, M.A., A potential role of mesenchymal stem cells derived from human umbilical cord blood in ameliorating psoriasis-like skin lesion in the rats. *Folia Morphol.*, 81, 3, 614–31, 2022.

124. Johnson-Huang, L.M., Lowes, M.A., Krueger, J.G., Putting together the psoriasis puzzle: an update on developing targeted therapies. *Dis. Models Mech.*, 5, 4, 423–33, 2012 Jul 1.
125. Yao, D., Ye, S., He, Z., Huang, Y., Deng, J., Wen, Z., Chen, X., Li, H., Han, Q., Deng, H., Zhao, R.C., Adipose-derived mesenchymal stem cells (AD-MSCs) in the treatment for psoriasis: results of a single-arm pilot trial. *Ann. Transl. Med.*, 9, 22, 1653, 2021 Nov.
126. Chandran, N.S., Bhupendrabhai, M.N., Tan, T.T., Zhang, B., Lim, S.K., Choo, A.B., Lai, R.C., A Phase 1, Open-label Study to determine Safety and Tolerability of the Topical Application of Mesenchymal stem/stromal cell (MSC) Exosome Ointment to Treat Psoriasis in Healthy Volunteers. *Cytotherapy*, 27, 633–641, 2025 Jan 23.
127. Meybodi, M.A., Nilforoushzadeh, M.A., KhandanDezfully, N., Mansouri, P., The safety and efficacy of adipose tissue-derived exosomes in treating mild to moderate plaque psoriasis: A clinical study. *Life Sci.*, 353, 122915, 2024 Sep 15.
128. De Jesus, M.M., Santiago, J.S., Trinidad, C.V., See, M.E., Semon, K.R., Fernandez Jr., M.O., Chung, F.S., Autologous adipose-derived mesenchymal stromal cells for the treatment of psoriasis vulgaris and psoriatic arthritis: a case report. *Cell Transplant.*, 25, 11, 2063–9, 2016 Nov.
129. Ahn, H., Lee, S.Y., Jung, W.J., Pi, J., Lee, K.H., Psoriasis treatment using minimally manipulated umbilical cord-derived mesenchymal stem cells: A case report. *World J. Clin. Cases*, 9, 23, 6798, 2021 Aug 8.
130. Wang, S.G., Hsu, N.C., Wang, S.M., Wang, F.N., Successful treatment of plaque psoriasis with allogeneic gingival mesenchymal stem cells: a case study. *Case Rep. Dermatological Medicine*, 2020, 1, 4617520, 2020.
131. Seetharaman, R., Mahmood, A., Kshatriya, P., Patel, D., Srivastava, A., Mesenchymal stem cell conditioned media ameliorate psoriasis vulgaris: a case study. *Case Rep. Dermatol. Med.*, 2019, 1, 8309103, 2019.
132. Comella, K., Parlo, M., Daly, R., Dominessy, K., First-in-man intravenous implantation of stromal vascular fraction in psoriasis: a case study. *Int. Med. Case Rep. J.*, 11, 59–64, 2018 Mar 21.
133. Wei, L., Yan, W., Shah, W., Zhang, Z., Wang, M., Liu, B., Xue, Z., Cao, Y., Hou, X., Zhang, K., Yan, B., Wang, X., Advancements and challenges in stem cell transplantation for regenerative medicine. *Heliyon*, *10*, 16, e35836, 2024, https://doi.org/10.1016/j.heliyon.2024.e35836.
134. Bahari, M., Mokhtari, H., Yeganeh, F., Stem Cell Therapy, the Market, the Opportunities and the Threat. *Int. J. Mol. Cell. Med.*, *12*, 3, 310–319, 2023, https://doi.org/10.22088/IJMCM.BUMS.12.3.310.
135. Aly, R.M., Current state of stem cell-based therapies: an overview. *Stem Cell Invest.*, *7*, 8, 2020.
136. Hentze, J.A., *et al.*, Regulatory Barriers to Stem Cell Research and Therapy. *J. Clin. Invest.*, 128, 2, 451–453, 2018.

137. Anderson, R.D., *et al.*, Standardization in Stem Cell Therapy and Its Impact on Clinical Applications. *Regener. Med.*, 13, 3, 353–361, 2018.
138. Chandra, S.M., *et al.*, Clinical Application of Stem Cell Therapy: Safety, Efficacy, and Regulatory Challenges. *Stem Cells Transl. Med.*, 7, 10, 745–753, 2018.
139. Carbone, A., *et al.*, Ethical Considerations in Stem Cell-Based Therapies. *Cell Stem Cell*, 22, 1, 1–8, 2018.
140. Lee, M.M.H.M., *et al.*, Navigating Regulatory Challenges in Stem Cell-Based Therapies. *Cell Stem Cell*, 23, 1, 2–5, 2018.
141. Xu, M.C., *et al.*, The Economic Impact of Stem Cell-Based Therapies: Opportunities and Challenges. *J. Stem Cell Res. Ther.*, 5, 4, 234–238, 2018.
142. DiGirolamo, A.G., *et al.*, Economic Considerations in Stem Cell Therapy: A Growing Challenge for Health Care Systems. *Regener. Med.*, 13, 1, 47–56, 2018.
143. Lyu, T.K.S., *et al.*, Personalized Approaches in Stem Cell-Based Therapy. *Trends Biotechnol.*, 36, 7, 652–662, 2018.
144. Yamanaka, S., Induced Pluripotent Stem Cells: Past, Present, and Future. *Cell Stem Cell*, 28, 4, 510–518, 2021.
145. Chen, W.M., *et al.*, Advances in Stem Cell-Based Therapies: Insights from Clinical Trials and Technological Innovations. *J. Regener. Med.*, 10, 3, 255–263, 2022.
146. Zhang, H., *et al.*, Ethical implications of stem cell-based therapies, PubMed, 2018, https://pubmed.ncbi.nlm.nih.gov/30271953/.
147. Pang, W., *et al.*, Genetic stability in stem cells: Implications for therapeutic application, PubMed, 2020, https://pubmed.ncbi.nlm.nih.gov/31901028/.
148. Ge, S., *et al.*, Variability in stem cell products: Implications for manufacturing. PubMed, 2019, https://pubmed.ncbi.nlm.nih.gov/31541747/.
149. Liu, X., *et al.*, Immune rejection in allogenic stem cell therapy: Challenges and solutions, PubMed, 2020, https://pubmed.ncbi.nlm.nih.gov/32185649/.
150. Xu, T., *et al.*, Challenges in stem cell production and scalability, PubMed, 2021, https://pubmed.ncbi.nlm.nih.gov/33525641/.
151. Zhao, L., *et al.*, Global regulatory challenges in stem cell therapies, PubMed, 2021, https://pubmed.ncbi.nlm.nih.gov/34114431/.
152. Chen, F., *et al.*, Financial barriers to stem cell research: A global perspective, PubMed, 2019, https://pubmed.ncbi.nlm.nih.gov/30755778/.
153. Shin, H., *et al.*, Long-term safety and efficacy of stem cell therapies: A critical review, PubMed, 2021, https://pubmed.ncbi.nlm.nih.gov/33559269/.
154. Tan, Y., *et al.*, Standardization of stem cell differentiation protocols, PubMed, 2020, https://pubmed.ncbi.nlm.nih.gov/32313655/.
155. Ammar, A., *et al.*, Tumorigenesis in stem cell therapies: A growing concern, PubMed, 2020, https://pubmed.ncbi.nlm.nih.gov/32686550/.
156. Gao, Y., *et al.*, Infrastructure challenges for stem cell therapies in developing countries, PubMed, 2020, https://pubmed.ncbi.nlm.nih.gov/33207006/.

157. Shao, W., *et al.*, Social and cultural factors influencing stem cell therapy adoption, PubMed, 2018, https://pubmed.ncbi.nlm.nih.gov/30159958/.
158. Lee, J., *et al.*, Ethical and regulatory frameworks in stem cell research, PubMed, 2021, https://pubmed.ncbi.nlm.nih.gov/33201490/.
159. https://clinicaltrials.ucsf.edu/psoriasis.
160. Galli, S., *et al.*, Regulation of stem cell therapies in the U.S.: Current status and future directions. *Regener. Med.*, 14, 4, 235–245, 2019.
161. Turner, L. and Knoepfler, P., Selling Stem Cells in the USA: Assessing the Direct-to-Consumer Industry. *Cell Stem Cell*, 22, 1, 1–4, 2018.
162. Alvarez, L., *et al.*, The role of the European Medicines Agency in regulating stem cell therapies. *Stem Cells Transl. Med.*, 10, 5, 630–635, 2021.
163. Patel, S., *et al.*, Regulatory frameworks in stem cell therapy: The Indian scenario. *J. Clin. Diagn. Res.*, 13, 5, LE01–LE05, 2020.
164. Kaur, G., *et al.*, Stem Cell Therapy: Regulatory and Ethical Challenges in India. *Indian J. Med. Ethics*, 5, 2, 72–78, 2020.
165. Kurtzberg, J., Ethical and regulatory issues in stem cell transplantation. *J. Pediatr.*, 222, 15–20, 2020.
166. Burt, R., *et al.*, Regulatory challenges in stem cell research: Perspectives from the ISSCR. *Regener. Med.*, 14, 6, 1–4, 2019.
167. Hu, J., *et al.*, Regulatory challenges in gene-editing and stem cell therapies. *Nat. Biotechnol.*, 39, 1, 23–28, 2021.
168. Shao, W., *et al.*, Global regulatory challenges in stem cell therapy. *Regener. Med.*, 14, 2, 133–141, 2019.
169. Eshghi, P., *et al.*, Navigating the global regulatory landscape of stem cell therapy. *Stem Cells Dev.*, 29, 1, 25–34, 2020.
170. McHugh, M., *et al.*, Stem Cell Therapy in Clinical Trials: Regulatory and Ethical Issues. *J. Clin. Trials*, 10, 2, 135–142, 2020.

4

Therapeutic Modalities of Stem Cells: Skin Cancer Treatment

Dipanjan Karati[1]*, Soumi Das[2] and Swarupananda Mukherjee[3]

[1]*Deparment of Pharmaceutical Chemistry, School of Pharmacy, Techno India University, West Bengal, Kolkata, India*
[2]*Department of Pharmacy Practice, ISF College of Pharmacy, Moga, Punjab, India*
[3]*Department of Pharmaceutics, NSHM College of Pharmaceutical Technology, NSHM Knowledge Campus, Kolkata - Group of Institutions, Kolkata, West Bengal, India*

Abstract

Stem cells are adaptable, unspecialized cells that may rebuild and repair damaged tissues, making them a potential target for future therapeutics. Their ability to differentiate varies from pluripotent to unipotent, allowing them to perform a variety of roles. The skin contains a wide range of stem cells that play an important part in its normal healing and regeneration processes. Interest in studying and applying skin stem cells for therapeutic purposes is growing among scientists and medical experts. The skin contains hematopoietic stem cells, which support the immune system and regulate various skin lymphocyte types; melanocyte stem cells preserve the natural color of the skin and hair; sebaceous gland stem cells support the skin's barrier; dermal mesenchymal stem cells support the regenerating dermal fibroblasts; and epidermal and follicular stem cells encourage the growth of new skin cells, aiding in wound healing. For international healthcare systems, skin cancer, one of the most common cancers in the world, continues to pose a serious problem. Various people have unsatisfactory results and a variety of negative consequences, even though there are diverse therapies available. The search for more potent treatment approaches has focused on stem cells, which are well-known for their exceptional capacity for differentiation and regeneration. These qualities make them potentially revolutionary in the treatment of skin cancer. This thorough chapter explores the many facets of skin stem cells, including their many kinds and functions in tissue regeneration and repair.

**Corresponding author*: karatibabai@gmail.com

Sumel Ashique, Sanjay Dey, Biplab Debnath, Mohhammad Ramzan and Swarupananda Mukherjee (eds.) Stem Cell Therapy in Dermatological Disorders, (79–108) © 2025 Scrivener Publishing LLC

Keywords: Skin cancer, stem cell, immunomodulation, melanoma, organoids

4.1 Introduction

One of the main causes of death worldwide is cancer, and the incidence of skin cancer is steadily rising in nations where these tumors are common [1]. With more and more exposure to ultraviolet (UV) radiation, skin cancers are the most frequently diagnosed cancers among Caucasians worldwide, and their incidence is increasing constantly. An imbalance toward either too much cell growth and survival in the epidermis or too little apoptosis is a hallmark of skin cancer. Although UV radiation is the major cause of skin cancer, viruses, food and chemical mutagens, and genetic predisposition are other causes [2]. Skin cancer can be prevented by reducing or eliminating these causes. By inhibiting blood supply to a tumor (antiangiogenesis), skin cancers can be readily eliminated, causing a decrease in tumor formation with increased survival of a patient. Many cancer cells lack functional apoptosis machinery or find means to circumvent it, allowing excessive cell proliferation to occur [3]. So, the final common pathway on which antineoplastic agents act is on the process of apoptosis. Currently, besides other types of treatments for skin cancer are cryosurgery, radiation therapy, chemotherapy, and surgical excision [4]. Although 5-fluorouracil and imiquimod are used in topical chemotherapy for superficial basal cell carcinoma (BCC) and squamous cell carcinoma (SCC) *in situ*, imiquimod is the only topically approved medication for cutaneous malignant melanoma. Making a selection is never simple because every treatment option has advantages and disadvantages [5]. The optimal course of action is determined by a variety of factors, including the cancer's location, the patient's condition, and the views of both the patient and the doctor. In a similar sense, stem cell treatment is a new approach that has attracted a lot of attention recently. The 2007 Nobel Prize in Physiology or Medicine was to Mario R. Capecchi, Sir Martin J. Evans, and Oliver Smithies for their understanding of the fundamentals of using embryonic stem cells (ESCs) to modify specific genes in mice. "Stem cell–derived products" are covered under the New Drugs and Clinical Trials Rules' new drug section because stem cells have been utilized for a variety of medical applications. The successful use of stem cell treatment in certain refractory illnesses has led to an expansion of the therapeutic arsenal in dermatology. Reviewing the state of stem cell therapy in dermatology is our goal in this work [6].

4.2 Stem Cells in Skin Cancer: A Brief Overview

4.2.1 Classification and Characteristics of Stem Cells

ESCs, mesenchymal stem cells (MSCs), and induced pluripotent stem cells (iPSCs) are among the several stem cells that are identified in skin cancer; each comes from a particular location and has a unique role in tumor biology. Each of these three stem cell types has a distinct function in the development, spread, and management of cancer.

ESCs: They are pluripotent cells that were produced from the blastocyst's inner cell mass. Despite not being the cause of skin cancer, ESCs' capacity to differentiate into distinct cell types and their ability to do so have helped to clarify a lot of the biology of tumors [7]. Studies state that to maintain their state of undifferentiating and proliferative, cancer cells, including skin cancers, may colonize pathways similar to ESCs like Oct4, Sox2, and Nanog. This implies the cruciality of ESC-like signaling with respect to maintaining the features of cancer stem cells (CSCs) involved in skin malignancies [8].

MSCs: Adipocytes, chondrocytes, osteoblasts, and other mesodermal lineages can all be produced from multipotent stromal cells. In the tumor microenvironment of skin malignancies, MSCs are essential. They are drawn to tumor locations, where they engage in interactions with cancer cells to stimulate immune modulation, invasion, and angiogenesis [9]. MSCs can release growth factors and cytokines that support cancer stemness and treatment resistance. This property of MSCs makes them a potential therapeutic target and a tool in regenerative medicine for reconstructive surgery after cancer treatment, considering their immunomodulatory and plasticity properties [10].

iPSCs: The introduction of Oct4, Sox2, Klf4, and c-Myc converted the somatic cells to become pluripotent. Currently, iPSCs have emerged as a strong model for studying the biological mechanism of cancer, especially skin cancer [11]. This technique of converting the somatic cancer cells into iPSCs has explained mechanisms of carcinogenesis and stemness of the cancer cells. Furthermore, iPSCs also serve as a system for tailored treatment and drug screening. Their use is not risk-free, though, as the reprogramming procedure can reactivate carcinogenic pathways, making their therapeutic use more difficult [12].

4.2.2 Tumorigenic Potential of CSCs in Skin Cancer

CSCs are thought to be the origin of skin cancer because they have the ability to create, initiate, and aid in the spread of malignancies. Because of their ability to differentiate and self-renew, CSCs—a subpopulation of cancer cells with stem-like characteristics—can display the range of cell types found in the tumor [13]. Often identified in SCC and BCC by markers such as CD44, CD133, and ALDH1, CSCs are also correlated with increased tumor-forming potential. For these cells to retain their stemness and proliferative potential, dysregulated signaling pathways such as hedgehog, Wnt/βx-catenin, and notch are necessary (Figure 4.1) [14]. Melanoma is aggressive and extremely invasive due to the great tumorigenic potential of CSCs detected by ABCB5, CD271, and nestin [15]. These cells' enhanced invasion and motility promote metastasis. The epithelial-to-mesenchymal transition and the stimulation of pathways such as transforming growth factor β are responsible for this [16].

The resistance to conventional therapies is one of the main components of CSC tumorigenicity. Some of the mechanisms that make CSCs resistant to radiation and chemotherapy include quiescence, overexpression of drug-efflux pumps, and effective DNA repair. In addition, CSCs interact with the tumor microenvironment to recruit immune and stromal cells

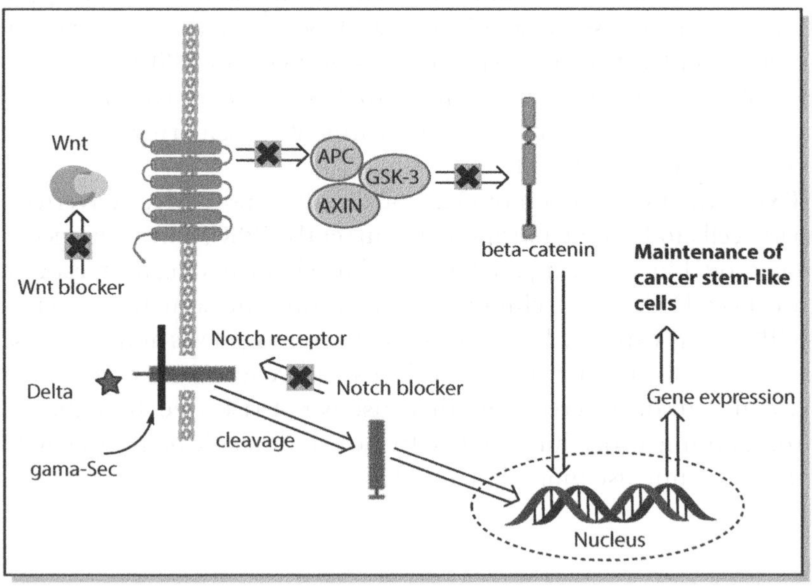

Figure 4.1 Stemness-related pathways functional for targeting CSCs.

to create a niche of support [17]. The activation of hypoxia-inducible factors (HIFs) under hypoxic conditions within the tumor niche enhances cellular stemness, promotes angiogenesis, and increases resistance to therapy [18]. It is the ensemble of these properties that allow CSCs to support tumor progression, evade therapy, and facilitate metastasis and recurrence. Understanding the tumorigenic capacity of CSCs in skin cancer will be imperative for the construction of more successful therapy schemes [19].

4.3 Therapeutic Modalities of Stem Cells in Skin Cancer Treatment

There are several therapeutic approaches for skin cancer treatment by stem cell therapy (Table 4.1).

4.3.1 Cell Replacement Therapy

In regenerative medicine, cell replacement therapy is one of the future strategies that can treat tissue dysfunction by replacing healthy stem cells with the dysfunctional or malignant cell population. The goal of this type of treatment for skin cancer is to replace the malignant cells with healthy and functioning stem cells, thereby regenerating healthy skin tissue [20]. Most promising at this point, although in its experimental stage, is the application of this in treating skin cancer and mending damage from the disease or its treatments [13]. In order to repair the skin and restore normal tissue architecture, healthy stem cells—such as MSCs, iPSCs, and tissue-specific epidermal stem cells—develop into keratinocytes, melanocytes, or other cell types. Particularly in nonmelanoma skin cancers such as SCC and BCC, stem cells can be used to restore function and regenerate damaged areas with minimal scarring following surgical, irradiation, or chemotherapy removal of the malignant tissue [21].

However, because melanoma is an aggressive and metastatic disease, cell replacement therapy is more challenging [22]. Recent developments in the utilization of stem cells as anticancer drugs' carriers provide a dual strategy that targets the illness while encouraging tissue regeneration. Engineered MSCs have altered cells to fix tissue, together with therapeutic delivery directly at the tumor, using immune-modulating compounds and proapoptotic drugs, for instance [23]. There is still such a thing as an unknown level of certainty; they will not ever grow into malignancies,

Table 4.1 Therapeutic modality of stem cells in skin cancer therapy.

Therapeutic modality	Stem cell type	Mechanism of action	Advantages	References
MSCs	Bone marrow-derived MSCs, adipose-derived MSCs	Tumor- targeting, secretion of antitumor cytokines, immune modulation	Antitumor properties, easy isolation	[34]
SCC	Arise from stem cells within the squamous epithelium	Ability to self-renew, rapidly proliferate, and drive tumor growth, leading to metastasis and resistance to therapies	Chemoresistance, metastasis potential, biomarker identification	[35]
BCC	Follicular basal keratinocyte stem cell	It mainly arises from a disturbance in the hedgehog signaling pathway, typically due to mutations in the "patched" (PTCH1) gene, which results in unchecked growth of basal cells in the skin	Slow growth, low metastasis rate, high cure rate	[36]
iPSCs	Reprogrammed somatic cell	Generation of patient specific immune cells for therapy	Personalized treatment, ethical advantage over embryonic stem cell	[37]
CSC	Targeting CSC markers	Inhibition of self-renewal and differentiation pathways	Prevents tumor relapse and metastasis	[38]

will not assist in establishing a malignancy, and do not ensure cancer. Such high and careful source identification of stem cell populations combined with genetic modifications could minimize such adverse effects during any preclinical phase [24]. Moreover, the tumor microenvironment that mainly favors the survival and proliferation of CSCs is altered to stop recurrence.

Effective delivery methods, such as hydrogel scaffolds, bioengineered skin grafts, or direct injection of stem cells into the afflicted area, are also required for the successful clinical application of cell replacement treatment in skin cancer to ensure proper integration and survival [25]. Another important aspect, especially in allogeneic stem cell transplantation, is overcoming immunological rejection. Despite these challenges, cell replacement therapy is a novel approach to cancer treatment that will eliminate the disease while restoring function, reducing scarring, regenerating healthy skin tissue, and improving quality of life in patients. This approach should not be used without further research, which will improve procedures, ensure safety, and establish clinical efficacy as a standard of care [26].

4.3.2 Immunomodulation

Another strategy for strengthening the immune system to aid in the fight against cancer, particularly against skin malignancies such as melanoma, BCC, and SCC, is new immunomodulatory stimulation involving stem cells [27]. Immune response regulation primarily relies on MSCs and iPSCs. It could do it either by enhancing its activity against cancer cells or suppressing hyperactivated immune responses that could hinder anticancer treatment.

The immunomodulatory properties of MSCs are particularly well-documented. They can influence both innate and adaptive immunity by releasing growth factors and cytokines that regulate immune cell function. For example, MSCs have been shown to suppress myeloid-derived suppressor cells (MDSCs) and regulatory T cells, which are often recruited by cancers to evade immune responses. MSCs help enhance antitumor immune responses by modulating these immune cells that support tumor growth [28]. Additionally, MSCs have the ability to activate cytotoxic T cells and natural killer (NK) cells, which are essential for locating and eliminating cancer cells. It is even possible to enhance immune responses to cancer through the delivery of genetically modified MSCs with immune-stimulatory factors such as interleukins or interferons directly into the tumor microenvironment [28].

iPSCs also assist in immunomodulation in the treatment of cancer. They can be differentiated into NK or dendritic cells to be used in adoptive cell therapy, which is a strategy that uses immune cells derived from stem cells targeting cancer to boost the immune system [29]. Through iPSCs, one can generate patient-specific immune cells that are genetically engineered to recognize and destroy specific cancer cells for improved accuracy and effectiveness in the treatment. One of the most immunogenic skin malignancies, melanoma, holds significant promise for treatment through stem cell–based immunomodulation. The immunosuppressive microenvironment that tumor cells frequently produce makes it challenging for the immune system to mount a successful defense [30]. By modifying the tumor microenvironment to promote immune activation, stem cells can reverse this. To improve the immune system's capacity to "see" tumor cells, MSCs can be modified to release chemicals that improve antigen presentation or proinflammatory cytokines [31]. There remain challenges despite all these advancements. The dual immunomodulatory function of stem cells, allowing either stimulation or inhibition of the immune system, has to be controlled to prevent unanticipated events such as tumor growth or autoimmune reactions. Another complexity is in the tumor microenvironment, containing immunosuppressive cells and immunological checkpoints, which can restrict the effectiveness of the treatment using stem cells [32]. In conclusion, immunomodulation based on stem cells is one of the potential approaches for cancer immunotherapy. This strategy may improve the capacity of the immune system to fight cancer, especially in immunogenic skin tumors such as melanoma, through the use of the special qualities of MSCs and iPSCs to modulate immune responses. Research is still being conducted on how to optimize these tactics for enhanced safety, accuracy, and therapeutic results [33].

4.3.3 Differentiation Therapy

A novel strategy known as differentiation therapy aims to inhibit the development of skin cancer by forcing CSCs or stem cells to undergo differentiation and become mature noncancerous cells [39]. This therapy restores tissue function to normal levels in addition to eliminating the cause of tumor formation and drug resistance through the promotion of CSC differentiation. Melanoma and other types of skin cancers, including BCC and SCC, have recently been responsive to differentiation therapy [40].

CSCs sustain the proliferation and survival of skin cancer tumors because of their stem-like features, including self-renewal and resistance to traditional therapy [41]. These features can be blocked through the promotion of differentiation, which makes the cells less hostile or hostile and susceptible to therapy [42]. For example, in the case of promoting terminal differentiation in CSCs, signaling pathways such as Wnt/β-catenin, hedgehog, and notch are often aberrant and could be targeted. For example, agents targeting these pathways, including retinoic acid, vitamin D analogs, or small molecule inhibitors, have been promising in preclinical studies.

Stem cells, specifically MSCs and iPSCs, are also essential for differentiation therapy. It is possible to engineer or condition MSCs to release factors that cause CSCs to differentiate, so turning them into nonproliferative cells.

Alternatively, immune cells from iPSCs are uniquely individual-specific [43]. The best candidates among them may be cytotoxic T cells and dendritic cells. Once found with a target cell population in cancer tissue, they would induce its programmed cell death, providing the intrinsic, autonomous function that might significantly fight against a neoplasm [44].

The highly plastic and invasive nature of melanoma cells poses special challenges for differentiation therapy. However, it has been shown that differentiation-inducing drugs targeting melanoma CSCs can reduce the aggressiveness of the tumor and make cancer cells more sensitive to conventional therapies such as immunotherapy and chemotherapy. Similarly, differentiation in BCC and SCC can prevent cancer cells from proliferating and lower the likelihood of recurrence [45].

Although differentiation therapy is a promising treatment for skin cancer, there are still challenges to be overcome. The efficiency of differentiation techniques may be limited by the tumor microenvironment, which often promotes stemness and resistance to therapy [46]. Moreover, it is essential to achieve exact control over differentiation without compromising normal tissue regeneration or producing unintended consequences. Novel drug delivery technologies such as hydrogel-based scaffolds or nanocarriers are currently being explored to improve targeted drug delivery that promotes differentiation [47].

Differentiation therapy aims to rewire skin CSCs into specialized, non-cancerous cells, hence reducing their capacity to cause tumors [48]. This method lowers the chance of recurrence and improves the effectiveness of current treatments in addition to interfering with tumor growth. To fully understand the promise of differentiation-inducing techniques in the fight against skin cancer, more investigation and clinical studies are required [49].

4.3.4 Gene Therapy

Gene therapy using stem cells is a novel approach to skin cancer treatment as it delivers therapeutic genes to target cancer cells or alters the tumor microenvironment [50]. Given their low immunogenicity, tendency to home to tumor sites, and ease of genetic manipulation, stem cells in general, but more specifically MSCs and iPSCs, are the perfect vectors for gene therapy. This method decreases systemic damage but allows for a direct attack on cancer cells [51].

For skin cancers such as melanoma, BCC, and SCC, stem cells can be genetically modified to produce genes that promote immune responses, inhibit tumor growth, or cause tumor cell death. MSCs are frequently used as vectors for delivering genes that cause cancer cells to undergo selective apoptosis, such as when p53 or tumor necrosis factor–related apoptosis-inducing ligand (TRAIL) are expressed [52]. Following their migration to the tumor site, these altered MSCs would release therapeutic gene products that would cause cancerous cells to undergo apoptosis while protecting healthy tissue.

Increasing the immune system's capacity to combat skin cancer is another intriguing use. It is possible to modify stem cells to release immune-stimulatory cytokines such as interleukin 2 (IL-2) or interferon γ (IFN-γ), which will increase the activation of cytotoxic T cells and NK cells against the tumor [53]. By using anti–cytotoxic T-lymphocyte-associated protein 4 (CTLA-4) or anti–programmed cell death 1 (anti–PD-1) antibodies to deliver checkpoint inhibitors intratumor or intratissue compartments, MSCs can also circumvent a frequent immune checkpoint evasion mechanism for skin cancer, especially melanoma.

Stem cell–based gene therapy can also be used to treat therapy-resistant cancers. For instance, genes that target drug-efflux pumps or enhance DNA repair pathways can be delivered by stem cells to reverse resistance to radiation or chemotherapy. Furthermore, the capacity of iPSCs to differentiate into a range of immune cells, such as NK cells and dendritic cells, enables the development of tailored immune therapy according to each patient's particular cancer profile [54].

On the other hand, stem cell–based gene therapy remains a promising technology despite its potential shortcomings, especially immunological rejection, insertional mutagenesis, and off-target consequences. In order to avoid these outcomes, it is important that the expression of therapeutic genes must occur only at the tumor site and not in tissues where they are not needed [55]. This will be achieved with further advancements in gene-editing technologies, such as the breakthrough CRISPR-Cas9. Furthermore, the efficacy of stem cell–based gene therapy can be improved by optimizing the delivery techniques, including hydrogel-based scaffolds, viral vectors, or nanoparticles [56].

Using stem cells as gene therapy vectors can overcome the problem of therapy resistance in this adaptable and focused strategy for skin cancer treatment. Through gene delivery to the tumor locations, this treatment holds the potential to boost immune responses, stop tumor growth, and reverse therapy resistance [57]. Although the step will dramatically change the approach to the treatment of skin cancer and transform patient outcomes, it is still bound by numerous dangers and limitations that require further study [58].

4.3.5 Tumor-Targeted Therapy

One of the most innovative treatments for skin cancer is tumor-targeted therapy, where stem cells are used to carry anticancer drugs directly to tumors [59]. This approach aims to reduce damage to surrounding healthy tissues while maximizing the accuracy and efficiency of cancer treatment. Stem cells are perfect delivery vehicles for anticancer drugs because they can be genetically altered, made to travel to tumor areas, and taken up and distribute therapeutic chemicals [60]. This is especially true for tumor-homing stem cells, iPSCs, and MSCs.

Using stem cells, tumor-targeted therapy takes use of the special characteristics of the tumor microenvironment in skin cancers such as melanoma, BCC, and SCC. Signals emitted by tumors often attract specific types of stem cells. For example, it has been shown that MSCs migrate specifically to the locations of tumors by secreting chemokines such as CCL5, which guide the cells to the malignant areas [61]. By taking advantage of their inherent homing potential, stem cells can be modified to deliver anticancer medications straight to the tumor. This would ensure a more effective and localized therapeutic effect while reducing systemic side effects [62].

The most popular strategy is to genetically modify stem cells to produce genes encoding cytotoxic substances, such as proapoptotic proteins like TRAIL or p53, cytokines such as interleukin 2 or tumor necrosis factor, or enzymes that transform nontoxic prodrugs into toxic substances at the tumor site [63]. Modified stem cells, after being injected into the patient, may go directly to the tumor and release therapeutic chemicals, killing cancer cells specifically. This targeted therapy reduces side effects of traditional treatment such as chemotherapy that hurt both healthy and malignant cells [64].

Using the delivery of stem cells as immune modulators to build a tumor-killing capability inside the body is another approach that could improve defenses. For instance, MSCs or iPSCs can be used as the carriers of immunologically stimulating cytokines such as IFN-γ or granulocyte-macrophage colony-stimulating factor for T cytotoxic cells and NK cells at the tumor site [65]. Besides this, the checkpoint inhibitors, for instance, anti–PD-1 or anti–CTLA-4, can be expressed by the stem cells to avoid the strategies of immune evasion used by the malignancies including melanoma [66].

A treatment option with great promise in the context of melanoma, a cancer that is notoriously metastatic and also extremely drug-resistant, lies in stem cell–targeted therapies. Antiangiogenic drugs such as endostatin and angiostatin, for example, may be delivered through MSCs in order to starve the tumor of its blood supply, shutting it down both in terms of its growth and spreading [67]. Another avenue that can be opened by targeting the drug efflux pumps or improving DNA repair pathways in stem cells is medicines reversing resistance to radiation or chemotherapy so that cancer cells are again more responsive to conventional treatments [68].

Although it holds much promise, stem cell–based tumor-targeted therapy is also associated with many disadvantages. For instance, ensuring that the stem cells target the tumor exclusively and efficiently without harming the normal tissues is one of the main problems [69]. Additionally, to prevent hazards such as insertional mutagenesis, unintended immunological responses, or tumor development, the stability and safety of genetically modified stem cells need to be strictly evaluated. Furthermore, the delivery methods are critical and need to be optimized to ensure the stem cells migrate effectively and deliver therapeutic chemicals at the tumor site [70]. Although it holds much promise, stem cell–based tumor-targeted therapy is also associated with many disadvantages. For instance, ensuring that the stem cells target the tumor exclusively and efficiently without harming the normal tissues is one of the main problems [71]. Additionally, to prevent hazards such as insertional mutagenesis, unintended immunological responses, or tumor development, the stability and safety of genetically modified stem cells need to be strictly evaluated. Furthermore, the delivery methods are critical and need to be optimized to ensure the stem cells migrate effectively and deliver therapeutic chemicals at the tumor site [72].

A potential and extremely specialized therapeutic approach is tumor-targeted therapy, which uses stem cells to deliver the anticancer drugs directly to skin cancers. This strategy takes advantage of innate homing abilities of stem cells and modifies these cells to carry therapeutic molecules, which increases treatment efficacy while reducing unwanted side effects. If more study is conducted on the safety and delivery concerns of stem cell–based tumor-targeted therapy, patients with skin cancer may experience a considerable improvement in their prognosis and quality of life [73].

4.4 Emerging Trends and Advanced Technologies in Stem Cell Therapy for Skin Cancer

Over the last few years, stem cell therapy has dramatically progressed, and advances in technology and trends continue to widen its possibilities in the treatment of skin cancer. As scientists delve deeper into understanding the principles behind CSCs and the intricacies of skin cancer, novel therapeutic approaches continue to be considered [74]. These advanced techniques, such as gene editing, organoid models, three-dimensional (3D) bioprinting, and combination therapies, are all designed to enhance the effectiveness, precision, and safety of stem cell–based skin cancer treatments.

4.4.1 Gene-Edited Stem Cells for Precision Therapy

With precision therapy tailored to the genetic makeup of particular tumors, gene-edited stem cells are an interesting and rapidly developing area in the treatment of skin cancer. By carefully modifying the genetic material of stem cells with state-of-the-art gene-editing tools such as CRISPR-Cas9, TALENs, and zinc finger nucleases, scientists may target specific mutations that cause skin cancer [75]. These alterations are common in melanoma, basal cell carcinoma (BCC), and squamous cell carcinoma (SCC), and gene-edited stem cells can potentially reverse them, offering a more targeted therapeutic strategy. For instance, mutations that occur within essential oncogenes in melanoma, BRAF, NRAS, and p53, respectively, can be corrected using the editing of stem cells to substitute or repair the dysfunctional genes responsible for this disease, thus ceasing tumor progression and preventing recurrence [76].

Gene-edited stem cells can make drug delivery more precise besides directly targeting genetic mutations. Scientists can ensure that treatment drugs are delivered precisely to the tumor site by modifying the stem cells to produce specific proteins or receptors [77]. Stem cells, such as MSCs or iPSCs, can be engineered to precisely home into cancer cells, increasing treatment efficacy while reducing adverse effects [78]. To further improve therapy outcomes, these modified stem cells can be used for the direct delivery of proapoptotic molecules, immune-modulating cytokines, or anticancer drugs to the tumor [79].

Using gene-edited stem cells in conjunction with immunotherapy holds additional promise [80]. These cells can be altered to produce immunoactivating cytokines such as IL-2 or IFN-γ or to express immune checkpoint inhibitors such as anti–PD-1 or anti–CTLA-4, which can boost the body's defenses against skin cancer [81]. By releasing therapeutic molecules at the tumor site, these altered stem cells will function as targeted delivery vehicles, improving the immune system's capacity to recognize and eliminate cancer cells, providing a more potent and durable anticancer response [82].

Gene-edited stem cells also hold promise to be very powerful for breaking resistance to the traditional therapies, which include radiation and chemotherapy. Scientists can make stem cells produce genes that suppress resistance mechanisms, such as improving DNA repair or blocking efflux pumps that remove drugs from the cancer cells. Moreover, this technique could make the cancers that were previously resistant to treatment-sensitive again [83].

Because gene-edited stem cells can be tailored to fit the specific genetic signature of each patient's skin cancer, personalized therapy is another critical benefit of this technology [84]. To identify which treatment will be most effective, patient-derived iPSCs—the genetic changes that are present in a patient's tumor—can be genetically modified to test multiple therapeutic approaches in a dish. This method allows for a more tailored treatment based on the specifics of each patient's cancer [85].

Although highly promising, the gene-edited stem cells are applied carefully so as not to bring along hazards. Gene editing is an innovative tool, but accuracy increases while its mutation rates become minimal. It may even get lesser risks when other mechanisms are formed in advance to combat those consequences with mechanisms called safety switches or suicide genes [86].

Gene-edited stem cells are a significant breakthrough in precision treatment for skin cancer. Gene-edited stem cells hold much promise for skin cancer therapy by bringing well-focused therapeutic approaches that could address alterations at the genetic level, strengthen immunological responses, and overcome resistance [87]. Although there is still much work to be done to optimize these treatments, the continued development of gene-editing technology may well provide skin cancer patients with safer, more targeted, and more effective options [88].

4.4.2 Organoids and 3D Culture Models in Skin Cancer Research

Organoids and 3D culture models are more accurate *in vitro* reconstructions of human tissue and tumors compared with conventional two-dimensional (2D) cell cultures, making them powerful tools in the study of skin cancer [89]. These advanced models are revolutionizing the way skin cancer is studied and opening up new avenues for testing stem cell therapy. Organoids are 3D constructs made from stem cells that resemble actual organs or tumors in both structure and function [90]. By growing organoids from patient-derived cells, researchers may more accurately mimic several skin cancers, such as melanoma, BCC, and SCC. This offers a platform for testing new stem cell–based treatments and enables the study of cancer biology in a setting that is more physiologically appropriate [91].

In comparison to conventional 2D cultures, 3D culture models, which are the growth of cells in a 3D matrix, also depict the tumor microenvironment more closely. Such models simulate the complicated relationships between cancer cells, stromal cells, and the extracellular matrix and yield valuable information on how treatment resistance might occur, tumor growth, and metastasis [92]. Before using gene-edited stem cells, MSCs, or iPSCs in clinical trials, this technique is highly helpful in assessing their safety and efficacy in treating skin cancer. Researchers can determine how the treatments based on stem cells would affect the cancerous cells along with the surrounding tissues by using stem cells to form 3D models [93].

The ability to use organoids and 3D culture models for high-throughput drug screening is another significant benefit. By using these models, scientists are able to test a variety of possible treatments, such as stem cell–based therapies, and see how they impact tumor development, invasion, and the response to immunotherapy or chemotherapy [94]. These models can also help find biomarkers for the stratification of patients in clinical settings and reveal new therapeutic targets by more accurately mimicking the milieu of skin cancer. Development of personalized therapy for skin cancers is significantly facilitated by 3D culture models [95]. Tumor models created from individual patients can allow researchers to determine how specific tumors will respond to specific therapies. In this regard, stem cells may be used to develop patient-specific models by selecting the therapy best suited for the distinct genetic makeup of a patient's tumor. Such a strategy can decrease unnecessary side effects and enhance the efficacy of treatments against skin cancers [96].

Organoids and 3D culture models are some of the best leading ways for skin cancer research. These are advanced platforms in which cancer biology can be studied and stem cell–based treatments evaluated. New approaches for treatment and enhanced stem cell–based treatments for skin cancer are made possible through these models, which ensure more precise and individualized drug development. As these technologies advance, they will likely have a significant impact on how precision medicine and skin cancer treatment are developed in the future [97].

4.4.3 Artificial Intelligence and Stem Cell Therapy Optimization

Artificial intelligence (AI) is now playing an increasingly important role in optimizing stem cell therapies for skin cancer, revolutionizing clinical and research applications. From the development of drugs and treatment optimization to patient classification and tracking, AI can significantly improve the accuracy, efficacy, and personalization of stem cell–based treatments for skin cancer through machine learning algorithms [98].

Accelerating discovery and development for novel treatment approaches is one of the most promising uses of AI in stem cell therapy. AI-enabled systems can scrutinize large datasets, including genomic, proteomic, and clinical data, to reveal novel treatment targets, biomarkers, and potential medication candidates in skin cancer [99]. When combined with stem cells, AI can assist in the development of gene-editing methods or the improvement of differentiation protocols to generate the specific cell types required for targeted therapy. By predicting the most effective drug combinations, AI can assist in developing more customized treatment regimens that are appropriate for the genetic and molecular features of each patient's skin cancer [100].

AI is also being used to refine the optimization process of stem cell differentiation in order to treat skin cancer. For regenerative purposes, programmed stem cells (MSCs or iPSCs) can develop into cell types associated with skin, including melanocytes and keratinocytes [101]. The signals as well as elements in the microenvironment that stimulate stem cells into specific lineages can then be optimized utilizing AI systems that guide the development of differentiation protocols [102]. These algorithms with the ability to predict the condition for ideal differentiations of the stem cells advance the uniformity and caliber in the cell type used for such treatment of skin cancers [103].

AI is also making possible more personalized treatment choices by better stratifying patients in the clinic. AI can assist in determining which patients, based on patient-specific information such as genetic mutations, tumor characteristics, and medication responses, are most likely to benefit from stem cell–based therapies [104]. With medicines tailored to each patient's unique genetic makeup, there are always improved outcomes and fewer side effects. AI can also help estimate the chance of tumor spread or recurrence, which helps doctors better monitor patients and modify treatment plans [105].

AI is transforming the tracking of stem cell–based treatments for skin cancer. More accurate and detailed analyses of tumor growth or response to treatment can be provided with AI-powered imaging methods, such as magnetic resonance imaging scans or deep learning–based analysis of skin biopsies [106]. Such techniques can also identify early indicators of adverse effects or challenges, such as tumor recurrence or problems associated with stem cells, in order to facilitate immediate action. In order to optimize stem cell–based therapies in clinical practice, AI models need to be able to continuously assess patient data and provide real-time insights into treatment efficacy and safety [107].

AI is helping to better develop preclinical models for skin cancer studies. AI can mimic the potential *in vivo* behavior of stem cell therapies by combining data from animal research, 3D organoid models, and patient records [108]. By reducing the necessity of extensive animal testing and speeding up the clinical trial approach, these models can help lead the way toward the formulation of safer and more effective therapies. AI-driven predictions are also improving the design of clinical trials on how various stem cell–based therapies interact with the tumor microenvironment, such that treatments can be adjusted even before human testing [109].

4.5 Challenges and Limitations of Stem Cell–Based Therapies

Despite its vast potential, stem cell therapy has several disadvantages and limitations that should be taken into consideration with caution. One significant disadvantage is the possibility of tumorigenesis, in which an unchecked rise in the quantity of transplanted stem cells could result in the development of malignancies [110]. To guarantee their safety, stem cells must pass stringent preclinical testing prior to being utilized in clinical settings. Because the recipient's immune system rejects the transplanted cells because it believes they are alien, the immune response also presents a problem. It is still difficult to develop strategies for improved engraftment and decreased immune rejection [111].

More significantly, one of the main obstacles is the difficulty of accurately regulating the development of stem cells into the required cell types. Safety and reliability in therapeutic outcome are also a cause for concern because of differences in techniques for differentiation and possible off-target effects. The logistics of producing enough quality-controlled stem cells for widespread clinical use in a cost-effective and scalable manner present another set of difficulties for the industry in reaching its objectives [112].

The debate over stem cell therapy, and more especially the use of ESCs, centers on ethical concerns. Discussions over the early human embryo's ethical standing arose from the killing of embryos to obtain ESCs [113]. In light of these concerns, researchers are looking at alternative sources of pluripotent stem cells, such as iPSCs, which are reprogrammed from adult cells and do not have the same ethical concerns as ESCs.

Regulatory frameworks are needed to navigate the ethical dilemmas raised by various stem cell treatments [114].

Regulations pertaining to the clinical use of stem cells vary by nation and can be either permissive or restrictive. It is still difficult for regulatory bodies to strike a balance between patient safety and innovation. Dynamic regulatory frameworks that can be adjusted for scientific advances will be required due to the dynamic nature of stem cell research and therapies [115]. There are still discussions in this field, mostly around the commercialization of stem cell therapies. Questions raised have to do with access, cost, and justice in distributing treatments to individuals in need [116].

Further, ethical complexity is added by the concern regarding early marketing of unproven stem cell therapies and an openness about how much science there is currently. In a nutshell, addressing the problems and ethical dilemmas surrounding stem cell therapy necessitates a multipronged strategy that includes in-depth scientific investigation, open communication, and flexible laws. As science advances, full potential of stem cell therapies will be met by striking the appropriate balance between innovation and responsibility [117].

4.6 Preclinical and Clinical Studies: Insights and Outcomes

The preclinical and clinical research conducted on skin cancer has provided vital information regarding the possibility of new treatments, such as stem cell–based therapy. Animal models have been used in preclinical studies to evaluate the safety, efficacy, and molecular pathways of stem cells in the treatment of skin cancer [118]. The ability of stem cells to target and repair damaged skin tissues, promote wound healing, or deliver targeted medications directly to cancers is often the focus of these investigations. The potential of stem cells to differentiate into various skin cell types and promote skin regeneration after chemotherapy or radiation therapy has also been investigated [119].

It is still challenging to translate these promising preclinical model results into the clinic. In order to assess the potential of stem cell therapy for patients with human skin cancer, initial-stage clinical trials have just started [120]. The results of these trials have been mixed; some have reported increased skin regeneration and preliminary signs of tumor reduction, whereas others have raised concerns over the risk of immunological rejection or carcinogenesis [121]. Most importantly, it can be very tough now to draw any ultimate conclusions because majority of the studies with clinical trials and cases are of such small size in which there were differences in a patient's reactions. Additionally, because skin cancer's multifactorial, heterogeneous nature comprised many genes and subtypes, tailored intervention and treatment plans were necessary to really maximize the utility of treatment provided [122].

4.7 Future Directions and Perspectives

The new findings and emerging technologies may revolutionize the vast range of medical specialties with the aid of stem cell therapy. Incorporation of stem cell therapy into precision medicine techniques promises better therapeutic results with fewer side effects, opening up a new era of personalized treatment based on the genetic makeup of an individual [123]. These integration effects are expected to be driven forward by genomics advancements and utilization of patient-specific stem cells [124]. Improvements in the modulating immune methods associated with treatments based on stem cells and further resolution of such issues as graft-versus-host reactions and immunological rejections are to be included in subsequent studies. It is aimed for better compatibility with treatments using stem cells by new techniques, such as immunomodulatory chemicals and modified stem cells [125].

Gene-editing technologies such as CRISPR-Cas9 are being developed, and their development will enhance the precision and safety of treatments to be provided using stem cells. The technique enables researchers to alter particular genes in stem cells, enabling the use of these cells for targeted therapeutic interventions and cellular-level therapy of inherited disorders. The combination of stem cell therapy with bioengineering is a significant area of study [126]. By combining stem cells with cutting-edge biomaterials, functional tissues and organs with exceptional structural and functional qualities can be produced. New pathways to tissue-specific regeneration and transplantation are made possible by bioengineered structures. These significant avenues highlight the multifaceted character of upcoming developments in stem cell therapy, combining bioengineering, gene editing, precision medicine, and immune modulation to drive the field toward groundbreaking discoveries [127].

Recent advancements in stem cell therapy have revealed a promising path that has the potential to fundamentally change modern medicine. This subject is now at the forefront of biomedical research since advances in science, innovation, and clinical trial applications have prompted the study of stem cells in a range of medical specializations [128]. The historical progression of medical research from the foundational ideas established by early 20th-century pioneers to groundbreaking discoveries such as the isolation of ESCs and the finding of iPSCs marks a significant advancement in the field [129].

Regeneration processes of embryonic, adult, induced pluripotent, and perinatal stem cells provide previously unimaginable avenues for therapeutic interventions. Stem cells are intimately associated with development, tissue repair, and regeneration due to their remarkable capacity to self-renew and specialize into a wide variety of cell types [130]. Their wide range of uses, including tissue damage, diabetes, spinal cord injuries, cardiovascular diseases, and neurodegenerative disorders, is creating new treatment options for crippling illnesses. However, as the field evolves, it is essential to address the key issues and complexities [131]. Deep research and creative solutions are needed to address problems such as immunological rejection, carcinogenesis, and the precise regulation of stem cell function. Because of its complexity, stem cell therapy necessitates a thorough comprehension of its historical foundations, present state, and potential future directions [132].

4.8 Conclusion

The use of stem cell–based therapies offers tremendous potential in the treatment of skin cancer, giving new and unique modes such as cell replacement therapy, immunomodulation, differentiation therapy, gene therapy, and tumor-targeted strategies. By exploiting some other unique properties such as regenerative ability by stem cells or even genetically engineered, these therapies can bring around considerable challenges that traditional treatments pose, including resistance to therapies, limited precision, and adverse side effects. Despite promising preclinical and early clinical studies, the road ahead still presents challenges related to immune rejection, tumorigenesis, and regulatory complexities. Advanced technologies in gene editing, organoid models, and AI further speed progress toward precision and personalized medicine. Continued research, interdisciplinary collaboration, and the resolution of ethical and logistical hurdles are likely to help stem cell–based therapies revolutionize skin cancer treatment and improve patient outcomes and quality of life.

AI disclosure statement: The author confirms that no artificial intelligence (AI)-generated content or AI-assisted writing tools were used in the preparation of this chapter (including summaries and the abstract). All text, analysis, and conclusions are the result of the author's original work.

References

1. Roky, A.H., Islam, M.M., Ahasan, A.M.F., Mostaq, M.S., Mahmud, M.Z., Amin, M.N., *et al.*, Overview of skin cancer types and prevalence rates across continents. *Cancer Pathog. Ther.*, 2, E01–E36, 2024.
2. De Gruijl, F., Skin cancer and solar UV radiation. *Eur. J. Cancer*, 35, 14, 2003–9, 1999.
3. Kubota, Y., Tumor angiogenesis and anti-angiogenic therapy. *Keio J. Med.*, 61, 2, 47–56, 2012.
4. Vafopoulou, P. and Kourti, M., Anti-angiogenic drugs in cancer therapeutics: A review of the latest preclinical and clinical studies of anti-angiogenic agents with anticancer potential. *J. Cancer Metastasis Treat.*, 8, 18, 2022.
5. Pérez, C.E.A., Ariza-Gómez, S.A., Morales-Naranjo, S.D., Cure rate of low-risk superficial basal cell carcinoma treated with topical 5-fluorouracil. *Rev. Asoc. Colomb. Dermatol. Cir. Dermatol.*, 32, 3, 253–61, 2024.
6. Burgess, R., *Stem cells: a short course*, John Wiley & Sons, 2015.
7. Fan, Y., Min, Z., Alsolami, S., Ma, Z., Zhang, E., Chen, W., *et al.*, Generation of human blastocyst-like structures from pluripotent stem cells. *Cell Discov.*, 7, 1, 81, 2021.
8. Kagawa, H., Javali, A., Khoei, H.H., Sommer, T.M., Sestini, G., Novatchkova, M., *et al.*, Human blastoids model blastocyst development and implantation. *Nature*, 601, 7894, 600–5, 2022.
9. Prajwal, G.S., Jeyaraman, N., Kanth, V.K., Jeyaraman, M., Muthu, S., Rajendran, S.N.S., *et al.*, Lineage differentiation potential of different sources of mesenchymal stem cells for osteoarthritis knee. *Pharmaceuticals*, 15, 4, 386, 2022.
10. Xu, Q., Hou, W., Zhao, B., Fan, P., Wang, S., Wang, L., *et al.*, Mesenchymal stem cells lineage and their role in disease development. *Mol. Med.*, 30, 1, 207, 2024.
11. Aguirre, M., Escobar, M., Forero Amézquita, S., Cubillos, D., Rincón, C., Vanegas, P., *et al.*, Application of the Yamanaka transcription factors Oct4, Sox2, Klf4, and c-Myc from the laboratory to the clinic. *Genes*, 14, 9, 1697, 2023.
12. Park, S., Gwon, Y., Khan, S.A., Jang, K.-J., Kim, J., Engineering considerations of iPSC-based personalized medicine. *Biomater. Res.*, 27, 1, 67, 2023.

13. Khan, N.H., Mir, M., Qian, L., Baloch, M., Khan, M.F.A., Ngowi, E.E., *et al.*, Skin cancer biology and barriers to treatment: Recent applications of polymeric micro/nanostructures. *J. Adv. Res.*, 36, 223–47, 2022.
14. Sahoo, O.S., Pethusamy, K., Srivastava, T.P., Talukdar, J., Alqahtani, M.S., Abbas, M., *et al.*, The metabolic addiction of cancer stem cells. *Front. Oncol.*, 12, 955892, 2022.
15. Mu, X., Zhou, Y., Yu, Y., Zhang, M., Liu, J., The roles of cancer stem cells and therapeutic implications in melanoma. *Front. Immunol.*, 15, 1486680, 2024.
16. Lin, Y.-T. and Wu, K.-J., Epigenetic regulation of epithelial-mesenchymal transition: focusing on hypoxia and TGF-β signaling. *J. Biomed. Sci.*, 27, 1, 39, 2020.
17. Nallasamy, P., Nimmakayala, R.K., Parte, S., Are, A.C., Batra, S.K., Ponnusamy, M.P., Tumor microenvironment enriches the stemness features: the architectural event of therapy resistance and metastasis. *Mol. Cancer*, 21, 1, 225, 2022.
18. Zhou, H.-M., Zhang, J.-G., Zhang, X., Li, Q., Targeting cancer stem cells for reversing therapy resistance: mechanism, signaling, and prospective agents. *Signal Transduct. Target. Ther.*, 6, 1, 62, 2021.
19. Zhong, C., Wang, G., Guo, M., Zhu, N., Chen, X., Yan, Y., *et al.*, The role of tumor stem cells in colorectal cancer drug resistance. *Cancer Control*, 31, 10732748241274196, 2024.
20. Hoang, D.M., Pham, P.T., Bach, T.Q., Ngo, A.T., Nguyen, Q.T., Phan, T.T., *et al.*, Stem cell-based therapy for human diseases. *Signal Transduct. Target. Ther.*, 7, 1, 1–41, 2022.
21. Mannino, G., Russo, C., Maugeri, G., Musumeci, G., Vicario, N., Tibullo, D., *et al.*, Adult stem cell niches for tissue homeostasis. *J. Cell. Physiol.*, 237, 1, 239–57, 2022.
22. Switzer, B., Puzanov, I., Skitzki, J.J., Hamad, L., Ernstoff, M.S., Managing metastatic melanoma in 2022: a clinical review. *JCO Oncol. Pract.*, 18, 5, 335–51, 2022.
23. Soltantoyeh, T., Akbari, B., Karimi, A., Mahmoodi Chalbatani, G., Ghahri-Saremi, N., Hadjati, J., *et al.*, Chimeric antigen receptor (CAR) T cell therapy for metastatic melanoma: challenges and road ahead. *Cells*, 10, 6, 1450, 2021.
24. Shajari, N., Baradaran, B., Tohidkia, M.R., Nasiri, H., Sepehri, M., Setayesh, S., *et al.*, Advancements in Melanoma Therapies: From Surgery to Immunotherapy. *Curr. Treat. Options Oncol.*, 25, 8, 1073–88, 2024.
25. Boukani, L.M., Khosroshahi, R.F., Kh, S.A., Rashtbar, M., Khosroshahi, A.F., Statistical study of clinical trials with stem cells and their function in skin wound. *Cell Tissue Res.*, 393, 2, 217–28, 2023.
26. Garg, A., Garg, S., Adlak, P., Kori, M.L., Lodhi, S., Stem Cells and Regenerative Strategies for Wound Healing: Therapeutic and Clinical Implications. *Current Pharmacol. Rep.*, 10, 2, 121–44, 2024.

27. Chiang, E., Stafford, H., Buell, J., Ramesh, U., Amit, M., Nagarajan, P., *et al.*, Review of the tumor microenvironment in basal and squamous cell carcinoma. *Cancers*, 15, 9, 2453, 2023.
28. Li, H., Dai, H., Li, J., Immunomodulatory properties of mesenchymal stromal/stem cells: The link with metabolism. *J. Adv. Res.*, 45, 15–29, 2023.
29. Huang, Y., Wu, Q., Tam, P.K.H., Immunomodulatory mechanisms of mesenchymal stem cells and their potential clinical applications. *Int. J. Mol. Sci.*, 23, 17, 10023, 2022.
30. Zhou, C. and Bai, X.-Y., Strategies for the induction of anti-inflammatory mesenchymal stem cells and their application in the treatment of immune-related nephropathy. *Front. Med.*, 9, 891065, 2022.
31. Refaie, A.F., Elbassiouny, B.L., Kloc, M., Sabek, O.M., Khater, S.M., Ismail, A.M., *et al.*, From mesenchymal stromal/stem cells to insulin-producing cells: immunological considerations. *Front. Immunol.*, 12, 690623, 2021.
32. Yeo, G.E.C., Ng, M.H., Nordin, F.B., Law, J.X., Potential of mesenchymal stem cells in the rejuvenation of the aging immune system. *Int. J. Mol. Sci.*, 22, 11, 5749, 2021.
33. Shen, Z., Huang, W., Liu, J., Tian, J., Wang, S., Rui, K., Effects of mesenchymal stem cell-derived exosomes on autoimmune diseases. *Front. Immunol.*, 12, 749192, 2021.
34. Zhou, L.N., Wang, J.C., Zilundu, P.L., Wang, Y.Q., Guo, W.P., Zhang, S.X., Luo, H., Zhou, J.H., Deng, R.D., Chen, D.F., A comparison of the use of adipose-derived and bone marrow-derived stem cells for peripheral nerve regeneration *in vitro* and *in vivo*. *Stem Cell Res. Ther.*, 11, 1–6, 2020 Dec.
35. Acharya, S.K., Shai, S., Choon, Y.F., Gunardi, I., Hartanto, F.K., Kadir, K., Roychoudhury, A., Amtha, R., Vincent-Chong, V.K., Cancer Stem Cells in Oral Squamous Cell Carcinoma: A Narrative Review on Experimental Characteristics and Methodological Challenges. *Biomedicines*, 12, 9, 2111, 2024 Sep 16.
36. Hoashi, T., Kanda, N., Saeki, H., Molecular mechanisms and targeted therapies of advanced basal cell carcinoma. *Int. J. Mol. Sci.*, 23, 19, 11968, 2022 Oct 8.
37. Weber, E.W., Maus, M.V., Mackall, C.L., The emerging landscape of immune cell therapies. *Cell*, 181, 1, 46–62, 2020 Apr 2.
38. Limonta, P., Chiaramonte, R., Casati, L., Unveiling the dynamic interplay between cancer stem cells and the tumor microenvironment in melanoma: implications for novel therapeutic strategies. *Cancers*, 16, 16, 2861, 2024 Aug 16.
39. Ramesh, S., Selvakumar, P., Ameer, M.Y., Lian, S., Abdullah Alzarooni, A.I.M., Ojha, S., *et al.*, State-of-the-art therapeutic strategies for targeting cancer stem cells in prostate cancer. *Front. Oncol.*, 13, 1059441, 2023.
40. Maucort, C., Di Giorgio, A., Azoulay, S., Duca, M., Differentiation of Cancer Stem Cells by Using Synthetic Small Molecules: Toward New Therapeutic Strategies against Therapy Resistance. *ChemMedChem*, 16, 1, 14–29, 2021.

41. Liu, Q., Guo, Z., Li, G., Zhang, Y., Liu, X., Li, B., *et al.*, Cancer stem cells and their niche in cancer progression and therapy. *Cancer Cell Int.*, 23, 1, 305, 2023.
42. Strobl, M.A., Gallaher, J., West, J., Robertson-Tessi, M., Maini, P.K., Anderson, A.R., Spatial structure impacts adaptive therapy by shaping intra-tumoral competition. *Commun. Med.*, 2, 1, 46, 2022.
43. Merkens, L., Sailer, V., Lessel, D., Janzen, E., Greimeier, S., Kirfel, J., *et al.*, Aggressive variants of prostate cancer: underlying mechanisms of neuroendocrine transdifferentiation. *J. Exp. Clin. Cancer Res.*, 41, 1, 46, 2022.
44. Wu, W., Klockow, J.L., Zhang, M., Lafortune, F., Chang, E., Jin, L., *et al.*, Glioblastoma multiforme (GBM): An overview of current therapies and mechanisms of resistance. *Pharmacol. Res.*, 171, 105780, 2021.
45. Ng, M.F., Simmons, J.L., Boyle, G.M., Heterogeneity in melanoma. *Cancers*, 14, 12, 3030, 2022.
46. Baghban, R., Roshangar, L., Jahanban-Esfahlan, R., Seidi, K., Ebrahimi-Kalan, A., Jaymand, M., *et al.*, Tumor microenvironment complexity and therapeutic implications at a glance. *Cell Commun. Signal.*, 18, 1–19, 2020.
47. Xiao, Y. and Yu, D., Tumor microenvironment as a therapeutic target in cancer. *Pharmacol. Ther.*, 221, 107753, 2021.
48. Faghihkhorasani, A., Dalvand, A., Derafsh, E., Tavakoli, F., Younis, N.K., Yasamineh, S., *et al.*, The role of oncolytic virotherapy and viral oncogenes in the cancer stem cells: a review of virus in cancer stem cells. *Cancer Cell Int.*, 23, 1, 250, 2023.
49. Correnti, M., Binatti, E., Gammella, E., Invernizzi, P., Recalcati, S., The emerging role of tumor microenvironmental stimuli in regulating metabolic rewiring of liver cancer stem cells. *Cancers*, 15, 1, 5, 2022.
50. Rahimi, A., Esmaeili, Y., Dana, N., Dabiri, A., Rahimmanesh, I., Jandaghian, S., *et al.*, A comprehensive review on novel targeted therapy methods and nanotechnology-based gene delivery systems in melanoma. *Eur. J. Pharm. Sci.*, 187, 106476, 2023.
51. Sher, E.K., Kalić, A., Džidić-Krivić, A., Zećo, M.B., Pinjić, E., Sher, F., Cellular therapeutic potential of genetically engineered stem cells in cancer treatment. *Biotechnol. Genet. Eng. Rev.*, 40, 4, 4062–97, 2024.
52. Wunderlich, K., Suppa, M., Gandini, S., Lipski, J., White, J., Del Marmol, V., Risk Factors and Innovations in Risk Assessment for Melanoma, Basal Cell Carcinoma, and Squamous Cell Carcinoma. *Cancers*, 16, 5, 1016, 2024.
53. Chmiel, P., Kłosińska, M., Forma, A., Pelc, Z., Gęca, K., Skórzewska, M., Novel approaches in non-melanoma skin cancers—a focus on hedgehog pathway in basal cell carcinoma (BCC). *Cells*, 11, 20, 3210, 2022.
54. Chen, L., Yang, F., Chen, S., Tai, J., Mechanisms on chemotherapy resistance of colorectal cancer stem cells and research progress of reverse transformation: A mini-review. *Front. Med.*, 9, 995882, 2022.

55. Abusalah, M.A.H., Abd Rahman, E.N.S.E., Choudhary, O.P., Evolving trends in stem cell therapy: an emerging and promising approach against various diseases. *Int. J. Surg.*, 110, 11, 6862–8, 2024.
56. Zhu, Y., Advances in CRISPR/Cas9. *Biomed Res. Int.*, 2022, 1, 9978571, 2022.
57. Kumari, P., Alam, M.A., Dhoundiyal, S., Kaur, A., Yadav, S., Treatment Considerations to Overcome the Barriers Associated with Skin Cancer Targeting. *Curr. Cancer Ther. Rev.*, 20, 6, 529–45, 2024.
58. de Oliveira, B.S.A., de Assis, A.C.C., Souza, N.M., Ferreira, L.F.R., Soriano, R.N., Bilal, M., *et al.*, Nanotherapeutic approach to tackle chemotherapeutic resistance of cancer stem cells. *Life Sci.*, 279, 119667, 2021.
59. Desai, V.M., Kumbhar, P., Kadam, A.Y., Swarup, J., Priya, S., Jain, A., *et al.*, Exploring the therapeutic modalities of targeted treatment approach for skin carcinoma: cutting-edge strategies and key insights. *Expert Opin. Drug Deliv.*, 21, 8, 1213–33, 2024.
60. Zeng, L., Gowda, B.J., Ahmed, M.G., Abourehab, M.A., Chen, Z.-S., Zhang, C., *et al.*, Advancements in nanoparticle-based treatment approaches for skin cancer therapy. *Mol. Cancer*, 22, 1, 10, 2023.
61. Xu, M. and Li, S., Nano-drug delivery system targeting tumor microenvironment: a prospective strategy for melanoma treatment. *Cancer Lett.*, 574, 216397, 2023.
62. García-Hevia, L., Soltani, R., González, J., Chaloin, O., Ménard-Moyon, C., Bianco, A., *et al.*, Carbon nanotubes targeted to the tumor microenvironment inhibit metastasis in a preclinical model of melanoma. *Bioact. Mater.*, 34, 237–47, 2024.
63. Han, X., Alu, A., Liu, H., Shi, Y., Wei, X., Cai, L., *et al.*, Biomaterial-assisted biotherapy: A brief review of biomaterials used in drug delivery, vaccine development, gene therapy, and stem cell therapy. *Bioact. Mater.*, 17, 29–48, 2022.
64. Zhou, H., He, Y., Xiong, W., Jing, S., Duan, X., Huang, Z., *et al.*, MSC based gene delivery methods and strategies improve the therapeutic efficacy of neurological diseases. *Bioact. Mater.*, 23, 409–37, 2023.
65. Jiang, J., Huang, Y., Zeng, Z., Zhao, C., Harnessing engineered immune cells and bacteria as drug carriers for cancer immunotherapy. *ACS Nano*, 17, 2, 843–84, 2023.
66. Nel, A.E., Mei, K.-C., Liao, Y.-P., Liu, X., Multifunctional lipid bilayer nanocarriers for cancer immunotherapy in heterogeneous tumor microenvironments, combining immunogenic cell death stimuli with immune modulatory drugs. *ACS Nano*, 16, 4, 5184–232, 2022.
67. Mir, G.J., Manzoor, I., Au, I., SA, G., Hamid, R., Cancer Stem Cells as Promising Nanotherapeutic Targets, in: *Personalized and Precision Nanomedicine for Cancer Treatment*, pp. 27–68, Springer, 2024.
68. Singh, D. and Prasad, S., A Pioneer Review on Lactoferrin-Conjugated Extracellular Nanovesicles for Targeting Cellular Melanoma: Recent

Advancements and Future Prospects. *ASSAY Drug Dev. Technol.*, 23, 2, 55–69, 2024.

69. Liu, H., Deng, S., Han, L., Ren, Y., Gu, J., He, L., *et al.*, Mesenchymal stem cells, exosomes and exosome-mimics as smart drug carriers for targeted cancer therapy. *Colloids Surf., B*, 209, 112163, 2022.
70. He, W., Li, Q., Lu, Y., Ju, D., Gu, Y., Zhao, K., *et al.*, Cancer treatment evolution from traditional methods to stem cells and gene therapy. *Curr. Gene Ther.*, 22, 5, 368–85, 2022.
71. Khosravi, N., Pishavar, E., Baradaran, B., Oroojalian, F., Mokhtarzadeh, A., Stem cell membrane, stem cell-derived exosomes and hybrid stem cell camouflaged nanoparticles: A promising biomimetic nanoplatforms for cancer theranostics. *J. Control. Release*, 348, 706–22, 2022.
72. Taeb, S., Rostamzadeh, D., Amini, S.M., Rahmati, M., Golshekan, M., Abedinzadeh, M., *et al.*, Revolutionizing Cancer Treatment: Harnessing the Power of Mesenchymal Stem Cells for Precise Targeted Therapy in the Tumor Microenvironment. *Curr. Top. Med. Chem.*, 25, 3, 243–262, 2025.
73. Fan, L., Wei, A., Gao, Z., Mu, X., Current progress of mesenchymal stem cell membrane-camouflaged nanoparticles for targeted therapy. *Biomed. Pharmacother.*, 161, 114451, 2023.
74. Zeng, C.-W., Advancing spinal cord injury treatment through stem cell therapy: A comprehensive review of cell types, challenges, and emerging technologies in regenerative medicine. *Int. J. Mol. Sci.*, 24, 18, 14349, 2023.
75. Chang, A.Y., CRISPR Genome Surgery in Stem Cells and Disease Tissues, in: *Genome engineering with CRISPR/Cas9, ZFNs, and TALENs*, pp. 39–45, Academic Press, 2022.
76. Vlašić, I., Horvat, A., Tadijan, A., Slade, N., p53 family in resistance to targeted therapy of melanoma. *Int. J. Mol. Sci.*, 24, 1, 65, 2022.
77. Bayarsaikhan, D., Bayarsaikhan, G., Lee, B., Recent advances in stem cells and gene editing: Drug discovery and therapeutics. *Prog. Mol. Biol. Transl. Sci.*, 181, 231–69, 2021.
78. Wan, T. and Ping, Y., Delivery of genome-editing biomacromolecules for treatment of lung genetic disorders. *Adv. Drug Deliv. Rev.*, 168, 196–216, 2021.
79. Hasanzadeh, A., Ebadati, A., Dastanpour, L., Aref, A.R., Sahandi Zangabad, P., Kalbasi, A., *et al.*, Applications of innovation technologies for personalized cancer medicine: stem cells and gene-editing tools. *ACS Pharmacol. Transl. Sci.*, 6, 12, 1758–79, 2023.
80. Kanaya, N., Kitamura, Y., Lopez Vazquez, M., Franco, A., Chen, K.-S., van Schaik, T.A., *et al.*, Gene-edited and-engineered stem cell platform drives immunotherapy for brain metastatic melanomas. *Sci. Transl. Med.*, 15, 698, eade8732, 2023.
81. Guo, S., Yao, Y., Tang, Y., Xin, Z., Wu, D., Ni, C., *et al.*, Radiation-induced tumor immune microenvironments and potential targets for combination therapy. *Signal Transduct. Target. Ther.*, 8, 1, 205, 2023.

82. Shukla, S., Dalai, P., Agrawal-Rajput, R., Metabolic crosstalk: Extracellular ATP and the tumor microenvironment in cancer progression and therapy. *Cell. Signal.*, 121, 111281, 2024.
83. Wellhausen, N., Baek, J., Gill, S.I., June, C.H., Enhancing cellular immunotherapies in cancer by engineering selective therapeutic resistance. *Nat. Rev. Cancer*, 24, 9, 614–28, 2024.
84. Neumayer, G., Torkelson, J.L., Li, S., McCarthy, K., Zhen, H.H., Vangipuram, M., *et al.*, A scalable and cGMP-compatible autologous organotypic cell therapy for Dystrophic Epidermolysis Bullosa. *Nat. Commun.*, 15, 1, 5834, 2024.
85. Pang, L.K., Huang, M.-F., Gingold, J.A., Pena, M., Zhao, R., Lee, D.-F., Progress and possibilities for patient-derived iPSCs and genetically engineered stem cells in cancer modeling and targeted therapies, in: *iPSCs-State of the Science*, pp. 247–88, 2022.
86. Dashnau, J.L., Xue, Q., Nelson, M., Law, E., Cao, L., Hei, D., A risk-based approach for cell line development, manufacturing and characterization of genetically engineered, induced pluripotent stem cell–derived allogeneic cell therapies. *Cytotherapy*, 25, 1, 1–13, 2023.
87. Brooks, I.R., Sheriff, A., Moran, D., Wang, J., Jacków, J., Challenges of gene editing therapies for genodermatoses. *Int. J. Mol. Sci.*, 24, 3, 2298, 2023.
88. Zorina, A., Zorin, V., Isaev, A., Kudlay, D., Manturova, N., Ustugov, A., *et al.*, Current Status of Biomedical Products for Gene and Cell Therapy of Recessive Dystrophic Epidermolysis Bullosa. *Int. J. Mol. Sci.*, 25, 19, 10270, 2024.
89. Habanjar, O., Diab-Assaf, M., Caldefie-Chezet, F., Delort, L., 3D cell culture systems: tumor application, advantages, and disadvantages. *Int. J. Mol. Sci.*, 22, 22, 12200, 2021.
90. Barbosa, M.A., Xavier, C.P., Pereira, R.F., Petrikaitė, V., Vasconcelos, M.H., 3D cell culture models as recapitulators of the tumor microenvironment for the screening of anti-cancer drugs. *Cancers*, 14, 1, 190, 2021.
91. Cacciamali, A., Villa, R., Dotti, S., 3D cell cultures: Evolution of an ancient tool for new applications. *Front. Physiol.*, 13, 836480, 2022.
92. Esposito, A., Ferraresi, A., Vallino, L., Garavaglia, B., Dhanasekaran, D.N., Isidoro, C., Three-Dimensional *In Vitro* Cell Cultures as a Feasible and Promising Alternative to Two-Dimensional and Animal Models in Cancer Research. *Int. J. Biol. Sci.*, 20, 13, 5293, 2024.
93. Antonelli, F., 3D cell models in radiobiology: improving the predictive value of *in vitro* research. *Int. J. Mol. Sci.*, 24, 13, 10620, 2023.
94. Urzì, O., Gasparro, R., Costanzo, E., De Luca, A., Giavaresi, G., Fontana, S., *et al.*, Three-dimensional cell cultures: the bridge between *in vitro* and *in vivo* models. *Int. J. Mol. Sci.*, 24, 15, 12046, 2023.
95. Wood, L.D. and Ewald, A.J., Organoids in cancer research: a review for pathologist-scientists. *J. Pathol.*, 254, 4, 395–404, 2021.
96. Zhao, Z., Chen, X., Dowbaj, A.M., Sljukic, A., Bratlie, K., Lin, L., *et al.*, Organoids. *Nat. Rev. Methods Primers*, 2, 1, 94, 2022.

97. Pennarossa, G., Arcuri, S., De Iorio, T., Gandolfi, F., Brevini, T.A., Current advances in 3D tissue and organ reconstruction. *Int. J. Mol. Sci.*, 22, 2, 830, 2021.
98. Kim, M. and Hong, S., Integrating artificial intelligence to biomedical science: New applications for innovative stem cell research and drug development. *Technologies*, 12, 7, 95, 2024.
99. Gharibshahian, M., Torkashvand, M., Bavisi, M., Aldaghi, N., Alizadeh, A., Recent advances in artificial intelligent strategies for tissue engineering and regenerative medicine. *Skin Res. Technol.*, 30, 9, e70016, 2024.
100. Mukherjee, S., Yadav, G., Kumar, R., Recent trends in stem cell-based therapies and applications of artificial intelligence in regenerative medicine. *World J. Stem Cells*, 13, 6, 521, 2021.
101. Saeed, W., Shahbaz, E., Maqsood, Q., Ali, S.W., Mahnoor, M., Cutaneous oncology: Strategies for melanoma prevention, diagnosis, and therapy. *Cancer Control*, 31, 10732748241274978, 2024.
102. Guerrisi, A., Falcone, I., Valenti, F., Rao, M., Gallo, E., Ungania, S., *et al.*, Artificial intelligence and advanced melanoma: treatment management implications. *Cells*, 11, 24, 3965, 2022.
103. Vera, J., Lai, X., Baur, A., Erdmann, M., Gupta, S., Guttà, C., *et al.*, Melanoma 2.0. Skin cancer as a paradigm for emerging diagnostic technologies, computational modelling and artificial intelligence. *Briefings Bioinf.*, 23, 6, bbac433, 2022.
104. Abubakar, M., Exploring the pivotal association of AI in cancer stem cells detection and treatment. *Proc. Anticancer Res.*, 8, 5, 52–63, 2024.
105. Xu, Y., Su, G.-H., Ma, D., Xiao, Y., Shao, Z.-M., Jiang, Y.-Z., Technological advances in cancer immunity: from immunogenomics to single-cell analysis and artificial intelligence. *Signal Transduct. Target. Ther.*, 6, 1, 312, 2021.
106. Chadaga, K., Prabhu, S., Sampathila, N., Chadaga, R., A machine learning and explainable artificial intelligence approach for predicting the efficacy of hematopoietic stem cell transplant in pediatric patients. *Healthc. Anal.*, 3, 100170, 2023.
107. Elemento, O., Leslie, C., Lundin, J., Tourassi, G., Artificial intelligence in cancer research, diagnosis and therapy. *Nat. Rev. Cancer*, 21, 12, 747–52, 2021.
108. Salybekov, A.A., Wolfien, M., Kobayashi, S., Steinhoff, G., Asahara, T., Personalized cell therapy for patients with peripheral arterial diseases in the context of genetic alterations: artificial intelligence-based responder and non-responder prediction. *Cells*, 10, 12, 3266, 2021.
109. Sebastian, A.M. and Peter, D., Artificial intelligence in cancer research: trends, challenges and future directions. *Life*, 12, 12, 1991, 2022.
110. Attia, N., Mashal, M., Puras, G., Pedraz, J.L., Mesenchymal stem cells as a gene delivery tool: Promise, problems, and prospects. *Pharmaceutics*, 13, 6, 843, 2021.
111. Sivandzade, F. and Cucullo, L., Regenerative stem cell therapy for neurodegenerative diseases: an overview. *Int. J. Mol. Sci.*, 22, 4, 2153, 2021.

112. Chentsova, O., Mesenchymal stem cells (MSC) in breast cancer treatment: Advantages and challenges over traditional therapies, Hochschule Rhein-Waal, 2023.
113. Choudhery, M.S., Arif, T., Mahmood, R., Harris, D.T., Stem cell-based acellular therapy: insight into biogenesis, bioengineering and therapeutic applications of exosomes. *Biomolecules*, 14, 7, 792, 2024.
114. Riha, S.M., Maarof, M., Fauzi, M.B., Synergistic effect of biomaterial and stem cell for skin tissue engineering in cutaneous wound healing: A concise review. *Polymers*, 13, 10, 1546, 2021.
115. Lukomskyj, A.O., Rao, N., Yan, L., Pye, J.S., Li, H., Wang, B., *et al.*, Stem cell-based tissue engineering for the treatment of burn wounds: a systematic review of preclinical studies. *Stem Cell Rev. Rep.*, 18, 6, 1926–55, 2022.
116. Voisin, A., Pénaguin, A., Gaillard, A., Leveziel, N., Stem cell therapy in retinal diseases. *Neural Regener. Res.*, 18, 7, 1478–85, 2023.
117. Abdul Kareem, N., Aijaz, A., Jeschke, M.G., Stem cell therapy for burns: story so far. *Biol. Targets Ther.*, 15, 379–97, 2021.
118. Li, Y., Jiang, Q.-L., Van der Merwe, L., Lou, D.-H., Lin, C., Preclinical efficacy of stem cell therapy for skin flap: a systematic review and meta-analysis. *Stem Cell Res. Ther.*, 12, 1–12, 2021.
119. Attia, N., Mashal, M., Pemminati, S., Omole, A., Edmondson, C., Jones, W., *et al.*, Cell-based therapy for the treatment of glioblastoma: an update from preclinical to clinical studies. *Cells*, 11, 1, 116, 2021.
120. Qian, H., Shan, Y., Gong, R., Lin, D., Zhang, M., Wang, C., *et al.*, Mechanism of action and therapeutic effects of oxidative stress and stem cell-based materials in skin aging: Current evidence and future perspectives. *Front. Bioeng. Biotechnol.*, 10, 1082403, 2023.
121. Mousaei Ghasroldasht, M., Seok, J., Park, H.-S., Liakath Ali, F.B., Al-Hendy, A., Stem cell therapy: from idea to clinical practice. *Int. J. Mol. Sci.*, 23, 5, 2850, 2022.
122. Chou, Y., Alfarafisa, N.M., Ikezawa, M., Khairani, A.F., Progress in the development of stem cell-derived cell-free therapies for skin aging. *Clin. Cosmet. Invest. Dermatol.*, 16, 3383–406, 2023.
123. Ratnayake, P., Udalamaththa, V., Samaratunga, U., Seneviratne, J., Udagama, P., Therapeutic Potential of Skin Stem Cells and Cells of Skin Origin: Effects of Botanical Drugs Derived from Traditional Medicine. *Stem Cell Rev. Rep.*, 18, 6, 1986–2001, 2022.
124. Tran, D.-K., Phuong, T.N.T., Bui, N.-L., Singh, V., Looi, Q.H., Koh, B., *et al.*, Exploring the potential of stem cell-based therapy for aesthetic and plastic surgery. *IEEE Rev. Biomed. Eng.*, 16, 386–402, 2021.
125. Shalhout, S.Z., Kaufman, H.L., Emerick, K.S., Miller, D.M., Immunotherapy for nonmelanoma skin cancer: facts and hopes. *Clin. Cancer Res.*, 28, 11, 2211–20, 2022.
126. Aldoghachi, A.F., Chong, Z.X., Yeap, S.K., Cheong, S.K., Ho, W.Y., Ong, A.H.K., Stem cells for cancer therapy: translating the uncertainties and

possibilities of stem cell properties into opportunities for effective cancer therapy. *Int. J. Mol. Sci.*, 24, 2, 1012, 2023.

127. Wu, S., Sun, S., Fu, W., Yang, Z., Yao, H., Zhang, Z., The role and prospects of mesenchymal stem cells in skin repair and regeneration. *Biomedicines*, 12, 4, 743, 2024.
128. Raoufinia, R., Rahimi, H.R., Keyhanvar, N., Moghbeli, M., Abdyazdani, N., Rostami, M., *et al.*, Advances in Treatments for Epidermolysis Bullosa (EB): emphasis on Stem Cell-Based Therapy. *Stem Cell Rev. Rep.*, 20, 1–13, 2024.
129. Sowa, Y., Mazda, O., Tsuge, I., Inafuku, N., Kishida, T., Morimoto, N., Roles of adipose-derived stem cells in cell-based therapy: current status and future scope—a narrative review. *Dig. Med. Res.*, 5, 1–17, 2022.
130. Ebrahimi, A., Ahmadi, H., Ghasrodashti, Z.P., Tanideh, N., Shahriarirad, R., Erfani, A., *et al.*, Therapeutic effects of stem cells in different body systems, a novel method that is yet to gain trust: A comprehensive review. *Bosn. J. Basic Med. Sci.*, 21, 6, 672, 2021.
131. Wang, M., Xu, X., Lei, X., Tan, J., Xie, H., Mesenchymal stem cell-based therapy for burn wound healing. *Burns Trauma*, 9, tkab002, 2021.
132. Yang, P., Zhang, S., Yan, T., Li, F., Zhang, S., The therapeutic application of stem cells and their derived exosomes in the treatment of radiation-induced skin injury. *Radiat. Res.*, 199, 2, 182–201, 2023.

5

Therapeutic Advancement of Stem Cell Therapy in Atopic Dermatitis

Kishanta Kumar Pradhan[1], Rojalin Samanta[2], Reechik Bandyopadhyay[3], Arnab Seth[3] and Chowdhury Mobaswar Hossain[4*]

[1]Department of Pharmaceutical Sciences and Technology, Birla Institute of Technology, Mesra, Ranchi, India
[2]Department of Pharmacy, Usha Martin University, Ranchi, Jharkhand, India
[3]Department of Pharmaceutical Sciences, Bharat Technology, Uluberia, Howrah, India
[4]Department of Pharmaceutical Technology, Maulana Abul Kalam Azad University of Technology, West Bengal, Nadia, India

Abstract

Atopic dermatitis (AD) is an unrelenting seditious dermal condition provoked by immunological, ecological, and hereditary conditions that can result in mutilation and sensitivity to the epidermis. Disparity among T_H2 (T helper 2) and T_H1 cells and the genetic modifications put into the pathogenesis of AD lead to persistent inflammation and disease development. Stem cell treatment has come forth due to its resurgence and immune-modulation characteristics. Different types of stem cells, *vis.* mesenchymal, induced pluripotent, and embryonic stem cells, provide promising therapeutic activity using immune inflection, control of cytokines, and tissue recovery. Notwithstanding heartening consequences in preclinical and early clinical studies that facilitate future translational research, numerous tribulations endure, including regulatory barriers, ethical predicaments, and immunological refusals. Stem cell treatments have been improved through modern strategies such as gene-edited stem cell and exosome-based therapy. This chapter put forward an inclusive study of stem cell treatment, research aspects, ethical inferences, and further prospects in AD.

**Corresponding author*: drcmhossain@gmail.com

Sumel Ashique, Sanjay Dey, Biplab Debnath, Mohhammad Ramzan and Swarupananda Mukherjee (eds.) Stem Cell Therapy in Dermatological Disorders, (109–140) © 2025 Scrivener Publishing LLC

Keywords: Atopic dermatitis (AD), pathogenesis, stem cell, immune-modulation, cytokine, exosome

5.1 Background of Atopic Dermatitis

Definition

Atopic dermatitis (AD) is one of the most common chronic inflammatory skin diseases. It affects a significant percentage of individual, compromising their skin barrier and resulting in skin inflammation and allergic sensitization [1].

Etiology

AD is a multifactorial condition involving environmental and genetic components, causing immune system and epidermis abnormalities. Asthma, AD, and allergic rhinoconjunctivitis make up the atopic triad. If a parent is atopic, the process of "atopic march" occurs, with 50% of offspring developing symptoms. Genetic changes such as filaggrin mutations can also predispose patients to other conditions. Food hypersensitivity can also worsen AD [2, 3].

Risk factors: AD is a challenging genetic disease influenced by multiple interactions between gene–gene and gene–environment. It is more likely in monozygotic twins (concordance rate of 75%) than in dizygotic twins (30%). However, environmental and developmental factors also contribute to susceptibility [4].

Genetics: Multiple genes, especially those that encode essential components of the immune system and epidermal structure proteins, have been linked to AD. The high correlation between AD and mutations in the filaggrin gene on chromosome 1 is a recent and intriguing genetic finding [5]. The most potent known genetic risk factor for AD is the filaggrin gene. About 50% of all AD sufferers had genetic changes in the filaggrin gene, compared to 10% of those from Western ethnicities. Mutations in the filaggrin gene cause the filaggrin protein to malfunction, compromising the skin barrier's integrity. Dry skin with cracks and an increased incidence of eczema are the clinical manifestations of such deficits. These mutations are absent in all AD sufferers, and other genetic variations have also been implicated [6]. AD develops through the interaction of all these genetic variations and risk factors related to the environment and development.

Environment: There is the notion that various environmental risk factors cause AD, but only a few are consistently accepted. For instance, there is strong testimony that our Western lifestyle has contributed to some of the reported rises in eczema manifestation over the past few years. However, this has neither been directly linked to environmental risk factors nor resulted in effective preventive measures [7]. Many support the hygiene theory to elucidate the sharp rise in the prevalence of eczema [8]. The hygiene theory suggests that early childhood exposure to infections such as hepatitis A and tuberculosis increases susceptibility to atopic diseases. Children in traditional farming environments with unpasteurized cow's milk and livestock quarters have a lower risk of developing the disease. Disease development is positively correlated with breastfeeding duration and the high social position of parents. These observations may support the hygiene hypothesis or the theory that eczema occurs in genetically susceptible individuals [9–12].

Epidemiology

Statistics on AD incidence and prevalence are rising, with the majority of cases happening in the first few years of life. With a steady plateau in Europe and North America and an increase in Asia, adult AD is rising. In children, the point prevalence ranged from 0% to 18.2%, whereas the 1-year prevalence ranged from 4.1% to 22.7%. The highest incidence of AD occurred during infancy and early childhood. In adults, the 1-year adult prevalence was 4.9% in the United States, 3.5% in Canada, and 4.4% in Europe [13, 14]. Both the 1-year prevalence and lifetime prevalence of diagnosed AD were higher in females than in males, except in the United Kingdom and the United States.

5.2 Stem Cell Biology and Mechanisms

Understanding stem cells, their traits, and their uses in science and medicine is the focus of the fascinating and dynamic field of stem cell biology research. The ability of stem cells to distinguish right into a wide type of specialized cells, consisting of muscle, nerve, or blood cells, sets them aside from other cells and may be quite beneficial in regenerative medication [15]. There are different types of stem cells, such as embryonic stem cells (ESCs), mesenchymal stem cells (MSCs), adult stem cells (somatic stem cells), induced pluripotent stem cells (iPSCs), perinatal stem cells, tissue-specific stem cells, and cancer stem cells.

MSCs: MSCs can multipotently differentiate into several cell types, such as osteoblasts, adipocytes, and chondrocytes. They are primarily localized in bone marrow, adipose tissue, and umbilical cord blood [16].

Key characteristics of MSCs:

Multipotency: MSCs can differentiate into various types of cells depending on the signals they receive, and this characteristic makes them very versatile in regenerative medicine and tissue repair [17].

Self-renewal: These cells can divide and replicate for long periods, sustaining the pool of stem cells in the body. One of the hallmarks of stem cells, including MSCs, is their ability to self-renew. It describes the capacity of stem cells to divide and generate both identical stem cells (replenishing the original stem cell population) and differentiated cells (these will become specialized cell types) [18].

In MSCs, the self-renewal process usually occurs through asymmetric cell division. Whenever an MSC undergoes cell division, one of the two daughter cells retains the characteristics of a stem cell (self-renewal). In contrast, the other cell commits to differentiate, resulting in a specialized cell type, such as a bone cell or cartilage cell [19, 20].

In vivo, MSCs are located within the stem cell niche, a specialized microenvironment that modulates stem cell activity. MSCs located in the bone marrow can be recruited into the injury sites following damage to bone tissue. They self-renew to replenish their pool and differentiate [21].

Immunomodulatory properties: MSCs will secrete several diverse classes of bioactive molecules, among them cytokines, which are small protein hormones working to regulate immune responses, as well as growth factors. Depending on the circumstances, those molecules can be either suppressors or enhancers of immune cell activity [22].

Cytokine secretion: MSCs secrete many bioactive molecules, including cytokines (small proteins regulating immune responses) and growth factors. Depending on the conditions, such molecules may suppress or aught immune cell activity [22].

Major cytokines and factors released by MSCs include:

TGF-β (transforming growth factor β): It plays a significant role in mediating anti-inflammation and inhibiting the activation of immune proinflammatory cells such as T cells and macrophages [23].

IL-10 (interleukin 10): This particular cytokine has such a strong anti-inflammatory effect that it counters the activities of inflammatory immune cells [23].

PGE2 (prostaglandin E2): This offers T cells any proinflammatory type of immune cell to suppress caps and thereby expand anti-inflammatory Tregs (regulatory T cells) [24].

HGF (hepatocyte growth factor): Inhibits inflammation and mounts some tissue-repairing activity [24].

MSC's influence on immunity includes direct cell-to-cell interactions: MSCs can also directly interact with immune cells through surface molecules on them. This interaction regulates T cell, B cell, natural killer (NK) cell, and macrophage activity and functions [25].

T-cell suppression: MSCs can prevent the proliferation and activation of T cells, especially the proinflammatory T_H1 and T_H17 types. They do this by expressing molecules such as PD-L1 (programmed death ligand 1), which bind to receptors on T cells and induce tolerance (i.e., the immune system does not attack the MSCs or the tissue they are helping to repair) [26].

Induction of Tregs: MSCs have been shown to promote the expansion of Tregs, a subset of T cells that suppress immune responses and prevent autoimmune reactions [27].

Macrophage polarization: MSCs can "reprogram" macrophages and favor their differentiation toward an anti-inflammatory phenotype, also known as the M2 type. The slow-healing macrophage, M2 (instead of the proinflammatory M1, a marker of chronic inflammation), is essential for wound healing and tissue repair [27].

Role of Treg induction: MSC has been demonstrated to support the proliferation of Tregs, which are T-cell subtypes capable of suppressing immunological reactions and protecting against autoimmunity.

Applications:

Biomechanics: MSCs could be used to treat mechanical damage. For example, they cure cardiac tissue, cartilage, and bone damage [28].

Immunotherapy and inflammation: MSCs have the potential for chronic inflammatory disorders, including autoimmune ailments, because of their immunomodulatory properties [29].

Cell-based therapies: MSCs are being investigated for potential application in stem cell therapies, which could replace diseased or damaged cells in various tissues [28, 29].

iPSCs: A type of stem cell known as iPSC can be created by reprogramming somatic cells, such as skin or blood cells, to become pluripotent again and differentiate into any cell in the human body. This vital discovery has opened the door for drug development, disease models, and regenerative medicine [28, 29].

Creation of iPSCs:

The process

Adult somatic cells are reprogrammed to return to a pluripotent state similar to ESCs to produce iPSCs. Shinya Yamanaka initially achieved this in 2006 by using the Yamanaka factors, a combination of only four genes (OCT4, SOX2, KLF4, and c-MYC) to trigger this reprogramming. These elements have been utilized to differentiate differentiated cells into a pluripotent phenotype and are crucial for maintaining pluripotency in ESCs [30].

iPSCs are produced through the following general process:

Somatic cell isolation: The individuals usually collect skin fibroblasts, blood cells, or other easily accessible cells [30, 31].

Introduction of reprogramming factors: The Yamanaka factors (or other variants) are introduced into the somatic cells using viral vectors or nonviral methods [30, 31].

Reprogramming: The somatic cells eventually become iPSCs as they progressively shed their specialized traits and recover the capacity to differentiate into any cell [30, 31].

Selection and expansion: Since the reprogrammed cells can differentiate into any cell, they are selected and multiplied to form more prominent clusters [30, 31].

Key characteristics of iPSCs

Pluripotency: iPSCs can differentiate into almost every cell type in the body, including blood, muscle, heart, and neuron cells. This trait is comparable to ESCs, which can differentiate into various cell types [32].
Self-renewal: Under the proper growth conditions, iPSCs can divide and reproduce endlessly, just like normal stem cells, and over time, they can develop into any form of cell [32].
Drug screening: Pharmaceutical firms can test the safety and effectiveness of new medications on human-derived cells, which may have a higher predictive value than animal models, thanks to iPSCs' ability to specialize in particular cell types pertinent to drug testing [33].

Applications of iPSCs:
iPSCs are being explored for a wide range of applications, particularly in regenerative medicine and therapeutic development:

Regenerative medicine: iPSCs can potentially regenerate damaged tissues or even entire organs. For instance, iPSCs could generate dopaminergic neurons for treating Parkinson disease, cardiac cells for heart disease, or pancreatic beta cells for diabetes [34, 35].
Tissue engineering and transplantation: iPSCs can be differentiated into specific cell types needed for tissue repair or organ regeneration. For example, iPSC-derived blood cells could be used for blood transfusions, or iPSC-derived skin cells might help with burn or wound healing [34, 35].
Gene therapy: iPSCs present a possible gene treatment approach. To treat a patient's ailment, iPSCs can be utilized to fix a genetic mutation *in vitro*. The patient can then get the repaired cells transplanted [36].
ESCs: Pluripotent stem cells, often called ESCs, are distinguished by their high developmental capacity, which means they can specialize in any cell type in the body. These cells are usually harvested from the inner cell mass of a blastocyst, a very early stage of embryonic development, around 4 to 5 days after fertilization. Pluripotent stem cells distinguish themselves by their high developmental capacity [37].

Indefinite division and self-renewal are two ways that pluripotent stem cells might produce more stem cells while maintaining their pluripotency.

ESC development: Scientists typically extract ESCs from the inner cell mass of a blastocyst, a very young embryo composed of roughly 50 to 150 cells. This phase happens roughly 4 to 5 days after fertilization. The inner cell mass will become the embryo; the outer layer of cells (the trophoblast) will help form the placenta. At this stage, the embryo has not yet differentiated into specific tissues or organs. The cells in the inner cell mass remain pluripotent; they can potentially become any cell type in the body. When ESCs are cultured in the laboratory, pluripotency is conserved, and these cells can continue to divide indefinitely to produce more ESCs without differentiating. This capacity for indefinite division is very important for long-term studies [38].

Properties of ESCs:

Pluripotency: As discussed, ESCs can differentiate into almost any type of cell in the body. These cells can turn into neurons, muscle cells, blood cells, or other cell types, which provides enormous potential for creating cells for regenerative medicine [39, 40].

Self-renewal: ESCs can proliferate indefinitely in a controlled environment, which means they can be maintained and expanded in a laboratory over long periods. However, this property is well controlled because, if not regulated, ESCs may form tumors (teratomas) *in vivo* [39, 40].

Teratoma formation: Their teratoma formation capability can readily identify pluripotent cells—benign tumors containing tissues originating from all three germ layers (ectoderm, mesoderm, and endoderm). This capability is utilized as a quality assay in stem cell research, verifying the pluripotency of ESCs [39, 40].

Applications of ESCs:

Regenerative medicine: Given developing medications for many diseases and injuries, potentially delivering success through cell therapy, ESCs become a source for producing

specific cell types enough to replace or regenerate damaged tissues and organs. Specific diseases include the following:
Neurogenic ailments: ESCs could create dopaminergic neurons to treat ailments such as Parkinson disease due to the specific degeneration of some cerebral cells. This holds promise for other neurodegenerative disorder-type treatments such as Alzheimer disease and spinal injuries [41].
Heart disease: ESCs could generate cardiomyocytes replacement for damaged tissue in heart failure during myocardial infarction. The research seeks to regenerate heart tissue and metabolism-hemodynamic function.
Liver, kidney, and lung diseases: The prospect of artificially functional ESCs—they could create liver, kidney, and lung cells that might help cure organ dysfunction. This could tend to supplant the need for organ transplantation in the future.
Drug development and disease modeling: ESCs serve immensely in the investigation of drugs. They can model diseases using differentiation into specific cell types resembling the disease state. A few examples include the following:

Mechanisms of action in immune modulation and skin repair

Stem cells, particularly MSCs, ESCs, and iPSCs, have portrayed immense possibilities for immune modulation and skin repair. The mechanisms of action implicated in these processes are complex and include different pathways that determine inflammation, tissue regeneration, and immune response [42].

Stem cells in immune modulation: Stem cells, especially MSCs, can modulate the immune response, acting on autoimmune diseases, inflammatory conditions, and even transplant rejection [43].

The mechanisms of immune modulation by stem cells are multiple in nature

Immunosuppressive effects of stem cells: MSCs may be isolated from bone marrow, adipose tissue, or the umbilical cord. They possess potent immunosuppressive properties that operate through a few mechanisms:
Secreting soluble factors: MSCs release soluble factors (cytokines, growth factors, and exosomes), which contribute

to the suppression of immune cell activation. These include TGF-β, IL-10, indoleamine 2,3-dioxygenase, and PGE2. These molecules exert their effect by inhibiting the activity of T cells, B cells, dendritic cells, and NK cells. Among these factors, TGF-β is hugely recognized for promoting immune tolerance and inhibiting autoimmune responses [44].

Induction of Tregs: MSCs stimulate the expansion and activation of Tregs important for immune homeostasis and preventing excessive inflammation. Tregs primarily inhibit effector T cells responsible for autoimmune responses and the promotion of chronic inflammation [44].

Inhibition of dendritic cells: MSCs interact with dendritic cells, which are important antigen-presenting cells triggering immune responses. MSCs might inhibit the maturation of dendritic cells, hence their ability to activate T cells and initiate immune responses [45].

Tolerance and autoimmune disease treatment: MSCs have been studied as potential treatments for autoimmune diseases (such as rheumatoid arthritis, multiple sclerosis, and lupus). Stem cell therapy is said to promote immune tolerance and reduce the chance of an overactive immune response that attacks body tissues, thus being a suitable therapeutic modality for the management of such conditions. In organ transplantation, MSCs are used in a strategy to prevent the grafts from rejection by reducing the activity of the immune system on the transplanted tissue, which adopts particular importance in the case of highly demanding long-term immunosuppressive drugs, which could also radioactively exert side effects [45].

Stem cells in skin repair: Skin stem cells are important for skin repair and regeneration after injury, burn, or chronic wounds. Skin is a thin but complex organ comprising three layers (epidermis, dermis, and hypodermis) undergoing different but coordinated repair processes [46].

Dermis regeneration (dermal stem cells): The dermis provides structural support and houses important skin components such as collagen, elastin, and blood vessels. Fibroblast is a type of cell that plays an important role in producing collagen and extracellular matrix (ECM) proteins. When the skin is injured, stem cells in the dermis (mostly MSCs that can differentiate into fibroblasts) cause ECM formation and collagen deposition, allowing the dermis to repair [47].

Wound healing

Inflammation: When the skin is injured, the immune system is activated and secretes inflammatory mediators. MSCs have anti-inflammatory properties that help control inflammation, making it less chronic and making the healing process more effective [48].

Dissemination: As part of the healing process, stem cells proliferate and differentiate into the desired skin type. For example, keratinocytes (epidermal stem cells) migrate to the injury site, covering the damaged area and forming a new epithelial layer. In the dermis, fibroblasts (derived from MSCs) contribute to the deposition of ECM, helping the skin repair process. This is important for wound healing because new blood vessels carry oxygen and nutrients to the damaged tissue. MSCs enable growth factors such as vascular endothelial growth factor (VEGF) to promote angiogenesis. This process can take several months, during which stem cells help the dermis regenerate and restore skin function [49].

Stem cells for burns and skin care

The mechanisms of action of stem cells in preventing and repairing skin diseases are important in terms of their potential in regenerative medicine. In immunomodulation, stem cells (especially MSCs) reduce inflammation, strengthen immunity, and help treat autoimmune diseases and organ transplants. In skin repair, stem cells play an important role in forming the epidermis and dermis, supporting wound healing, and restoring skin function after injury or disease. We are promising in the hospital. Despite the difficulties such as managing differences, preventing tumor formation, and ensuring safety and efficacy, we hope to achieve this goal. Research in these areas is ongoing, and stem cell therapies will become an important part of regenerative medicine in the future [50].

5.3 Types and Pathophysiology of AD

It has several types based on age, symptoms, and triggers.

Types based on age

1. *Infantile AD:* Occurs in neonates and young children (0–2 years of age). It typically impacts the facial region, scalp, and extensor surfaces (upper and lower limbs). Skin appears red, weepy, and crusty due to pruritic behavior. It is frequently correlated with hypersensitive reactions to dietary substances.
2. *Childhood AD:* Childhood AD affects children aged 2 to 12 years, causing dry, thickened, and intensely itchy skin on elbows, knees, wrists, and neck. It may improve with age but can persist.
3. *Adolescent and adult AD:* Stress, weather changes, and allergens trigger adolescent and adult AD. It commonly affects the hands, feet, eyelids, and neck.

Types based on clinical features

1. *Acute AD*: Acute AD is characterized by red, swollen, and blistered skin, often caused by allergens or irritants.
2. *Chronic AD*: It is a long-lasting form of AD with dry, thickened, and cracked skin. Skin appears leathery due to constant scratching.
3. *Nummular (discoid) eczema*: Presents as coin-shaped, scaly, itchy patches. It is often found on the arms, legs, and torso. It is more common in adults.
4. *Dyshidrotic eczema*: Sweat, stress, or allergens can trigger dyshidrotic eczema. It presents small, itchy blisters on the hands and feet.
5. *Contact AD*: Caused by irritants or allergens such as soaps, detergents, and metals. It often affects the face, hands, or areas of direct contact.

Types based on severity

1. *Intrinsic AD (nonallergic type):* Intrinsic AD has normal immunoglobulin E (IgE) levels and no specific allergen trigger.

2. *Extrinsic AD (allergic type):* Extrinsic AD is strongly associated with allergies and asthma, resulting in elevated IgE levels [51].

Pathophysiology of AD:
AD, also known as eczema, is an inflammatory skin disease that usually affects the skin's surface, causing dryness, irritation, itching, and recurring rashes. It usually begins in childhood and can persist into adulthood. A complex mix of genetic, immunological, and environmental variables contributes to the particular etiology of AD [52].

Genetic factors and skin barrier dysfunction
Genetic alterations that compromise the integrity of the skin barrier have been linked to AD. The skin's outermost layer, known as the epidermal barrier, is crucial in keeping allergies, irritants, and microbes out of the body. Because it aids in skin connection formation and moisture retention, the barrier function is significant. Filaggrin gene mutations (like the FLG mutation) in AD make it more difficult for allergens, irritants, and microbes to penetrate the skin, impairing the skin's capacity to retain moisture and defend against environmental factors, which can result in scarring. An immunological reaction brings on the pain, redness, and itching associated with AD [53].

The key features of the immune system in AD are as follows:

T helper 2–biased immune system: T helper 2 (T_H2) cells are overactive in the early stages of atopic dermatitis. T_H2 cells produce cytokines such as IL-4, IL-5, and IL-13, which drive the inflammatory response in the skin. These cytokines stimulate the activity of other immune cells, including mast cells and eosinophils, leading to the release of inflammatory chemicals (e.g., histamine, prostaglandins). The production of inflammatory cytokines is associated with allergic reactions and may contribute to the progression of AD. This leads to the development of skin conditions such as erythema (redness), eczema (itchy, scaly skin), and lithification (thickened, scaly skin) [53].

Altered skin microbiome: The skin microbiome (the community of bacteria living on the skin) plays an important role in skin care and disease prevention. In people with AD, the normal balance of the microbiota is disrupted, leading to the onset of disease [54].

Environmental triggers: Many environmental factors can trigger or worsen AD symptoms, especially in people with a genetic predisposition or a history of skin problems. Environmental triggers include:

- **Allergens:** Pollen, dust mites, cat dander, and mold can be environmental allergens that induce flare-ups. These allergens can activate the immune system and cause an allergic reaction when they pass the compromised skin barrier [54].
- **Irritants:** Certain substances, such as soaps, detergents, fragrances, and chemical exposures, can irritate the skin, causing it to become inflamed and itchy [54].
- **Weather conditions:** Cold, dry air can exacerbate barrier dysfunction, whereas hot and humid temperatures may induce sweating, further irritating the skin [54].
- **Stress:** Researchers have linked psychological stress to the onset and worsening of AD flare-ups. Stress can impact the immune system and increase the release of proinflammatory cytokines [54].
- **Chronic and relapsing:** AD is a chronic, remitting disease. The skin will go through a period of flare-ups followed by a period of remission, in which the symptoms improve [54].

Systemic involvement

Eosinophilic inflammation: Systemic allergic conditions may lead to eosinophils (a type of white blood cell) in affected tissues, further contributing to inflammation and immune activation.

Elevated IgE levels: Many individuals with AD have elevated serum IgE levels, indicating a heightened allergic response. IgE is produced in response to allergens and contributes to the inflammatory processes observed in AD [55].

5.4 Different Stem Cell Application on AD

Various stem cell sources, including MSCs from umbilical cord blood, adipose tissue, and amniotic fluid, suppress IgE levels, reduce inflammation, and prevent immune cell activation. Stem cell–derived exosomes and hematopoietic stem cells (HSCs) further aid skin homeostasis by down-regulating immune reactions and promoting keratinocyte differentiation.

Modifying the immune system and reassuring skin rejuvenation, stem cell remedy has revealed promising AD management. Due to the regenerative and immunomodulatory efficacy and low side effects, MSCs have come forth as beneficial for AD. This aims to repair the skin barrier, resist immune rejoinders, and diminish inflammation, contrary to the traditional approaches, which only prevent inflammations. MSCs have also been proven for preclinical and clinical studies from different sources such as palatine tonsils (TMSCs), umbilical cord blood (hUCB-MSCs), amniotic fluid (AFSCs), and adipose tissue (AD-MSCs) [56, 57]. The hUCB-MSCs have established distinguished enhancement in AD patients, depressing seditious indicators, pruritus, and the sternness of eczema without producing severe adverse properties. Furthermore, by adapting cytokine countenance and immunological actions, exosomes derived from MSCs offer a cell-free healing approach [58, 59]. Supplementary factors that sustain tissue renaissance and scar-free healing of lesions comprise AFSCs and EPSCs (epidermal stem cells) [60, 61]. Although the treatments are propitious, additional clinical study is essential for determining long-term efficacy, safety, and personalized treatment plans. The applications of different stem cell therapy in AD is illustrated in Figure 5.1. The different types of stem cells that are applied in curing AD are enumerated below:

MSCs—a promising approach: Regulation of diverse immunological routes has made MSC-based treatment a generative approach toward AD. Conferring to several researches conducted on human and animal models, it has been found that

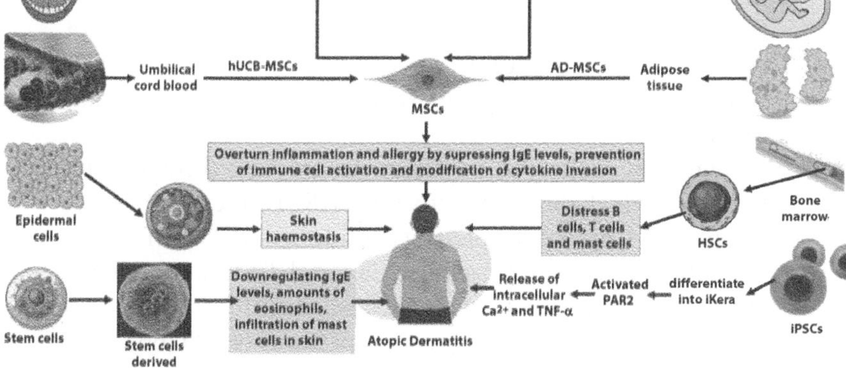

Figure 5.1 Illustration of how stem cell therapy helps treat AD by regulating immune responses and promoting skin healing.

MSCs can overturn inflammation and allergy by suppressing IgE levels, preventing immune cell activation, and modifying cytokine invention. MSCs can affect both adaptive and innate immunity *via* direct interactions. They can increase the production of IL-10 (anti-inflammatory cytokine) to downregulate allergic reactions by counteracting IL-3 and 4 [62]. In AD, AD-MSCs and hUCB-MSCs have proven prominent actions. Intravenous administration of AD-MSC upsurge of T_H1 and T_H2 responses, as well as downsurge IgE levels, MIP-2, and proinflammatory cytokines, leads to anti-inflammatory activity. hUCB-MSCs work through suppression of tumor necrosis factor α (TNF-α), hindering T_H2 differentiations and depressing IL-4 and IL-22 by concealing epidermal growth factor (EGF) [57]. TMSCs have also been shown as a probable therapy for AD as studied in the DNFB-induced AD animal models. It enhances TGF-β and downregulates IL-6, IL-1β, IL-4, and TNF-α [63, 64].

Epidermal stem cells and amniotic fluid stem cells: EPSCs are indispensable for skin hemostasis due to their self-regeneration and differentiation, and they are also used along with tissue engineering for application in the field of wound healing. AFSCs are flawless for clinical submissions and wound healing, irrespective of ethical issues. These can directly differentiate into the desired cell or boost neighboring cell dedifferentiation over transplantation [61].

HSCs: HSCs distress B cells, T cells, and mast cells, which assist in controlling immunity and subsidize AD. Patients may experience a new onset of AD due to allogeneic HSCT (HSC transplantation) because of GVHD (graft-versus-host disease) and alloimmunity. Recent approaches aim at HSC-based therapy, such as gene excision and immunity regulations [65].

Human iPSCs: The iPSCs differentiate into iKera (keratinocyte-like cells), which direct the activated PAR2 (protease-activated receptor 2). This is prompted by peptide ligand trypsin, which led to the release of intracellular Ca^{2+} and TNF-α. It has been found that iKera aids as an appreciated stage for transmission of drug targets in the inflammation pathway of AD [66].

Stem cell–derived exosome therapy: Exosomes derived from stem cells have been studied as an alternative option for AD treatment. These are nanoscale vesicles ranging from

30 to 200 nm and encompass lipids, proteins, and nucleic acids. By downregulating IgE levels, amounts of eosinophils, and infiltration of mast cells in the skin, these exosomes have been shown to reduce inflammation in AD. They also apply in skin abrasions due to their TNF-α, IL-4, IL-23, and IL-31 suppression ability. In contrast to full stem cell treatment, they have a low risk of teratoma formation [67].

5.5 Challenges and Ethical Considerations

Granting cellular rehabilitation and stem cell intercessions drew worldwide considerations; ethical problems, legal breaches, and dubious efficacy hinder their medical incorporation. The Sturner guideline is obligatory due to qualms about unfettered market maneuvers, misrepresentation, and the revelation of adverse properties. Because of the regenerative property, treatments based on SCs in dermatology grasp the potential for wound healing, skin disorders, and AD [68, 69]. Yet, there are questions about long-term efficiency and clinical reliability. Malfunctions of the dermal barrier and immunologic deregulation cause healing AD a bit more difficult. Emerging formularies based on nanotechnology and MSC-EV (mesenchymal stromal cell–derived extracellular vesicles) offer newer therapeutic opportunities by increasing immunomodulation and bioavailability. At the same time, traditional procedures aim only for symptomatic relief. Glitches, including erratic results, patient-specific alterability, and deprived clinical translation, still exist notwithstanding preclinical results [70, 71]. Table 5.1 shows the recent studies of stem cell therapy on AD.

Challenges in stem cell therapy for AD: Large-scale randomized controlled trials are obligatory to govern the usefulness and safety for long-term application of stem cell remedy in AD, even though early phase and preclinical measures give opportunistic results. The worth of treatment may be abolished if the receiver's immunity detects allogeneic stem cells as foreign bodies and produces a defense mechanism; hence, there might be a need for immunosuppressive agent application, which could have a negative impact. The rejuvenation potential of different stem cells, such as iPSCs, HSCs, and MSCs, diverges, making it grim to elect the most suitable one. If the pluripotent stem cells are not checked periodically, it may cause the development of teratoma,

Table 5.1 Recent studies of stem cell therapy on AD.

Mechanism of action	**Key findings**	**Reference**
Immunomodulation and anti-inflammatory effects	- hUCB-MSCs secrete TGF-β, which inhibits proinflammatory cytokines (e.g., TNF-α, IgE) - Reduces severity of AD symptoms in preclinical models - Modulates immune response and balances cytokine production	[79]
Restoration of skin barrier function	- Promotes filaggrin production, improving skin hydration and reducing transepidermal water loss - Enhances keratinocyte differentiation, maintaining skin barrier integrity	[80]
Mechanism of exosomes	- Exosomes derived from adipose tissue induce ceramide synthesis, improving skin hydration and barrier function - Reduces inflammatory cytokines (e.g., IL-4, IL-5, IL-13) - Enhances lamellar body formation	[81, 82]
Promotion of angiogenesis and tissue remodeling	- Releases VEGF to promote blood vessel formation. - Stimulates collagen production, improving skin elasticity and reducing scarring	[83]
Personalized medicine approaches	- Tailoring stem cell therapies based on genetic and environmental factors to enhance treatment efficacy	[79]
Regulatory pathways	- Focus on TGF-β and other cytokines released by stem cells - Understanding interactions between stem cells and immune cells for targeted therapies	[79]
Combination therapies	- Combining stem cell therapy with traditional immunomodulatory therapies to address multiple aspects of AD pathogenesis	[79, 81]

which enforces a safety issue. The iPSCs require more safety measures, and the applications of ESCs boon ethical queries [72, 73]. As there is an unavailability of proper standards for the growth, administration, and utilization of stem cells, therapeutic efficacy may be affected due to distinctions in differentiation methods and culture environments. The high expenditure on isolation, development, and regulatory endorsements makes it difficult for many patients to avail themselves of these services. It is also challenging to find a widespread stem cell–based therapy for AD, as the disorder involves skin barrier failure, immunity deregulation, and changes in the microbiome. Unpredictable results could be one of the consequences due to the impact of genetic diversity, environmental conditions, and distinctions of microenvironment. Even though MSCs and MSC-EV demonstrate positive outcomes, supplementary research is required to establish the mechanistic approach conveying skin renaissance and regulating immunity. To make the most of combination tactics, compatibility with systemic, topical, and nanotech-based medication must be evaluated [71, 74].

Ethical considerations in stem cell therapy for AD: Although tissue regenerative properties and immunomodulatory activity of stem cells make them a promising tool for AD, the origin of stem cells raises ethical queries. For instance, iPSCs circumvent the use of embryos (invented by Takahashi and Yamanaka in 2006) but raise inquiries regarding genetic alterations and risks of tumors. Adult SCs, like MSCs, although ethically favorable, show complications in obtainability and therapeutic constancy. ESCs encompass the obliteration of embryos, kindling moral debates [75, 76]. Informed agreement is essential to confirm patient safety, and unlicensed clinics that bid on untested treatment must be evaded. Strong governing directions are required for regimenting procedures and ceasing unethical activities. Clinical measures and authorized outlines are significantly impacted by the intricate moral considerations raised by SC study and treatment. Governing agencies such as US regulations leading SC clinics have been tautened by the Food and Drug Administration, and they now necessitate clinics to acquiesce medical stats and undergo opposite testing before offering treatments, confirming that SC therapies endure laborious clinical tribunals similar to other therapeutic measures [77, 78].

5.6 Recent Advancement of Stem Cell Therapy on AD

Recent advancements in stem cell therapy for AD have been significant, particularly focusing on the therapeutic potential of various types of stem cells, such as human umbilical cord blood–derived MSCs (hUCB-MSCs). Here is some key advancements:

1. **Immunomodulation and anti-inflammatory effects:** Different studies interpreted hUCB-MSC application in AD. One of the key mechanisms involves the secretion of TGF-β, which constrains the secretion of proinflammatory cytokines (e.g., TNF-α and IgE from mast cell and B cell). This mechanism contributes an effective approach to improve the symptoms related to AD. Recent research revealed that the administration of hUCB-MSCs could reduce the sternness of AD symptoms in preclinical models. The stem cells seem to modulate the immune response, promoting an equilibrium in cytokine production that mitigates the inflammatory processes responsible for skin lesions [79].
2. **Restoration of skin barrier function:** MSC contributes to the regeneration and remodeling of the skin barrier. Filaggrin is a key protein involved in skin barrier function, which is promoted by unregulated production of filaggrin. It improves the skin hydration and reduces the transepidermal water loss. Stem cell also enhances the differentiation of keratinocytes to maintain the structural integrity of the skin barrier [80].
3. **Mechanism of exosomes:** Exosomes have contained various bioactive molecules that have been derived from adipose tissues. These exosomes induce *de novo* synthesis of ceramides, which is essential for maintaining skin hydration, and barrier functionality in AD can prevent the epidermal disruption. The therapy has been observed to reduce levels of inflammatory cytokines (e.g., IL-4, IL-5, IL-13) associated with AD, thereby modifying systemic inflammation and enhancing the skin's repair processes. MSC-exosomes significantly improve the skin barrier's integrity by reducing transepidermal water loss

and enhanced hydration of the stratum corneum. The therapy promotes the lamellar body's formation, which secretes lipids that contribute to the skin's protective barrier [81]. The use of MSC-exosomes represents a cell-free therapeutic approach, which may reduce the risks associated with direct stem cell transplantation while providing effective treatment for AD patients. This approach could be particularly beneficial for patients who do not respond well to conventional therapies, presenting a novel avenue for managing AD [82].

4. **Promotion of angiogenesis and tissue remodeling:** Stem cells enhance wound healing and tissue remodeling in AD by releasing VEGF, which promote blood vessel formation. It also promotes the production of collagen, which improves skin elasticity and reduces scarring [83].
5. **Regulatory pathways:** In recent researches focuses on the understanding the regularity pathway that includes TGF-β and other cytokines released by stem cells. Deeper understanding of how stem cells interact with immune cells provides insights for developing targeted therapies to improve patient compliance and treatment outcome. There has been increasing interest in understanding the regulatory pathways that involve TGF-β and other cytokines released by stem cells. This research facilitates a deeper understanding of how stem cells interact with the immune system and provides insights for developing targeted therapies that can improve treatment outcomes for patients with AD.
6. **Personalized medicine approaches:** Advances in understanding the genetic and environmental factors contributing to AD are paving the way for personalized stem cell therapies. By tailoring treatments based on individual patient profiles, the efficacy of stem cell interventions may be significantly enhanced.
7. **Combination therapies:** Recent studies are exploring the use of stem cell therapy in combination with other treatments. This approach aims to enhance therapeutic efficacy by addressing multiple aspects of the disease pathogenesis simultaneously, such as the use of stem cells alongside traditional immunomodulatory therapies.

MSCs derived from patients with AD show increased levels of proinflammatory cytokines such as IL-6 and IL-13 compared to healthy MSCs, while the level of IL-4 remains unchanged. Additionally, AD-MSCs enhance the activation of T_H1 and T_H17 cells and promote the production of inflammatory cytokines, contrasting with the decreased production of T_H2 cytokines. This indicates a dysregulated immune response in AD-MSCs compared to their healthy counterparts [20].

MSCs derived from patients with AD exhibit higher expression levels of proinflammatory cytokines such as IL-6 and IL-13 compared to healthy MSCs. In contrast, the expression of IL-4 does not show significant differences between AD-MSCs and healthy MSCs. Additionally, AD-MSCs promote the activation of T_H1 and T_H17 cells while decreasing T_H2 cell activity, indicating a shift in the immune response [84].

5.7 Clinical Research Updates

Preclinical studies

Different preclinical studies have demonstrated that the efficacy of MSC therapy in animal models in subcutaneous administration of hUCB-MSCs significantly reduced the severity of AD-like symptoms in NC/Nga mice, a well-established model of AD. The treatment shows a reduction in epidermal thickness, mast cell infiltration, and serum IgE levels [79]. Similarly, Shin *et al.* (2020) showed that MSC-exosomes improved epidermal barrier function and reduced inflammation in a mouse model of oxazolone-induced dermatitis [79, 81].

Clinical studies:

MSC treatment for AD shows promising results in preclinical as well as in preliminary clinical trials. However, it is too early to ensure all aspects of safety and efficacy [85, 86]. Using hUCB-MSCs in moderate to severe atopic dermatitis shows a significant impact in phase I/IIa clinical trials, which is measured by the Eczema Area and Severity Index and the Scoring Atopic Dermatitis index [85]. Patients also reported reduced pruritus (itchy skin) and improved quality of life. The study concluded that hUCB-MSCs are well-tolerated and safe, with no reported serious adverse events.

Clinical research updates

Although the clinical trial findings are promising, MSC therapies still have several challenges in clinical trials, which include:

1. **Limited tangible outcomes**: Where numerous trials are involved in stem cell therapy for skin, only a few trials have demonstrated significant results on a benefit patient [86].
2. **Standardization issues**: The lack of standard protocol for the source, preparation, and dosing of MSCs leads to variability in treatment efficacy and effectiveness [86].
3. **Regulatory hurdles**: Regulatory aspects of cell-based therapies are very complex, which leads to delays in the approval and establishment of effective MSC therapy [86].
4. **Long-term efficacy and safety**: Long-term safety and efficacy need to be focused on, which has been insufficient in many studies. Research should be focused on the long-term effectiveness and potential adverse effects of MSC therapies [86].
5. **Patient variability**: Genetic background, the underlying cause of skin disorders, and concurrent patient treatment can also be influential factors for the complication and interpretation of the clinical trial data outcome [86].

Apart from the challenges mentioned previously, the source of stem cells has a critical impact on the study. MSCs derived from patients with AD show increased levels of proinflammatory cytokines compared to healthy MSCs. This indicates a dysregulated immune response in AD-MSCs compared to their healthy counterparts. To improve MSC treatments' efficacy in treating disorders such as AD and advance them in clinical trials, these issues must be addressed [87, 88].

Future Directions

Optimization of MSC therapy: Future research must focus on the optimization of MSC therapy for AD, which includes the identification of effective MSC sources (e.g., umbilical cord, adipose tissues, bone marrow), the optimum dosage, and suitable route of administration (e.g., intravenous, subcutaneous, and topical). To reduce the risk of live cell therapy, we must focus on the alternative use of exosomes. Exosomes hold great promise as a cell-free alternative, as they can be easily stored and be standardized [89, 90].

Combination therapies: Combining existing therapy with MSCs can improve and enhance the effect of treatment; such examples are topical corticosteroids or calcineurin inhibitors. Traditional treatment could provide immediate relief, whereas MSCs could reduce inflammation and restore the skin barrier. Future studies are needed to explore the potential synergistic effects of such combination therapies [91, 92].

Biomarker development: To predict the response to treatment and monitor disease progression, we need biomarkers that need further research. Identifying suitable biomarkers associated with MSC therapy could help personalize treatment and improve patient compliance [93].

Conclusion

AD remains a difficult disease to manage because of its multifactorial etiology and chronic relapsing nature. However, current research into skin physiology and drug development lays the foundation for more effective and targeted treatments. By treating the underlying dysregulation of the immune system and barrier dysfunction in the skin, future therapies could dramatically enhance the quality of life for individuals with AD. Ongoing coordination between dermatologists, researchers, and drug companies will be critical to bringing these advances into practice.

AI disclosure statement: The author confirms that no artificial intelligence (AI)-generated content or AI-assisted writing tools were used in the preparation of this chapter (including summaries and the abstract). All text, analysis, and conclusions are the result of the author's original work.

References

1. Paller, A., Jaworski, J.C., Simpson, E.L., Boguniewicz, M., Russell, J.J., Block, J.K., Tofte, S., Dunn, J.D., Feldman, S.R., Clark, A.R., Schwartz, G., Major comorbidities of atopic dermatitis: beyond allergic disorders. *Am. J. Clin. Dermatol.*, 19, 821–38, 2018 Dec.
2. Oliveira, A.D., Sodré, C.S., de Carvalho Ferreira, D., de Dios Abad, E., Saintive, S., Ribeiro, M., Cavalcante, F.S., Piciani, B., Gonçalves, L.S., Oral aspects identified in atopic dermatitis patients: a literature review. *Open Dent. J.*, 12, 424, 2018.
3. Berg, A.K., Nørgaard, K., Thyssen, J.P., Zachariae, C., Hommel, E., Rytter, K., Svensson, J., Skin problems associated with insulin pumps and sensors in

adults with type 1 diabetes: a cross-sectional study. *Diabetes Technol. Ther.*, 20, 7, 475–82, 2018 Jul 1.

4. Thomsen, S.F., Ulrik, C.S., Kyvik, K.O., Hjelmborg, J.V., Skadhauge, L.R., Steffensen, I., Backer, V., Importance of genetic factors in the etiology of atopic dermatitis: a twin study. *Allergy Asthma Proc.*, 28, 5, 535, 2007 Sep 1.
5. Palmer, C.N., Irvine, A.D., Terron-Kwiatkowski, A., Zhao, Y., Liao, H., Lee, S.P., Goudie, D.R., Sandilands, A., Campbell, L.E., Smith, F.J., O'Regan, G.M., Common loss-of-function variants of the epidermal barrier protein filaggrin are a major predisposing factor for atopic dermatitis. *Nat. Genet.*, 38, 4, 441–6, 2006 Apr 1.
6. Irvine, A.D., McLean, W.I., Leung, D.Y., Filaggrin mutations associated with skin and allergic diseases. *N. Engl. J. Med.*, 365, 14, 1315–27, 2011 Oct 6.
7. Douwes, J. and Pearce, N., Asthma and the westernization 'package'. *Int. J. Epidemiol.*, 31, 6, 1098–102, 2002 Dec 1.
8. Strachan, D.P., Hay fever, hygiene, and household size. *BMJ Br. Med. J.*, 299, 6710, 1259, 1989 Nov 11.
9. Bach, J.F., The effect of infections on susceptibility to autoimmune and allergic diseases. *N. Engl. J. Med.*, 347, 12, 911–20, 2002 Sep 19.
10. von Mutius, E., Maternal farm exposure/ingestion of unpasteurized cow's milk and allergic disease. *Curr. Opin. Gastroenterol.*, 28, 6, 570–6, 2012 Nov 1.
11. Hong, S., Choi, W.J., Kwon, H.J., Cho, Y.H., Yum, H.Y., Son, D.K., Effect of prolonged breast-feeding on risk of atopic dermatitis in early childhood. *Allergy Asthma Proc.*, 35, 1, 66–70, 2014 Jan 1.
12. Hammer-Helmich, L., Linneberg, A., Thomsen, S.F., Glümer, C., Association between parental socioeconomic position and prevalence of asthma, atopic eczema and hay fever in children. *Scand. J. Public Health*, 42, 2, 120–7, 2014 Mar.
13. Mortz, C.G., Andersen, K.E., Dellgren, C., Barington, T., Bindslev-Jensen, C., Atopic dermatitis from adolescence to adulthood in the TOACS cohort: prevalence, persistence and comorbidities. *Allergy*, 70, 7, 836–45, 2015 Jul.
14. Shreberk-Hassidim, R., Hassidim, A., Gronovich, Y., Dalal, A., Molho-Pessach, V., Zlotogorski, A., Atopic dermatitis in Israeli adolescents from 1998 to 2013: trends in time and association with migraine. *Pediatr. Dermatol.*, 34, 3, 247–52, 2017 May.
15. He, K., Zang, J., Ren, T., Feng, S., Liu, M., Zhang, X., Sun, W., Chu, J., Xu, D., Liu, F., Therapeutic Potential and Mechanisms of Mesenchymal Stem Cell and Mesenchymal Stem Cell-Derived Extracellular Vesicles in Atopic Dermatitis. *J. Inflamm. Res.*, 17, 5783–800, 2024 Dec 31.
16. Sierra-Sánchez, A., Montero-Vilchez, T., Quiñones-Vico, M.I., Sanchez-Diaz, M., Arias-Santiago, S., Current advanced therapies based on human mesenchymal stem cells for skin diseases. *Front. Cell Dev. Biol.*, 9, 643125, 2021 Mar 9.
17. Sah, S.K., Agrahari, G., Nguyen, C.T., Kim, Y.S., Kang, K.S., Kim, T.Y., Enhanced therapeutic effects of human mesenchymal stem cells transduced

with superoxide dismutase 3 in a murine atopic dermatitis-like skin inflammation model. *Allergy*, 73, 12, 2364–76, 2018 Dec.
18. Prodinger, C.M., Reichelt, J., Bauer, J.W., Laimer, M., Current and future perspectives of stem cell therapy in dermatology. *Ann. Dermatol.*, 29, 6, 667–87, 2017 Dec 1.
19. Kaiser, M. and Badiavas, E.V., Stem Cell Therapy in Telemedicine and Technological Advances in Dermatology, in: *Telemedicine and Technological Advances in Dermatology*, pp. 387–98, 2024 Nov 14.
20. Kader, H.A., Azeem, M., Jwayed, S.A., Al-Shehhi, A., Tabassum, A., Ayoub, M.A., Hetta, H.F., Waheed, Y., Iratni, R., Al-Dhaheri, A., Muhammad, K., Current insights into immunology and novel therapeutics of atopic dermatitis. *Cells*, 10, 6, 1392, 2021 Jun 4.
21. Yan, J., Zhang, L., Ye, J., Yan, S., Han, X., Han, Z., Effective remission of an elderly patient with refractory atopic dermatitis and foot ulcers following infusion with human umbilical cord□derived mesenchymal stem cells: A case report. *World Acad. Sci. J.*, 3, 1, 1, 2021 Jan 1.
22. Khandpur, S., Gupta, S., Gunaabalaji, D.R., Stem cell therapy in dermatology. *Indian J. Dermatol. Venereol. Leprol.*, 87, 6, 753–67, 2021 Oct 23.
23. Davies, O.G., Williams, S., Goldie, K., The therapeutic and commercial landscape of stem cell vesicles in regenerative dermatology. *J. Controlled Release*, 353, 1096–106, 2023 Jan 1.
24. Sugita, K. and Akdis, C.A., Recent developments and advances in atopic dermatitis and food allergy. *Allergol. Int.*, 69, 2, 204–14, 2020.
25. Hua, C., Chen, S., Cheng, H., Therapeutic potential of mesenchymal stem cells for refractory inflammatory and immune skin diseases. *Hum. Vaccines Immunother.*, 18, 6, 2144667, 2022 Nov 30.
26. Bhatt, D.M., Singh, A., Madke, B., Jangid, S.D., Ramya, T.S., Pharmacological Trends in the Management of Atopic Dermatitis: A Comprehensive Review. *Cureus*, 16, 7, e64302, 2024 Jul 11.
27. Hemrajani, C., Negi, P., Parashar, A., Gupta, G., Jha, N.K., Singh, S.K., Chellappan, D.K., Dua, K., Overcoming drug delivery barriers and challenges in topical therapy of atopic dermatitis: A nanotechnological perspective. *Biomed. Pharmacother.*, 147, 112633, 2022 Mar 1.
28. Li, H., Tian, Y., Xie, L., Liu, X., Huang, Z., Su, W., Mesenchymal stem cells in allergic diseases: Current status. *Allergol. Int.*, 69, 1, 35–45, 2020.
29. Ra, J.C., Kang, S.K., Shin, I.S., Park, H.G., Joo, S.A., Kim, J.G., Kang, B.C., Lee, Y.S., Nakama, K., Piao, M., Sohl, B., Stem cell treatment for patients with autoimmune disease by systemic infusion of culture-expanded autologous adipose tissue derived mesenchymal stem cells. *J. Transl. Med.*, 9, 1–1, 2011 Dec.
30. Ryu, B., Baek, J., Kim, H., Lee, J.H., Kim, J., Jeong, Y.H., Lee, S.G., Kang, K.R., Oh, M.S., Kim, E.Y., Kim, C.Y., Anti-inflammatory effects of M-MSCs in DNCB-induced atopic dermatitis mice. *Biomedicines*, 8, 10, 439, 2020 Oct 21.

31. Zibandeh, N., Genc, D., Ozgen, Z., Duran, Y., Goker, K., Baris, S., Ergun, T., Akkoc, T., Mesenchymal stem cells derived from human dental follicle modulate the aberrant immune response in atopic dermatitis. *Immunotherapy*, 13, 10, 825–40, 2021 Jul.
32. Legiawati, L., Suseno, L.S., Sitohang, I.B., Yusharyahya, S.N., Ardelia, A., Paramastri, K., Stem Cells as a therapeutic choice in dermatological disorders. *Curr. Stem Cell Res. Ther.*, 18, 8, 1069–75, 2023 Nov 1.
33. Hall, M.N., Rosenkrantz, W.S., Hong, J.H., Griffin, C.E., Mendelsohn, C.M., Evaluation of the potential use of adipose-derived mesenchymal stromal cells in the treatment of canine atopic dermatitis: a pilot study. *Vet. Ther.*, 11, 2, E1–14, 2010 Jan 1.
34. Myles, I.A., Castillo, C.R., Barbian, K.D., Kanakabandi, K., Virtaneva, K., Fitzmeyer, E., Paneru, M., Otaizo-Carrasquero, F., Myers, T.G., Markowitz, T.E., Moore, I.N., Therapeutic responses to Roseomonas mucosa in atopic dermatitis may involve lipid-mediated TNF-related epithelial repair. *Sci. Transl. Med.*, 12, 560, eaaz8631, 2020 Sep 9.
35. Nishimoto, R., Kodama, C., Yamashita, H., Hattori, F., Human Induced Pluripotent Stem Cell–Derived Keratinocyte-Like Cells for Research on Protease-Activated Receptor 2 in Nonhistaminergic Cascades of Atopic Dermatitis. *J. Pharmacol. Exp. Ther.*, 384, 2, 248–53, 2023 Feb 1.
36. Sideris, N., Paschou, E., Bakirtzi, K., Kiritsi, D., Papadimitriou, I., Tsentemeidou, A., Sotiriou, E., Vakirlis, E., New and upcoming topical treatments for atopic dermatitis: a review of the literature. *J. Clin. Med.*, 11, 17, 4974, 2022 Aug 24.
37. Salman, A. and Giménez-Arnau, A.M., Emerging systemic treatment options in atopic dermatitis. *Balk. Med. J.*, 41, 4, 239, 2024 Jul 5.
38. Kim, J.H., Kim, J.E., Kang, S.J., Yoon, J.K., Exosomes and Exosome-Mimetics for Atopic Dermatitis Therapy. *Tissue Eng. Regener. Med.*, 22, 1–6, 2025 Jan 20.
39. Hoffman, A.M. and Dow, S.W., Concise review: stem cell trials using companion animal disease models. *Stem Cells*, 34, 7, 1709–29, 2016 Jul 1.
40. Skayem, C. and Duong, T.A. (Eds.), *Latest Breakthroughs in the Treatment of Atopic Dermatitis*, Intech Open, 2024 Feb 28.
41. Zibandeh, N., Genc, D., Ozgen, Z., Duran, Y., Kasap, N., Goker, K., Baris, S., Ergun, T., Akkoc, T., Effect of dental follicle mesenchymal stem cell on Th1 and Th2 derived naive T cells in atopic dermatitis patients. *Clin. Exp. Health Sci.*, 9, 3, 220–7, 2019 Jul 1.
42. Zhang, L., Yang, X., Han, Z., Han, Z., Guo, T., Gao, X., Cai, H., Natural Killer Cells in Atopic Dermatitis Opening Doors to New Treatments, in: *Latest Breakthroughs in the Treatment of Atopic Dermatitis*, IntechOpen, 2023 Aug 17.
43. Song, S.J., Nam, Y., Rim, Y.A., Ju, J.H., Sohn, Y., Comparative analysis of regulations and studies on stem cell therapies: focusing on induced pluripotent

stem cell (iPSC)-based treatments. *Stem Cell Res. Ther.*, 15, 1, 447, 2024 Nov 21.

44. Miyagaki, T. and Sugaya, M., Recent advances in atopic dermatitis and psoriasis: genetic background, barrier function, and therapeutic targets. *J. Dermatol. Sci.*, 78, 2, 89–94, 2015 May 1.
45. Lee, B.C. and Kang, K.S., Functional enhancement strategies for immunomodulation of mesenchymal stem cells and their therapeutic application. *Stem Cell Res. Ther.*, 11, 1–0, 2020 Dec.
46. Bellei, B., Migliano, E., Picardo, M., Therapeutic potential of adipose tissue-derivatives in modern dermatology. *Exp. Dermatol.*, 31, 12, 1837–52, 2022 Dec.
47. Majewska, L., Dorosz, K., Kijowski, J., Efficacy of Rose Stem Cell-Derived Exosomes (RSCEs) in Skin Treatment: From Healing to Hyperpigmentation Management: Case Series and Review. *J. Cosmet. Dermatol.*, 24, 1, e16776, 2025 Jan.
48. Shukla, S., Mamale, K.B., Arya, R.K., Kaundal, R.K., Shukla, R., Therapeutic potential of microneedles-based delivery systems for the management of atopic dermatitis. *J. Drug Deliv. Sci. Technol.*, 84, 104493, 2023 Jun 1.
49. Ali, S., Ion, A., Orzan, O.A., Bălăceanu-Gurău, B., Emerging Treatments and New Vehicle Formulations for Atopic Dermatitis. *Pharmaceutics*, 16, 11, 1425, 2024 Nov 7.
50. Zuo, W., Yue, Z., Xu, S., Sun, C.H., Zou, X.F., Ma, J., Yan, H., Gu, X.W., Wang, M.Y., Emerging trends and research hotspots in the relationship between mast cells and atopic dermatitis based on the literature from 2001 to 2024: A bibliometric and visualized analysis. *Skin Res. Technol.*, 30, 9, e70053, 2024 Sep.
51. Tokura, Y., Extrinsic and intrinsic types of atopic dermatitis. *J. Dermatol. Sci.*, 58, 1, 1–7, 2010 Apr 1.
52. Eichenfield, L.F., Ahluwalia, J., Waldman, A., Borok, J., Udkoff, J., Boguniewicz, M., Current guidelines for the evaluation and management of atopic dermatitis–a comparison of the Joint Task Force Practice Parameter and American Academy of Dermatology Guidelines. *Alergol. Pol.-Polish J. Allergol.*, 4, 4, 158–68, 2017 Oct 1.
53. Marsella, R., New Treatments and Validations for Atopic Dermatitis in Humans after Comparative Approach with Canine Models, in: *Latest Breakthroughs in the Treatment of Atopic Dermatitis*, IntechOpen, 2023 May 18.
54. Vafaeian, A., Rajabi, F., Rezaei, N., The role of Toll-like receptors in the pathogenesis of atopic dermatitis. *Heliyon*, 11, 3, 2025 Jan 31.
55. Grajales, D.B., Sewdat, N., Leo, R., Kar, S., Unveiling abrocitinib: A thorough examination of the 2022 USFDA-approved treatment for atopic dermatitis (AD). *Med. Drug Discov.*, 20, 100161, 2023 Oct 10.
57. Daltro, S.R.T., Meira, C.S., Santos, I.P., Ribeiro dos Santos, R., Soares, M.B.P., Mesenchymal stem cells and atopic dermatitis: a review. *Front. Cell Dev. Biol.*, 8, 326, 2020.

58. Yang, J., Xiao, M., Ma, K., Li, H., Ran, M., Yang, S., Yang, Y., Fu, X., Yang, S., Therapeutic effects of mesenchymal stem cells and their derivatives in common skin inflammatory diseases: Atopic dermatitis and psoriasis. *Front. Immunol.*, 14, 1092668, 2023.
59. Kim, H.S., Yun, J.W., Shin, T.H., Lee, S.H., Lee, B.C., Yu, K.R., Seo, Y., Lee, S., Kang, T.W., Choi, S.W., Seo, K.W., Human umbilical cord blood mesenchymal stem cell-derived PGE2 and TGF-β1 alleviate atopic dermatitis by reducing mast cell degranulation. *Stem Cells*, 33, 4, 1254–1266, 2015.
60. Cho, B.S., Kim, J.O., Ha, D.H., Yi, Y.W., Exosomes derived from human adipose tissue-derived mesenchymal stem cells alleviate atopic dermatitis. *Stem Cell Res. Ther.*, 9, 1, 187, 2018.
61. Luo, H., Wang, Z., Qi, F., Wang, D., Applications of human amniotic fluid stem cells in wound healing. *Chin. Med. J.*, 135, 19, 2272–2281, 2022.
62. Yang, R., Yang, S., Zhao, J., Hu, X., Chen, X., Wang, J., Xie, J., Xiong, K., Progress in studies of epidermal stem cells and their application in skin tissue engineering. *Stem Cell Res. Ther.*, 11, 1–13, 2020.
63. Dias, I.E., Pinto, P.O., Barros, L.C., Viegas, C.A., Dias, I.R., Carvalho, P.P., Mesenchymal stem cells therapy in companion animals: useful for immune-mediated diseases? *BMC Vet. Res.*, 15, 358, 2019.
64. Jung, H., Son, G.M., Lee, J.J., Park, H.S., Therapeutic effects of tonsil-derived mesenchymal stem cells in an atopic dermatitis mouse model. *In Vivo*, 35, 2, 845–857, 2021.
65. Horino, S., Yamaguchi, Y., Miyabayashi, H., Aki, H., Nanjo, Y., Onuma, M., Rikiishi, T., Yabe, H., Imaizumi, M., Sato, A., Miura, K., Topical therapy and skin care for transplant-associated atopic dermatitis in children and adolescents. *Pediatr. Transplant.*, 28, 1, e14653, 2024.
66. Take, N., Nakahara, T., Kido-Nakahara, M., Furue, M., Three cases of adult-onset atopic dermatitis after hematopoietic stem cell transplantation. *Allergol. Int.*, 67, 4, 529–531, 2018.
67. Park, K.Y., Han, H.S., Park, J.W., Kwon, H.H., Park, G.H., Seo, S.J., Exosomes derived from human adipose tissue-derived mesenchymal stem cells for the treatment of dupilumab-related facial redness in patients with atopic dermatitis: A report of two cases. *J. Cosmet. Dermatol.*, 21, 2, 844–849, 2022.
68. Charitos, I.A., Ballini, A., Cantore, S., Boccellino, M., Di Domenico, M., Borsani, E., Nocini, R., Di Cosola, M., Santacroce, L., Bottalico, L., Stem cells: a historical review about biological, religious, and ethical issues. *Stem Cells Int.*, 1, 9978837, 2021.
69. Abdumutallibovna, M.M. and Orifovna, O.B., Stem Cell Therapy: Advances and Ethical Considerations. *Ethiop. Int. J. Multi. Res.*, 12, 03, 16–20, 2025.
70. Gatmaitan, J.G. and Lee, J.H., Challenges and future trends in atopic dermatitis. *Int. J. Mol. Sci.*, 24, 14, 11380, 2023.
71. Al-Masawa, M.E., Elfawy, L.A., Ng, C.Y., Ng, M.H., Law, J.X., Mesenchymal Stromal Cell-Derived Extracellular Vesicles in the Management of Atopic

Dermatitis: A Scoping Review of Therapeutic Opportunities and Challenges. *Int. J. Nanomed.*, 20, 2673–2693, 2025.

72. Volarevic, V., Markovic, B.S., Gazdic, M., Volarevic, A., Jovicic, N., Arsenijevic, N., Armstrong, L., Djonov, V., Lako, M., Stojkovic, M., Ethical and safety issues of stem cell-based therapy. *Int. J. Med. Sci.*, 15, 1, 36, 2018.
73. Moradi, S., Mahdizadeh, H., Šarić, T., Kim, J., Harati, J., Shahsavarani, H., Greber, B., Moore, J.B., Research and therapy with induced pluripotent stem cells (iPSCs): social, legal, and ethical considerations. *Stem Cell Res. Ther.*, 10, 1–13, 2019.
74. Chakraborty, D. and De, A., Stem-cell therapy in dermatology–Challenges and opportunities. *Indian J. Skin Allergy*, 3, 2, 93–105, 2024.
75. Turner, L., Direct-to-consumer marketing of stem cell interventions by Canadian businesses. *Regen. Med.*, 3, 6, 643–58, 2018.
76. King, N.M. and Perrin, J., Ethical issues in stem cell research and therapy. *Stem Cell Res. Ther.*, 5, 1–6, 2014.
77. Qiu, T. and Toumi, M., *Regenerative Medicine: Unlocking Patient Access and Commercial Potential*, CRC Press, 2023.
78. Nadel, L., The Future of Stem Cell Therapy Regulation under the FDA's Comprehensive Regenerative Medicine Policy Framework through a Public Health Lens. *Hous. J. Health L. Pol'y*, 21, 223, 2021.
79. Park, H.H., Lee, S., Yu, Y., Yoo, S.M., Baek, S.Y., Jung, N., Seo, K.W., Kang, K.S., TGF-β secreted by human umbilical cord blood-derived mesenchymal stem cells ameliorates atopic dermatitis by inhibiting secretion of TNF-α and IgE. *Stem Cells*, 38, 7, 904–16, 2020 Jul 1.
80. Lee, H.J., Lee, J.K., Lee, H., Carter, J.E., Chang, J.W., Oh, W., Yang, Y.S., Suh, J.G., Lee, B.H., Jin, H.K., Bae, J.S., Human umbilical cord blood-derived mesenchymal stem cells improve neuropathology and cognitive impairment in an Alzheimer's disease mouse model through modulation of neuroinflammation. *Neurobiol. Aging*, 33, 3, 588–602, 2012 Mar 1.
81. Shin, K.O., Ha, D.H., Kim, J.O., Crumrine, D.A., Meyer, J.M., Wakefield, J.S., Lee, Y., Kim, B., Kim, S., Kim, H.K., Lee, J., Exosomes from human adipose tissue-derived mesenchymal stem cells promote epidermal barrier repair by inducing de novo synthesis of ceramides in atopic dermatitis. *Cells*, 9, 3, 680, 2020 Mar 10.
82. Matsuda, K., Falkenberg, K.J., Woods, A.A., Choi, Y.S., Morrison, W.A., Dilley, R.J., Adipose-derived stem cells promote angiogenesis and tissue formation for *in vivo* tissue engineering. *Tissue Eng. Part A*, 19, 11–12, 1327–35, 2013 Jun 1.
83. Swain, H.N., Boyce, P.D., Bromet, B.A., Barozinksy, K., Hance, L., Shields, D., Olbricht, G.R., Semon, J.A., Mesenchymal stem cells in autoimmune disease: A systematic review and meta-analysis of pre-clinical studies. *Biochimie.*, 223, 54–73, 2024.

84. Jiang, W., Xu, J., Immune modulation by mesenchymal stem cells. *Cell Prolif.*, 53, 1, e12712. 2020.
85. Leung, D.Y., New insights into atopic dermatitis: role of skin barrier and immune dysregulation. *Allergol. Int.*, 62, 2, 151–61, 2013.
86. Kaur, G., Ramirez, A., Xie, C., Clark, D., Dong, C., Maki, C., Ramos, T., Izadyar, F., Najera, S.O., Harb, J., Hao, J., A double-blinded placebo-controlled evaluation of adipose-derived mesenchymal stem cells in treatment of canine atopic dermatitis. *Vet. Res. Commun.*, 46, 251–260, 2022 Feb 1.
87. Li, H., Zhang, Z., Zhang, H., Guo, Y., Yao, Z., Update on the pathogenesis and therapy of atopic dermatitis. *Clin. Rev. Allergy Immunol.*, 1–5, 2021 Dec 1.
88. Cotugno, N., Finocchi, A., Cagigi, A., Di Matteo, G., Chiriaco, M., Di Cesare, S., Rossi, P., Aiuti, A., Palma, P., Douagi, I., Defective B-cell proliferation and maintenance of long-term memory in patients with chronic granulomatous disease. *J. Allergy Clin. Immunol.*, 135, 3, 753–61, 2015 Mar 1.
89. Boguniewicz, M. and Leung, D.Y., Atopic dermatitis: a disease of altered skin barrier and immune dysregulation. *Immunol. Rev.*, 242, 1, 233–46, 2011 Jul.
90. Phinney, D.G. and Pittenger, M.F., Concise review: MSC-derived exosomes for cell-free therapy. *Stem Cells*, 35, 4, 851–8, 2017 Apr 1.
91. Lai, R.C., Yeo, R.W., Lim, S.K., Mesenchymal stem cell exosomes. *Semin. Cell Dev. Biol.*, 40, 82–88, 2015 Apr 1, Academic Press.
92. Shin, T.H., Kim, H.S., Choi, S.W., Kang, K.S., Mesenchymal stem cell therapy for inflammatory skin diseases: Clinical potential and mode of action. *Int. J. Mol. Sci.*, 18, 2, 244, 2017.
93. Arslan, F., Lai, R.C., Smeets, M.B., Akeroyd, L., Choo, A., Aguor, E.N., Timmers, L., van Rijen, H.V., Doevendans, P.A., Pasterkamp, G., Lim, S.K., Mesenchymal stem cell-derived exosomes increase ATP levels, decrease oxidative stress and activate PI3K/Akt pathway to enhance myocardial viability and prevent adverse remodeling after myocardial ischemia/reperfusion injury. *Stem Cell Res.*, 10, 3, 301–12, 2013 May 1.

6

Stem Cell Therapy: A Significant Therapeutic Approach in Wound Healing

Richa Dayaramani[1], Soma Jana[2], Bratati Bandyopadhyay[2], Amlan Bishal[2*] and Subhashis Debnath[3]

[1]Centre for Medical Devices, National Institute of Pharmaceutical Education and Research-Ahmedabad, Gandhinagar, Gujarat, India
[2]Department of Pharmaceutical Technology, Bharat Technology, Uluberia, Howrah, West Bengal, India
[3]Department of Pharmaceutics, Bharat Pharmaceutical Technology, Amtali, Agartala, Tripura, India

Abstract
Wound healing is a dynamic, complex biological process that comprises hemostasis, inflammation, proliferation, and remodeling of tissue. Despite the progress in traditional therapeutic regimens, nonhealing and chronic wounds remain a recalcitrant clinical issue, resulting in impaired quality of life and increased healthcare costs. Stem cell therapy has been hailed as a breakthrough therapy with its regenerative and tissue repair potential, anti-inflammatory action, and the ability to modulate the wound microenvironment. Stem cells, namely, mesenchymal stem cells (MSCs), induced pluripotent stem cells (iPSCs), and embryonic stem cells (ESCs), possess certain unique features such as self-renewal, differentiation, and paracrine signaling, which serve as the foundation for tissue regeneration and repair. The review highlights the mechanism of stem cell–treated wound healing, with focus on their roles in angiogenesis induction, extracellular matrix remodeling, immune modulation, and epithelialization. The recent developments in stem cell delivery systems, including hydrogels, scaffolds, and exosome-based therapy, are highlighted as the focus to enhance their therapeutic potential. The limitations of stem cell therapy, including immunogenicity, scalability, and regulation, are critically evaluated. Current trends, including gene editing and bioprinting, are highlighted for the potential to further enhance stem cell–based therapies. In conclusion, stem cell therapy has the potential to be a revolutionary treatment for

**Corresponding author*: bishalamlan31@gmail.com

Sumel Ashique, Sanjay Dey, Biplab Debnath, Mohhammad Ramzan and Swarupananda Mukherjee (eds.) Stem Cell Therapy in Dermatological Disorders, (141–162) © 2025 Scrivener Publishing LLC

acute and chronic wounds and opens the door to personalized and regenerative medicine. Further research and clinical advancements are required to overcome the current limitations and achieve the maximum therapeutic potential of stem cells in wound healing.

Keywords: Wound healing, steam cells, mesenchymal stem cells (MSCs), regenerative medicine, tissue regeneration, induced pluripotent stem cells (iPSCs), chronic wounds

6.1 Introduction

Wound healing is a critical biological process necessary for tissue regeneration and repair. Although superficial wounds heal autonomously, chronic wounds, trauma-related injuries, and severe wounds pose considerable difficulties due to compromised healing processes [1]. Chronic wounds, frequently linked to diabetes, vascular disorders, and infections, typically result in extended inflammation, heightened infection risk, and diminished quality of life. Likewise, severe wounds resulting from burns, complications of surgery, or traumatic injuries necessitate sophisticated therapeutic measures to rehabilitate functional tissue [2]. Conventional wound care techniques, such as dressings, antibiotics, as well as skin grafts, possess limitations in facilitating comprehensive regeneration. Under these conditions, stem cell treatment has evolved as a viable regenerative medicine enhancing tissue regeneration and wound healing [3]. Although superficial wounds heal autonomously, chronic wounds, trauma-related injuries, and severe wounds pose considerable difficulties due to compromised healing processes [4]. Chronic wounds, frequently linked to diabetes, vascular disorders, and infections, typically result in extended inflammation, heightened infection risk, and diminished quality of life. Likewise, severe wounds resulting from burns, complications of surgery, or traumatic injuries necessitate sophisticated therapeutic measures to rehabilitate functional tissue. Conventional wound care techniques, such as dressings, antibiotics, as well as skin grafts, possess limitations in facilitating comprehensive regeneration [5]. Impaired angiogenesis hinders tissue regeneration by diminishing the supply of blood to the wound, whereas chronic inflammation extends wound exposure and elevates the risk of infection. Furthermore, inadequate extracellular matrix (ECM) remodeling diminishes skin strength and elasticity, thereby obstructing the healing process. These factors lead to severe scar formation, affecting both the functionality and aesthetics of the skin [6]. Traditional wound-healing methods prioritize symptom management over rectifying the fundamental biological deficiencies. These encompass wound dressings (e.g., hydrocolloids, alginates) that maintain moisture but do not actively facilitate tissue regeneration. Antibiotic therapy, crucial for managing infections, may foster antibiotic resistance. Skin grafting, a conventional technique for treating deep wounds, encounters difficulties such as donor site morbidity, rejection, and scarring. Growth factor treatment is constrained by swift degradation and variable potency in chronic wounds. Regenerative methods, including stem cell therapy, have garnered interest for their capacity to improve wound healing by rectifying cellular deficiencies and facilitating tissue regeneration, owing to these limitations [7]. Stem cells, with self-renewal and differentiation potential, are crucial for regenerative medicine, specifically for wound healing. Mesenchymal stem cells (MSCs), derived from bone marrow, umbilical cord, or adipose tissue, facilitate tissue regeneration through paracrine mechanisms, releasing growth factors and cytokines that facilitate angiogenesis and inhibit inflammation. They increase ECM remodeling, increasing new skin tissue development, whereas differentiation capacity into skin fibroblasts, keratinocytes, and endothelial cells increases regeneration. Induced pluripotent stem cells (iPSCs) and epidermal stem cells are also being researched for personalized wound care treatments, presenting a hopeful option for future skin regeneration therapies [8]. Comprehensive preclinical and clinical research substantiates the stem cell efficacy in wound healing, illustrating their performance to promote tissue regeneration. Research involving animals has demonstrated that MSCs facilitate collagen deposition, re-epithelialization, and angiogenesis, thereby expediting wound closure. Clinical trials corroborate these findings, indicating enhanced recovery rates in ulcers associated with diabetes and burn injuries subsequent to stem cell therapy. Moreover, biomaterial scaffolds combined with stem cells facilitate regulated release and enhanced integration at wound sites, thereby optimizing therapeutic results. These advancements collectively underscore the stem cell–based methodologies in chronic wound treatment and regenerative medicine [9].

6.2 Physiology of Wound-Healing Process

Wound healing is a four-stage process with little overlap among the stages.

Hemostasis: Platelets are deposited onto the wound bed at the time of injury and release clotting factors, which initiate fibrin clot formation to control bleeding.

Inflammation: Immune cells such as neutrophils and macrophages flood into the wound within hours to destroy pathogens, dead cells, and debris and release cytokines and growth factors such as transforming growth factor β (TGF-β) and vascular endothelial growth factor (VEGF).

Proliferation: Fibroblasts multiply and produce ECM proteins such as collagen and keratinocytes and migrate to re-epithelialize the wound. VEGF supports angiogenesis to restore blood supply.

Remodeling (maturation): The final stage may take months, during which type III collagen is replaced by type I collagen to increase the tensile strength of the wound. Fibroblasts and myofibroblasts are responsible for remodeling of the ECM and wound contraction [10].

Cellular and molecular mechanisms

Wound-healing process requires many cells and signaling molecules to interact with each other to restore the integrity of tissue. Platelets trigger clotting and discharge platelet-derived growth factor to admit immune cells, and neutrophils play a role in the first defense against infection by discharging proteases to digest necrotic tissue. Macrophages play a dual function of clearing the debris and discharge proinflammatory cytokines tumor necrosis factor α and interleukin 1β (IL-1β) and the anti-inflammatory cytokine IL-10, to regulate inflammation and promote fibroblast growth. Fibroblasts facilitate the construction of the tissue by producing ECM components such as collagen and glycosaminoglycans. Endothelial tissues simultaneously facilitate angiogenesis by VEGF and fibroblast growth factor (FGF), thereby maintaining good blood supply to repair the tissue. Lastly, keratinocytes migrate from the surroundings of the wound, facilitating restoration of epithelium and establishment of the epidermis, thereby completing the complex process of wound healing [11].

Tissue repair and regeneration mechanism

Wound healing occurs by two principal mechanisms: regeneration and repair. Regeneration transpires in highly regenerative tissues such as the liver and skin, wherein lost tissue is substituted with identical, fully functional cells, reinstating the original structure and function. Conversely, repair is the primary mechanism in tissues with restricted regenerative ability, such as cardiac muscle, where healing entails fibrotic scarring. This process substitutes damaged tissue with collagen-dense scar tissue, which offers structural support but frequently lacks the original functionality, potentially resulting in enduring physiological deficits [12].

Factors affecting wound healing

Wound healing is induced by intrinsic and extrinsic factors. Intrinsic factors include genetics, age, diabetes, and immune deficiencies, which can slow down the body's natural repair processes. The aging process slows down cellular regeneration, whereas genetic factors and conditions such as diabetes can lead to delayed healing and increased susceptibility to infection. Immune deficiencies slow down the ability to fight infection and repair tissue. Extrinsic causes, such as poor nutrition, infection, smoking, and inadequate oxygenation, play a major role in the problem. Proper nutrition provides essential nutrients for cell growth, whereas infections prolong inflammation and inhibit tissue repair [13]. Smoking reduces blood flow and oxygen levels, both of which are necessary for healing, whereas inadequate oxygenation inhibits cellular metabolism, thus slowing down the healing process. Mechanical stress: Excessive tension or movement at the wound area can hinder healing [14].

Healing of wounds in different tissues

Different tissues exhibit distinct healing mechanisms. Skin heals through re-epithelialization and ECM remodeling, restoring barrier function and structural integrity. Mucosal tissues heal faster due to their high vascularity and abundance of stem cells, which promote rapid regeneration [15]. Bone undergoes both endochondral and intramembranous ossification to regenerate and restore its structural strength [16]. In contrast, nervous tissue has limited regenerative ability and primarily relies on glial cells for support and repair [17]. Muscle regeneration involves myoblasts and satellite cells, which contribute to muscle fiber repair and recovery following injury. Efficient wound healing depends on the coordinated activity of cellular as well as molecular mechanisms, which vary across different tissue types [18].

6.3 Biology of Stem Cells and their Role in Wound Healing

Wound healing is a complex biological process with multiple cellular and molecular pathways. Stem cells have been of interest due to their ability to self-renew and differentiate into various cell types, which can result in tissue regeneration [19].

Types of stem cells

Stem cells are classified based on their origin and differentiation potential:

- **Embryonic stem cells:** Embryonic stem cells (ESCs) are derived from the inner cell mass of the blastocyst and are pluripotent, the capacity to differentiate into the three germ layers—ectoderm, mesoderm, and endoderm. This special potential makes them valuable in regenerative medicine, disease modeling, and drug screening. However, their clinical application is limited due to ethical concerns of embryo destruction and tumorigenicity risk, as ESCs possess the capacity to form teratomas upon transplantation [20]. These challenges have led researchers to explore alternative stem cell sources, such as iPSCs, which offer similar pluripotency without the associated ethical dilemmas.
- **Adult stem cells:** Adult stem cells, or somatic or tissue-specific stem cells, are present in different tissues such as bone marrow, skin, and adipose tissues. They are unlike ESCs, which are multipotent but have significant differentiation potential and only to produce cell types within the lineage of their own tissue. Even though restricted, adult stem cells play a crucial role in tissue homeostasis and repair by replacing damaged or old cells. Adult stem cells are favored in regenerative medicine due to the absence of ethical considerations and less likelihood of tumor formation, and thus they are an ideal candidate for cell-based therapy and personalized medicine [21].
- **Induced pluripotent stem cells:** iPSCs are developed through the reprogramming of somatic cells with specific transcription factors, i.e., Oct4, Sox2, Klf4, and c-Myc. iPSCs are as pluripotent as ESCs and differentiate into any cell type that is derived from all three germ layers. Unlike ESCs, iPSCs avoid the ethical problem of embryo destruction. Their ability to be derived from the cells of a patient makes them highly hopeful for personalized medicine, regenerative therapy, and disease modeling. Despite their promise, problems such as genetic instability and tumor formation are under active investigation [22].
- **Mesenchymal stem cells:** MSCs are multipotent stem cells that are present in different sources such as bone marrow, adipose tissue, umbilical cord, and placenta. MSCs can be classified into osteoblasts, chondrocytes, and adipocytes and are thus used in bone, cartilage, and fat tissue formation.

MSCs also have immunomodulatory properties, which allow them to modulate the immune system and inhibit inflammation, thus their growing therapeutic application. Due to these characteristics, MSCs are widely researched for use in wound healing, tissue engineering, and regenerative medicine. Their ease of isolation, low immunogenicity, and autologous transplantation potential make them perfect candidates for clinical applications [23].

Stem cell–mediated healing signaling pathways

Stem cell–mediated repair is regulated by numerous molecular signaling pathways that are associated in cell growth, cell differentiation, and tissue regeneration. The Wnt/β-catenin pathway is central to cell proliferation, differentiation, and migration and specifically stimulates epidermal stem cells in wound healing. The TGF-β pathway regulates the deposition of the ECM fibrosis and differentiation of MSCs and wound contraction [24]. Notch signaling pathway controls stem cell fate determination and is required for dermal fibroblast activation and angiogenesis. Similarly, Hedgehog signaling pathway controls epithelial regeneration and activation of hair follicle stem cells, involved in skin repair [24]. Moreover, hypoxia-inducible factor 1α signaling cycle promotes stem cell survival under hypoxic conditions and angiogenesis by inducing VEGF expression to ensure a sufficient blood supply to the healing tissue. Together, these pathways regulate stem cell functions for efficient tissue healing and wound closure [25].

Mechanisms of stem cell action in tissue repair

Stem cells promote tissue repair by multifaceted mechanisms that involve paracrine signaling, immunomodulation, direct differentiation, and ECM remodeling. They release cytokines, growth factors, such as VEGF, FGF, and epidermal growth factor, and extracellular vesicles to promote tissue regeneration. Their immunomodulatory activity suppresses inflammation by regulating macrophage polarization, preventing excessive scarring and fibrosis. Stem cells also promote wound healing by direct differentiation into keratinocytes, fibroblasts, and endothelial cells, promoting skin regeneration and vascularization. Moreover, they regulate fibroblast activity in ECM remodeling, inducing collagen deposition and organization to increase tissue strength and structural integrity. Through these interrelated mechanisms, stem cells are actively engaged in promoting wound healing and enhancing tissue repair outcomes [26].

6.4 Stem Cell Therapy for Wound Healing and Histopathological Studies

Stem cell therapy significantly helps heal wounds through the utilization of different stem cell populations to induce tissue repair and regeneration. MSCs stimulate fibroblast proliferation and ECM deposition and inhibit inflammation and induce angiogenesis and play an important role in the repair of deep tissue. Epidermal stem cells, found in the basal epidermal layer, play roles in re-epithelialization and skin regeneration and help repair the skin barrier [27]. Hematopoietic stem cells (HSCs) support wound healing by differentiating into immune cells that regulate inflammation and infection control, as well as influencing macrophage polarization toward wound resolution [28]. Additionally, iPSCs offer a patient-specific source of stem cells for regenerative medicine, with the potential to generate functional skin substitutes and vascularized grafts. Collectively, these stem cell–based approaches provide promising therapeutic strategies for both cutaneous and deep tissue wound healing [29]. Histopathological studies of stem cell therapy in wound healing reveal significant improvements in tissue regeneration and repair. These evaluations demonstrate increased collagen synthesis and well-organized ECM deposition, which contribute to enhanced structural integrity and skin strength [30]. Additionally, stem cell therapy promotes neovascularization and granulation tissue formation, facilitating improved blood supply and accelerated healing. A notable reduction in inflammatory cell infiltration and fibrosis is also observed, indicating the immunomodulatory effects of stem cells in minimizing chronic inflammation and excessive scar formation [31]. These histological findings support the therapeutic potential of stem cells in promoting efficient and functionally superior wound healing.

6.5 Stem Cell Delivery Mechanisms for Wound Healing

Stem cell therapy offers a promising approach for wound healing by promoting tissue regeneration and reducing inflammation. Several delivery mechanisms ensure optimal stem cell survival, integration, and function at the wound site.

- ***Biomaterials as stem cell scaffolds:*** Biomaterials act as carriers for the facilitation of stem cell retention, viability, and differentiation. Hydrogels, nanofibers, and collagen matrices create a permissive environment, mimicking the ECM.

Scaffolds may be designed for growth factor delivery to enhance stem cell function and wound healing [32].

- ***Local vs. systemic delivery:*** Direct injection or application of stem cells at the wound site ensures higher cell concentration and faster tissue repair. Methods include hydrogels, microneedles, or spray-based applications. However, challenges include limited cell survival and immune responses. Intravenous or intra-arterial administration allows stem cells to home to injury sites *via* chemotactic signals. Although this method is less invasive, it may lead to lower efficiency due to cell trapping in nontarget organs [33].
- ***Bioengineered skin substitutes:*** Engineered skin substitutes integrate stem cells with biomaterials to replace lost tissue. Examples include cell-seeded scaffolds, 3D bioprinted skin, and decellularized matrices, which are new developments in skin tissue science for wound healing and transplantation. Cell-seeded scaffolds grow stem cells on biocompatible dermal scaffolds to promote tissue regeneration and enhance transplant integration. Layering bioinks composed of cells, growth factors, and biomaterials enables 3D bioprinting technology to precisely create multilayered skin structures that closely mimic the composition and properties of real skin [34]. Decellularized matrices, which are derived from real skin tissue, also offer a scaffold that is repopulated with patient-specific stem cells, lowering the risk of immunological rejection and promoting personalized tissue repair. These techniques hold great promise for improving the outcomes of severe burn injuries, chronic wounds, and skin transplants [35].

6.6 Clinical Uses of Stem Cell Therapy in Wound Healing

Stem cell therapy is a challenging solution to the augmentation of wound healing, where the traditional therapies are ineffective. Stem cells have regenerative capacity, regulating the inflammatory response and initiating angiogenesis, and induce tissue repair. Different stem cells, such as MSCs, endothelial progenitor cells, and iPSCs, have been used to treat acute wounds and chronic wounds, burns, and trauma injuries. Depending on the disease, these cells can be administered either by direct injection, biomaterial scaffolds, topical gels, or bioengineered skin grafts. The following table outlines the clinical use of stem cell treatment in wound healing.

6.7 Recent Developments and Technologies in Stem Cell Research for Wound Healing

Advancements in stem cell study have improved the efficacy and precision of regenerative medicine for wound healing. Emerging technologies such as exosome-based therapies, 3D bioprinting, and gene editing (CRISPR/Cas9) are revolutionizing the field by enhancing cell survival, tissue regeneration, and treatment personalization [36].

- ***Exosome-based therapies:*** Small extracellular vesicles, known as exosomes, contain proteins, lipids, and RNA, playing a crucial role in intercellular communication. In wound healing, these vesicles enhance angiogenesis and fibroblast proliferation, modulate the immune response to reduce inflammation, and stimulate collagen synthesis, accelerating tissue repair. A recent advancement in this field involves the development of exosome-loaded hydrogels, which enable the prolonged release of regenerative factors at the wound space, improving healing efficiency and offering a promising approach for advanced wound care therapies [37].
- ***3D bioprinting for wound healing*:** 3D bioprinting technology utilizes bioinks containing stem cells to fabricate layered, skin-like structures that closely mimic natural tissue. In wound healing, this innovation enables the creation of personalized skin grafts using patient-derived cells, allowing for the precise placement of keratinocytes, fibroblasts, and endothelial cells to induce functional skin growth, while reducing the risk of graft rejection. A recent breakthrough in this field includes the successful printing of vascularized skin substitutes, which enhance blood flow integration and improve graft survival, marking a significant advancement in regenerative medicine [34].
- ***Gene editing technologies (CRISPR/Cas9):*** CRISPR/Cas9 technology induces characterization of stem cell DNA to correct mutations or enhance regenerative properties, offering significant advancements in wound healing. By engineering stem cells with improved survival and differentiation capabilities, this

approach promotes efficient tissue regeneration while reducing fibrosis and excessive scarring through targeted gene modulation. Additionally, CRISPR-edited cells can be enhanced with antimicrobial properties to prevent wound infections. A recent breakthrough in this field demonstrated that MSCs overexpressed growth factors led to 50% faster wound closure in experimental models, highlighting the potential of gene-edited therapies in accelerating tissue repair [38].

6.8 Challenges in Stem Cell Treatment

- ***Regulatory and ethical aspects:*** Clinical application using stem cells needs certain ethical and safety considerations because the entire field of biomedical investigation and technique is changing so quickly these days. Stem cell investigators are working on various projects, such as addressing genetic disorders and creating novel stem cell–derived tissues from humans and biological substrates utilized in the study of genomes and regenerative medicine. Clinical studies and results show that stem cell–associated treatment has enormous therapeutic potential in developing the treatment of chronic, autoimmune, and genetic disorders [39].
- ***Tumorigenesis risk:*** Many clinical trials investigating the applications of stem cells for the cancer therapy and degenerative diseases, as well as repair of injured tissues, have begun in the rapidly evolving discipline. Stem cells therapy having a multilineage differentiation capacity during the production of progeny clutches the skill in the study of many degenerative as well as age-related diseases. The safe application of stem cell–based treatments still faces numerous challenges, despite its enormous potential. Administering pluripotent stem cells in humans may give rise to teratoma formation and when administering pluripotent ESCs in animal turns to teratocarcinoma (a malignant tumor formation). It is very difficult to evaluate the tumorigenesis risk of pluripotent cells because of having unviability of any conducted human trials with pluripotent ESC yet [40]. Certain circumstances may influence the tumorigenic tendency of iPSCs, which is associated with the risk of the pluripotent state. The focus formation assay evaluates the tumorigenicity of particular genes

in fibroblasts and shares a lot of similarities with the process used to create iPSCs from fibroblasts. A number of genes are forced to express themselves in somatic cells to prepare iPSCs, and genes may increase the tumorigenic tendency of iPSCs [41]. Helicobacter-induced gastric cancer may arise from undifferentiated stem cells, specifically hematopoietic stem cells (HSCs) and non-hematopoietic mesenchymal stem cells (MSCs), injected into specific tissues or infused intravenously. Notably, natural MSCs have been applied in juvenile leukemia, gastric cancer, and osteogenic sarcoma, and it has been suggested that grown-up somatic stem cells act as an important part in the pathogenesis of cancer that develops into tumorigenic cancer [42].

- ***Immune rejection and safety:*** The immune system guards the body against tumors and pathogens. The safety issues raised by stem cell therapies cannot be described by conventional analytical techniques designed for low-molecular-weight compounds or other biopharmaceuticals. The capacity to track cell distribution in the body is a unique problem because, after injection, the cells might become nearly identical to the host cells. An objective evaluation of the threat in the context of tumorigenicity or aberrant ectopic tissue creation requires the ability to track the therapeutic cells. When the stem cells are administered through intravenous route instead of locally, this is especially important due to the possibility of widespread diffusion. The ability to measure the biodistribution of injected cells adds to challenges because tracking the foreign cells would require the creation of such new technologies. It is necessary to exclude misplaced cells, which is currently not physically feasible. Removal of such uncontrolled cells is a more difficult task involving incorporation of a "self-destruct" mechanism engineered into the cells inducing apoptosis upon exposure to a specific stimulus. One of the largest problems with stem cell therapy is the tumorigenic potential. Injection of a cell with unlimited potential for renewal and the ability to differentiate into any human cell type is a safety burden not related to any other types of treatment. Therefore, there is a significant need for technological advancements in biomonitoring as well as the creation of new resources for the removal of administered cells that become inappropriately placed debated at the current time. Undifferentiated ESCs develop into teratomas when administered into immunocompromised animals, emphasizing how poorly this problem is being described. Further, if the cells produce genetic defects, these may have the tendency to develop teratocarcinomas. These would be extremely malignant, similar to the respective testicular germ cell tumors seen in young men. Another definite safety issue that needs to be met by stem cell therapy providers is immunogenicity [43]. Although ESCs from humans have been shown to have immune privilege, any foreign cell that is implanted into an individual will be suspected of immune surveillance. In addition to resolving the questions of effectiveness of stem cell therapies, Figure 6.1 show evidence of effective application of novel cell-based therapies will depend largely on our ability to resolve these significant safety questions, both preclinically and clinically. Every step of the way down the road to creating stem cell therapies must be rigorously tested, from the source of the cells used to expansion, manipulation, and preclinical testing through further engraftment into the host.

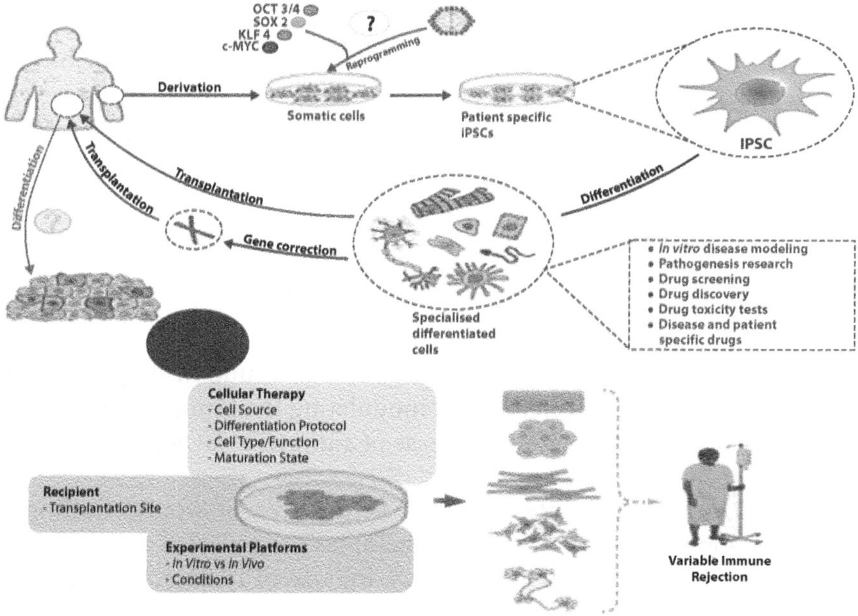

Figure 6.1 Variable immune rejection and safety concerns.

- ***Cost and accessibility:*** The cost of treatment varies greatly from \$5000 to \$50,000, depending on a number of factors, including the variety of stem cells administered, the quantity of cells, the characteristics of the cells, the laboratory's location, and the origin of the stem cells. For orthopedic conditions such as rotator cuff tendonitis, knee osteoarthritis, tennis elbow arthritis, and cartilage defects, the cost is naturally lower compared with intravenous cell therapy treatment for autoimmune, degenerative, or inflammatory conditions as the quantity of cells or dosage may be lesser, ranging between \$5000 and \$8000.

6.9 Potential Future Applications of Stem Cell–Derived Products in the Use of Stem Cells for Wound Healing

Personalized medicine approaches: Stem cell technology is a very important area of medical research. The development of iPSCs provides a potentially unlimited and personalized source of cells, eliminating the risk of immune rejection in therapeutic applications and opening new avenues for disease modeling and drug discovery.

Tissue engineering: Organoids, miniature 3D tissue models, enable accurate drug testing and disease modeling, offering insights beyond traditional methods.

HSC transplantation: Hematopoietic stem cell transplantation has been analyzed and utilized in clinical trials for the treatment of multiple sclerosis (MS) [5].

Wound healing: Stem cells play a crucial role in wound healing by stimulating cell growth and specialization, modulating the immune system, and combating infection through the release of antimicrobial agents.

Incorporation with nanotechnology: In the area of wound healing, a multiplication number of novel nanotherapies has revealed and is undergoing clinical research. Table 6.1 shows the use of nanotechnologies for wound healing applications.

Table 6.1 Use of nanotechnologies for wound-healing applications.

Type of technology	Specific characteristics	Applications	References
Nanomaterials			
Silver-coated nanomaterials	Antimicrobial and anti-inflammatory qualities, as well as improved wound-healing appearance	Diabetic ulcers and burn injuries	[44]
Copper nanoparticles	Improved rate of wound healing and antibacterial activity (biosynthesized copper nanoparticles)	Rat model of excisional wound	[45]
Cerium nanoparticles (nanoceria)	Antioxidant properties and quicker healing of wounds	Full-thickness wounds in mice	[46]
Nanosized bioactive glass particles	Promoting wound closure, angiogenesis, and cell proliferation	Human umbilical vein endothelial cells cultivated for an *in vitro* wound-healing assay	[47]
Zinc oxide nanomaterials	Infection prevention in wound healing, encouraging chemotaxis, angiogenesis, and cell division	Endothelial cell-based wound-healing test	[48]
Carbon-based nanomaterials	Anti-inflammatory and antioxidant qualities that influence cell division	Using primary human keratinocytes in an *in vitro* experiment	[49]
Nanomaterials for delivery of therapeutic agents			
Gold nanoparticles	Used for gene delivery; functionalized with antioxidants, antibiotics, and scavengers of reactive oxygen species	Diabetic wounds in mice	[50]

(*Continued*)

Table 6.1 Use of nanotechnologies for wound-healing applications. (*Continued*)

Type of technology	Specific characteristics	Applications	References
Nanomaterials for delivery of therapeutic agents			
Nanoparticles bearing nitric oxide	Speeding up wound closure, decreasing inflammation, and promoting collagen deposition, fibroblast proliferation, and blood vessels; antibiofilm activity	Wounds in immunodeficient, diabetic, and nonobese NOD-SCID mice; excisional wounds in mice infected with *Pseudomonas aeruginosa*	[51]
Nanomaterial-containing lipids (liposomes)	Distribution of phytodrugs with anti-inflammatory and antioxidant benefits	Full-thickness skin defect models *in vivo* and *in vitro*	[52]
Metal (silver) incorporated electrospun mats	Targeting chronic wound biofilms, polyvinyl alcohol reduced silver ions	Antimicrobial activity against *Escherichia coli* and *Staphylococcus aureus*	[53]
Scaffolds			
A hybrid nanofibrous scaffold made of poly (lactide-co-glycolic acid) (PLGA) and silk fibroin (SF)	Optimization of the PLGA/SF ratio is necessary for the adhesion and proliferation of L929 cells	Diabetic rat model of excisional wounds	[54]
Nanofibers of gelatine and poly-ε-caprolactone	Utilizing needleless electrospinning method to create the nanofibers; adhesion and proliferation of cells	Rats with full-thickness wounds	[55]
Porous scaffold made of fibrin, collagen, and fibrin	Matrices for fibroblast, keratinocyte, and epidermal cell motility	Skin regeneration	[56]

6.10 Conclusion

The drawbacks of conventional therapies are overcome by stem cell therapy, a novel therapy for wound-healing management. A number of interdependent stages, including inflammation, proliferation, hemostasis, and remodeling, constitute the wound-healing process. All these stages are important for the attainment of tissue function and integrity, and dysregulation results in chronic wounds, particularly in those patients also having a concomitant underlying disease such as diabetes. Biology of the stem cell is the focus here. Different types of stem cells, including epidermal stem cells, iPSCs, and MSCs, have been investigated for their use in healing, using a variety of mechanisms. These include the secretion of cytokines and growth factors to induce vascularization, modulation of inflammatory response, and regeneration of tissue. Despite the promises demonstrated by the stem cell therapies, there are still hurdles to be overcome. Problems related to regulatory hurdles, tumorigenesis, immunological rejection, and economic cost need to be overcome to maximize their clinical use. Recent advances with techniques such as 3D bioprinting and treatments with exosomes further expand the promise of harnessing the use of stem cells in facilitating wound healing. Overall, stem cell therapy promises to be mind-boggling for revolutionizing wound-healing medicine and maximizing the outcome of the patient with a chronic wound. Further studies need to be instituted to further mature these therapies with a focus toward making them safe, effective, and available to use in a clinical setting. This integration of the stem cell technology into the personalization of the medicine's regimen could bring even superior treatments and patient care into the future.

AI disclosure statement: The author acknowledges the use of AI-based tools for minor language editing or grammatical correction. However, all intellectual contributions, analyses, and conclusions presented in this chapter are entirely the author's own.

References

1. Rodrigues, M., Kosaric, N., Bonham, C.A., Gurtner, G.C., Wound Healing: A Cellular Perspective. *Physiol. Rev.*, 99, 665–706, 2019, https://doi.org/10.1152/physrev.00067.2017.
2. Sen, C.K., Human wound and its burden: updated 2020 compendium of estimates. *Adv. Wound Care*, 10, 5, 281–92, 2021 May 1.

3. Kolimi, P., Narala, S., Nyavanandi, D., Youssef, A.A., Dudhipala, N., Innovative treatment strategies to accelerate wound healing: trajectory and recent advancements. *Cells*, 11, 15, 2439, 2022 Jan.
4. Wallace, H.A., Basehore, B.M., Zito, P.M., Wound Healing Phases. [Updated 2023 Jun 12], in: *StatPearls* [Internet], StatPearls Publishing, Treasure Island (FL), 2025 Jan.
5. Gould, L., Abadir, P., Brem, H., Carter, M., Conner-Kerr, T., Davidson, J., DiPietro, L., Falanga, V., Fife, C., Gardner, S., Grice, E., Chronic wound repair and healing in older adults: current status and future research. *Wound Repair Regener.*, 23, 1, 1–3, 2015 Jan.
6. Jin, Y., Li, S., Yu, Q., Chen, T., Liu, D., Application of stem cells in regeneration medicine. *MedComm.*, 4, 4, e291, 2023 Aug.
7. Shariatzadeh, F.J., Currie, S., Logsetty, S., Spiwak, R., Liu, S., Enhancing wound healing and minimizing scarring: A comprehensive review of nanofiber technology in wound dressings. *Prog. Mater Sci.*, 6, 101350, 2024 Aug.
8. Maxson, S., Lopez, E.A., Yoo, D., Danilkovitch-Miagkova, A., LeRoux, M.A., Concise review: role of mesenchymal stem cells in wound repair. *Stem Cells Transl. Med.*, 1, 2, 142–9, 2012 Feb 1.
9. Wu, S., Zhou, Z., Li, Y., Jiang, J., Advancements in diabetic foot ulcer research: Focus on mesenchymal stem cells and their exosomes. *Heliyon*, 10, 17, e37031, 2024 Sep 15.
10. Schultz, G.S., Chin, G.A., Moldawer, L., Diegelmann, R.F., Principles of wound healing, in: *Mechanisms of vascular disease: A reference book for vascular specialists* [Internet], University of Adelaide Press, Adelaide (AU), 2011.
11. Mamun, A.A., Shao, C., Geng, P., Wang, S., Xiao, J., Recent advances in molecular mechanisms of skin wound healing and its treatments. *Front. Immunol.*, 15, 1395479, 2024 May 21.
12. Eming, S.A., Martin, P., Tomic-Canic, M., Wound repair and regeneration: mechanisms, signaling, and translation. *Sci. Transl. Med.*, 6, 265, 265sr6, 2014 Dec 3.
13. Guo, S.A. and DiPietro, L.A., Factors affecting wound healing. *J. Dent. Res.*, 89, 3, 219–29, 2010 Mar.
14. Jensen, J.A., Goodson, W.H., Hopf, H.W., Hunt, T.K., Cigarette smoking decreases tissue oxygen. *Arch. Surg.*, 126, 9, 1131–4, 1991 Sep 1.
15. Xue, M. and Jackson, C.J., Extracellular matrix reorganization during wound healing and its impact on abnormal scarring. *Adv. Wound Care*, 4, 3, 119–36, 2015 Mar 1.
16. Breeland, G., Sinkler, M.A., Menezes, R.G., Embryology, Bone Ossification. [Updated 2023 May 1], in: *StatPearls* [Internet], StatPearls Publishing, Treasure Island (FL), 2025 Jan.
17. Quan, L., Uyeda, A., Muramatsu, R., Central nervous system regeneration: the roles of glial cells in the potential molecular mechanism underlying remyelination. *Inflamm. Regen.*, 42, 1, 7, 2022 Mar 2.

18. Forcina, L., Cosentino, M., Musarò, A., Mechanisms regulating muscle regeneration: insights into the interrelated and time-dependent phases of tissue healing. *Cells*, 9, 5, 1297, 2020 May 22.
19. Yang, R., Liu, F., Wang, J., Chen, X., Xie, J., Xiong, K., Epidermal stem cells in wound healing and their clinical applications. *Stem Cell Res. Ther.*, 10, 1–4, 2019 Dec.
20. National Research Council (US) and Institute of Medicine (US) Committee on the Biological and Biomedical Applications of Stem Cell Research, *Stem Cells and the Future of Regenerative Medicine*, National Academies Press (US), Washington (DC), 2002.
21. Khuda-Bukhsh, A.R., Das, S., Samadder, A., Theories of stem cell aging, in: *Stem Cells and Aging*, pp. 1–7, Academic Press, 2021 Jan 1.
22. Al Abbar, A., Ngai, S.C., Nograles, N., Alhaji, S.Y., Abdullah, S., Induced pluripotent stem cells: Reprogramming platforms and applications in cell replacement therapy. *Biores. Open Access*, 9, 1, 121–36, 2020 Apr 1.
23. Orbay, H., Tobita, M., Mizuno, H., Mesenchymal stem cells isolated from adipose and other tissues: basic biological properties and clinical applications. *Stem Cells Int.*, 2012, 1, 461718, 2012.
24. Gumede, D.B., Abrahamse, H., Houreld, N.N., Targeting Wnt/β-catenin signaling and its interplay with TGF-β and Notch signaling pathways for the treatment of chronic wounds. *Cell Commun. Signaling*, 22, 1, 244, 2024 Apr 26.
25. Krock, B.L., Skuli, N., Simon, M.C., Hypoxia-induced angiogenesis: good and evil. *Genes Cancer*, 2, 12, 1117–33, 2011 Dec.
26. Baraniak, P.R. and McDevitt, T.C., Stem cell paracrine actions and tissue regeneration. *Regener. Med.*, 5, 1, 121–43, 2010 Jan 1.
27. Wu, S., Sun, S., Fu, W., Yang, Z., Yao, H., Zhang, Z., The role and prospects of mesenchymal stem cells in skin repair and regeneration. *Biomedicines*, 12, 4, 743, 2024 Mar 27.
28. Urao, N., Liu, J., Takahashi, K., Ganesh, G., Hematopoietic stem cells in wound healing response. *Adv. Wound Care*, 11, 11, 598–621, 2022 Nov 1.
29. Aboul-Soud, M.A., Alzahrani, A.J., Mahmoud, A., Induced pluripotent stem cells (iPSCs)—roles in regenerative therapies, disease modelling and drug screening. *Cells*, 10, 9, 2319, 2021 Sep 5.
30. Kanji, S. and Das, H., Advances of stem cell therapeutics in cutaneous wound healing and regeneration. *Mediators Inflamm.*, 2017, 1, 5217967, 2017.
31. Shi, W., Xin, Q., Yuan, R., Yuan, Y., Cong, W., Chen, K., Neovascularization: the main mechanism of MSCs in ischemic heart disease therapy. *Front. Cardiovasc. Med.*, 8, 633300, 2021 Jan 26.
32. Xu, Y., Chen, C., Hellwarth, P.B., Bao, X., Biomaterials for stem cell engineering and biomanufacturing. *Bioact. Mater.*, 4, 366–79, 2019 Dec 1.
33. Lin, X., Zhang, X., Wang, Y., Chen, W., Zhu, Z., Wang, S., Hydrogels and hydrogel-based drug delivery systems for promoting refractory wound

healing: Applications and prospects. *Int. J. Biol. Macromol.*, 285, 138098, 2024 Nov.
34. Weng, T., Zhang, W., Xia, Y., Wu, P., Yang, M., Jin, R., Xia, S., Wang, J., You, C., Han, C., Wang, X., 3D bioprinting for skin tissue engineering: Current status and perspectives. *J. Tissue Eng.*, 12, 20417314211028574, 2021 Jul.
35. Zhang, X., Chen, X., Hong, H., Hu, R., Liu, J., Liu, C., Decellularized extracellular matrix scaffolds: Recent trends and emerging strategies in tissue engineering. *Bioact. Mater.*, 10, 15–31, 2022 Apr 1.
36. Hussen, B.M., Taheri, M., Yashooa, R.K., Abdullah, G.H., Abdullah, S.R., Kheder, R.K., Mustafa, S.A., Revolutionizing medicine: recent developments and future prospects in stem-cell therapy. *Int. J. Surg.*, 110, 12, 8002–24, 2024 Dec 1.
37. Sun, T., Li, M., Liu, Q., Yu, A., Cheng, K., Ma, J., Murphy, S., McNutt, P.M., Zhang, Y., Insights into optimizing exosome therapies for acute skin wound healing and other tissue repair. *Front. Med.*, 18, 2, 258–84, 2024 Apr.
38. Abdelnour, S.A., Xie, L., Hassanin, A.A., Zuo, E., Lu, Y., The potential of CRISPR/Cas9 gene editing as a treatment strategy for inherited diseases. *Front. Cell Dev. Biol.*, 9, 699597, 2021 Dec 15.
39. World Health Organization, WHO considerations on regulatory convergence of cell and gene therapy products, World Health Organization, Geneva, pp. 1–22, 2022.
40. Hentze, H., Soong, P.L., Wang, S.T., Phillips, B.W., Putti, T.C., Dunn, N.R., Teratoma formation by human embryonic stem cells: evaluation of essential parameters for future safety studies. *Stem Cell Res.*, 2, 3, 198–210, 2009 May 1.
41. Cerneckis, J., Cai, H., Shi, Y., Induced pluripotent stem cells (iPSCs): molecular mechanisms of induction and applications. *Signal Transduct. Target Ther.*, 9, 1, 112, 2024 Apr 26.
42. Herberts, C.A., Kwa, M.S., Hermsen, H.P., Risk factors in the development of stem cell therapy. *J. Transl. Med.*, 9, 1–4, 2011 Dec.
43. Volarevic, V., Markovic, B.S., Gazdic, M., Volarevic, A., Jovicic, N., Arsenijevic, N., Armstrong, L., Djonov, V., Lako, M., Stojkovic, M., Ethical and safety issues of stem cell-based therapy. *Int. J. Med. Sci.*, 15, 1, 36, 2018 Jan 1.
44. Tian, J., Wong, K.K., Ho, C.M., Lok, C.N., Yu, W.Y., Che, C.M., Chiu, J.F., Tam, P.K., Topical delivery of silver nanoparticles promotes wound healing. *ChemMedChem: Chemistry Enabling Drug Discovery*, 2, 1, 129–36, 2007 Jan 15.
45. Tiwari, M., Narayanan, K., Thakar, M.B., Jagani, H.V., Venkata Rao, J., Biosynthesis and wound healing activity of copper nanoparticles. *IET Nanobiotechnol.*, 8, 4, 230–7, 2014 Dec.
46. Chigurupati, S., Mughal, M.R., Okun, E., Das, S., Kumar, A., McCaffery, M., Seal, S., Mattson, M.P., Effects of cerium oxide nanoparticles on the growth of

keratinocytes, fibroblasts and vascular endothelial cells in cutaneous wound healing. *Biomaterials*, 34, 9, 2194–201, 2013 Mar 1.
47. Mao, C., Chen, X., Miao, G., Lin, C., Angiogenesis stimulated by novel nanoscale bioactive glasses. *Biomed. Mater.*, 10, 2, 025005, 2015 Mar 25.
48. Barui, A.K., Veeriah, V., Mukherjee, S., Manna, J., Patel, A.K., Patra, S., Pal, K., Murali, S., Rana, R.K., Chatterjee, S., Patra, C.R., Zinc oxide nanoflowers make new blood vessels. *Nanoscale*, 4, 24, 7861–9, 2012.
49. Gao, J., Wang, H.L., Iyer, R., Suppression of Proinflammatory Cytokines in Functionalized Fullerene-Exposed Dermal Keratinocytes. *J. Nanomater.*, 2010, 1, 416408, 2010.
50. Chen, S.A., Chen, H.M., Yao, Y.D., Hung, C.F., Tu, C.S., Liang, Y.J., Topical treatment with anti-oxidants and Au nanoparticles promote healing of diabetic wound through receptor for advance glycation end-products. *Eur. J. Pharm. Sci.*, 47, 5, 875–83, 2012 Dec 18.
51. Blecher, K., Martinez, L.R., Tuckman-Vernon, C., Nacharaju, P., Schairer, D., Chouake, J., Friedman, J.M., Alfieri, A., Guha, C., Nosanchuk, J.D., Friedman, A.J., Nitric oxide-releasing nanoparticles accelerate wound healing in NOD-SCID mice. *Nanomed. Nanotechnol. Biol. Med.*, 8, 8, 1364–71, 2012 Nov 1.
52. Castangia, I., Nácher, A., Caddeo, C., Valenti, D., Fadda, A.M., Díez-Sales, O., Ruiz-Saurí, A., Manconi, M., Fabrication of quercetin and curcumin bionanovesicles for the prevention and rapid regeneration of full-thickness skin defects on mice. *Acta Biomater.*, 10, 3, 1292–300, 2014 Mar 1.
53. Nguyen, T.H., Kim, Y.H., Song, H.Y., Lee, B.T., Nano Ag loaded PVA nano-fibrous mats for skin applications. *J. Biomed. Mater. Res. Part B Appl. Biomater.*, 96, 2, 225–33, 2011 Feb.
54. Shahverdi, S., Hajimiri, M., Esfandiari, M.A., Larijani, B., Atyabi, F., Rajabiani, A., Dehpour, A.R., Gharehaghaji, A.A., Dinarvand, R., Fabrication and structure analysis of poly (lactide-co-glycolic acid)/silk fibroin hybrid scaffold for wound dressing applications. *Int. J. Pharm.*, 473, 1–2, 345–55, 2014 Oct 1.
55. Dubský, M., Kubinová, S., Širc, J., Voska, L., Zajíček, R., Zajícová, A., Lesný, P., Jirkovská, A., Michálek, J., Munzarová, M., Holáň, V., Nanofibers prepared by needleless electrospinning technology as scaffolds for wound healing. *J. Mater. Sci.: Mater. Med.*, 23, 931–41, 2012 Apr.
56. Han, C.M., Zhang, L.P., Sun, J.Z., Shi, H.F., Zhou, J., Gao, C.Y., Application of collagen-chitosan/fibrin glue asymmetric scaffolds in skin tissue engineering. *J. Zhejiang Univ. Sci. B*, 11, 7, 524–30, 2010 Jul.

7

Stem Cell Therapy: Skin Aging Blocking

Dipanjana Ash[1], Sohini Sen[1], Sneha Shaw[2], Ayon Dutta[3]*, Leena Kumari[4] and Swarupananda Mukherjee[5]

[1]*Department of Pharmaceutics, BCDA College of Pharmacy & Technology, Barasat, Kolkata, West Bengal, India*
[2]*Department of Pharmacology, BCDA College of Pharmacy & Technology, Barasat, Kolkata, West Bengal, India*
[3]*Department of Pharmaceutical Technology, Brainware University, Barasat, Kolkata, West Bengal, India*
[4]*Department of Pharmacy, Amity Institute of Pharmacy, Amity University, Kolkata, West Bengal, India*
[5]*Department of Pharmaceutics, NSHM College of Pharmaceutical Technology, NSHM Knowledge Campus, Kolkata - Group of Institutions, Kolkata, West Bengal, India*

Abstract

Aging, the time-related alteration and/or deterioration of structural integrity and functional capacity of organisms for survival and fertility, has been correlated with lowering of the number and regeneration ability of stem cells leading to senescent alopecia and other skin related manifestations such as decreased elasticity, wrinkles, relaxation, aberrant pigmentation, and wound healing abnormalities, graying of the hair along with the degradation of the pilosebaceous unit. The phenomenon of skin aging is becoming more noticeable globally as the birth rate continues to drop and human life expectancy continues to rise. Consequently, tackling the issues has emerged as a current area of study. No perfect remedies have been discovered as of yet. Stem cell therapies offer a wide range of potential possibilities in the field of rehabilitation healthcare to either delay or reverse aging, attributing to their immense biological properties, including flexibility, diverse differentiation, and self-regeneration ability. The chapter explores an overview of skin aging and different resources of stem cell with a special emphasis on the mechanism of stem cell therapies in the treatment of skin regeneration both preclinically and

**Corresponding author*: dutta8ayon7@gmail.com

Sumel Ashique, Sanjay Dey, Biplab Debnath, Mohhammad Ramzan and Swarupananda Mukherjee (eds.) Stem Cell Therapy in Dermatological Disorders, (163–194) © 2025 Scrivener Publishing LLC

clinically. In this regard, the chapter contributes an insightful perspective to scientists for the future clinical research on reversing or blocking skin aging through precise techniques.

Keywords: Aberrant pigmentation, self-regeneration, skin aging, stem cell

7.1 Introduction

In 21st century, it is anticipated that, in the ensuing decades, age-related stress would also rise significantly. With several established and emerging nations in particular heading toward severely aging populations in their civilizations, the world's population will continue to age. A research estimates that, by 2022, there will be 28.04 million people 60 years and older, or 19.8% of the total population; of these, there will be 209.78 million people 65 years and older, or 14.9% [1, 2]. In an effort to comprehend and mitigate the negative impacts of aging, contemporary civilizations are becoming more interested in the issue of aging as the population ages and living standards rise [3].

Skin aging is one of the most noticeable signs of aging, and women are especially concerned about it [3]. In 2023, the worldwide antiaging treatments industry was predicted to be valued around US $47 billion. The market is anticipated to achieve a compound annual growth rate of about 6% between 2023 and 2032 [4]. Skin aging, a combined impact of both endogenous and exogenous variables such as hormonal deficiencies, photoaging, chronological aging, and environmental factors, respectively, are of four types including intrinsic, extrinsic, photoaging, and hormonal skin aging [5]. In contrast to extrinsic aging, which can be brought on by excessive sunlight exposure and a variety of exogenous factors (pro-oxidant and antioxidant) on cell turnover *via* neuroendocrine immune biological response modifiers, intrinsic aging is characterized by finely wrinkled, uniform, unmarked, pale, less humid without flexibility skin. Extrinsically aged skin results from the loss of collagen type VII and fibril weakening between the superficial and epidermal layer of skin, whereas histological alterations in the basal cell layer of naturally aged skin result from internal factors that cause skin thinning and drooping. Photoaging, the resultant of three detrimental radiations of sunlight (visible: 44%, infrared: 52%–55%, and ultraviolet: 3%) primarily affects the face and neck. Reduced collagen formation, skin thickness with excessive moisture content, and epidermal barrier integrity are the main causes of hormonal aging [5–9]. Any layer of the skin may experience skin aging. Aging of the skin is often caused by two sources: internal and external influences. Every year, the synthesis of skin collagen and elastin declines by 1%, causing connective tissue to become disorganized. The dermal layer of the skin develops wrinkles as a result of the loss of collagen and elastin. Impaired visual activity, chronic arthritis, peeling skin, wrinkles, chronic inflammation, skin loosening, and hypothermia are the primary signs of aging skin [3, 5, 10, 11]. Numerous hypotheses or molecular mechanistic pathways have been brought up to explain aging, such as the theories of deoxyribonucleic acid (DNA), free radicals, membranes, Hayflick boundaries, telomerase, mitochondrial depletion, and neuroendocrine networking theories [5]. There are various treatment strategies to prevent aging including topically and systematically applied antiaging formulations (synthetic or natural, conventional, or nanotechnology-based liposomes, nanoemulsions, nanotopes, nanogems, metal oxide nanoparticles), hormone replacement therapy, and invasive techniques (chemical peelings, intense pulsed Light scattering, radiofrequency, injectable skin biostimulation and rejuvenation, ablative and/or nonablative laser photorejuvenation) [5, 12]. The conventional and nanocarrier-based therapies focus on only surface-level symptoms of aging [13].

However, stem cell therapies (SCTs) seeks to revitalize the skin at the cellular level, possibly targeting the most fundamental reasons of aging, including cell degeneration and decreased collagen formation [13, 14]. Despite being young, stem cells (SCs) (both embryonic and mesenchymal stem cells [MSCs]) have a remarkable ability to self-renew indefinitely and specialize into multiple cell lineages. Their distinct ability to serve as a storehouse for the development, restoration, rehabilitation, and rejuvenation of many tissues sets them apart from other kinds of cells [15]. The bioactive components (secretome and exosome) are released from SCs, and the interaction of SCs with the components demonstrates a promising therapeutic approach, ranging from treating aging-related degeneration to encouraging wound healing, *via* regulating immune system with fostering an anti-inflammatory environment [16–20].

The present chapter focuses on the concepts of SCs with their resources and the potential preclinical–clinical applications of SCTs with an especial emphasize on the molecular mechanistic pathways of SCs in the treatment of skin aging. The major existing obstacles of SCTs and mitigations of them are also explored in this chapter.

7.2 Stem Cell: An Overview

All phases of life contain SCs, a special type of cell population with the capacity to self-renew and specialize into several cell lineages. Although skin SCs are classified as somatic SCs, other kinds of skin SCs have been discovered in recent decades because of the diversity of cells of skin. With the aid of molecular mechanism, notable progress has recently been achieved in distinguishing between several kinds such as epidermal, follicular, melanocyte, sebaceous gland, mesenchymal, neural progenitor, and hematopoietic skin SCs [21, 22]. These cells are essential for the development of newborns and the healing process following an accident or illness because they are the source of certain cell types found in differentiated tissues and organs [23]. Although SCs' primary purpose in adults is rehabilitative and restorative, in neonates they serve to undergo differentiation and proliferation into the variety of cell types and lineages necessary for ongoing development [24].

Adult tissues as well as neonatal tissues such as the placenta or umbilical cord can be used to develop SCs. Although embryonic SCs have been used for research in the past, SCs from other sources have mainly supplanted them due to ethical issues [25]. Adipose tissue, trabecular bone, and bone marrow are common tissues from which adult oligopotent and unipotent SCs are extracted. The most popular location for obtaining non–neonatal-derived SCs has historically been bone marrow; nevertheless, this process is intrusive and unpleasant. Progenitor cells from peripheral blood have been used instead of bone marrow cells. However, this method was initially a less effective source of SCs and has its own problems and hazards. Additionally, it is now recognized that the proliferation and differentiation capacity of SCs varies according to their origin. It was shown that cells obtained from adipose tissue and umbilical Wharton jelly multiplied much faster than cells obtained from bone marrow and fetal origins [26, 27]. Induced pluripotent SCs (iPSCs) are a rapidly developing source of SCs that are now being used clinically as well. Unlike traditional sources, iPSCs are harvested using less invasive methods and are derived from somatic cells that have been reprogrammed back to a pluripotent state using reprogramming factors [28, 29]. The first bone marrow transplant was performed in 1956, marking the beginning of SCs' use in contemporary regenerative medicine in the 1950s. This discovery cleared the path for the currently available SC therapies and provided insight into the prospective treatments that might be developed in the future with more clinical procedure development and improvement [30, 31].

7.2.1 Different Sources for SCs

Stem cells have demonstrated significant capacity in skin restoration. MSCs are one of the most prevalent classes of SCs, which are vital for skin tissue regeneration and are subdivided into various categories, namely, bone marrow stem cells (BMSCs), adipose-derived stem cells (AdiSCs), amniotic membrane stem cells (AmnMSC), human umbilical cord MSCs (HumUCMSCs), and iPSCs [32]. The different sources of SCs used in the treatment of skin aging are described below:

7.2.1.1 Bone Marrow

Self-regenerating tissues need SCs to substitute for aging and damaged cells because of regular physiological processes. Hematopoietic stem cell (HSC)–derived myeloid and lymphoid lineage cells are both very short-lived cell types that need a steady supply of freshly differentiated repair cells [33]. The bone marrow contains HSCs, which serve as a source for the many blood cell types needed for regular immunological and biological activities. These cells are able to divide asymmetrically because they occupy a physiological niche. Stem cells that divide asymmetrically produce both a differentiated daughter cell and one identical daughter cell as part of their offspring [34, 35]. Certain microenvironments, exogenous signals, and developmental factors that the cell encounters direct the differentiation of these daughter cells into specialized cell types. This process enables bone marrow SC counts to remain mostly stable even as progeny continues to proliferate and differentiate [36, 37]. Because of its availability, early source recognition, and extensive research history, bone marrow has long been the most popular harvesting location for SC collection. Hospitalization or other problems are possible outcomes of the invasive and painful surgical process used to isolate SCs from bone marrow. When compared to other harvesting methods, patients also report higher levels of preprocedural anxiety and postprocedural discomfort. In contrast, bone marrow has been shown to be a more abundant source of cells than other harvesting methods, producing 18 times as many cells as blood vessel progenitor cell extracting techniques at first. However, as techniques became advanced, it was discovered that administering a cytokine rehabilitation to patients before peripheral bloodstream progenitor cell acquiring stimulated many of the targeted cells into the bloodstream, substantially boosting the technique's effectiveness and making it clinically feasible [38–41].

Under the right differentiation conditions, BMSCs have versatile maturation pattern and exceptional self-regenerating capabilities both *in vitro* and *in vivo*. MSCs aid in tissue healing through the substitution of differentiated cells and/or reorganization of the surroundings *via* releasing growth factors and chemokines. Therefore, the bone marrow–derived mesenchymal SC therapies support several scientific and clinical research projects for the management of numerous inherited and acquired illnesses. When administered systemically or locally, MSCs have been shown to accelerate the healing of cutaneous wounds. Since the antiapoptotic and antioxidative properties of the SCs have been demonstrated in animal studies, these cells might be taken into consideration as an alternate therapeutic approach for skin rejuvenation [42–47].

7.2.1.2 Adipose Tissue

Despite the fact that bone marrow SCs are now often used, problems with collecting methods and the possibility of poor cell yields have prompted researchers to look for other sources of MSCs from both animal and human adipose tissues [48]. Stromal vascular fraction, a heterogeneous adipocyte indicator, is developed following enzymatic breakdown of adipose tissue and the adipose tissue–derived SCs are present in it. The SC shows almost similar characteristics as that of MSCs [49–51]. Ease of harvesting, immunized accessibility attributing to the absence of significant histocompatibility II, and costimulatory molecular structure provide beneficial effects of adipose tissue–derived SCs over bone marrow–derived SCs [50, 51].

Although AdiSCs are multipotent cells, their functions are as regulatory cells rather than proliferating *in vivo*. They are distinguished by the surface expression of $CD34^+$, $CD44^+$, $CD31^-$, and $CD45^-$ while encouraging angiogenesis in dermis and have high capacity for proliferation *via* releasing a variety of growth factors such as endothelium, fibroblasts, and anti-inflammatory cytokines. AdiSCs promote tissue regeneration *via* encouraging the release of extracellular proteins (collagen and elastin) as well as metalloproteinases. The immune response's nonspecific pathways initiate the SCs [52–55]. Additionally, it has been discovered that MSCs derived from adipose tissue may reduce wrinkles triggered by UV radiation and protect photoaged epidermal fibroblasts from oxidative damage [56].

7.2.1.3 Umbilical Cord

Umbilical cord blood from donors may be harvested with no risk to the donor, and HumUCMSCs can be extracted from a number of sources, such as umbilical vein endothelial cells and umbilical cord lining, amnion, chorion, perivascular cells. Umbilical cord–derived SCs are significantly more accessible than bone marrow–derived SCs. There are three main kinds of SCs that may be detected in the umbilical cord blood such as hematopoietic, mesenchymal, and embryonic-like SCs [30].

Umbilical cord–derived SCTs can raise the ability of cells to scavenge free radicals, improve oxidative damage of skin, and have a favorable impact in lowering cell degeneration induced by oxidative stress. However, while the antioxidant benefits of the therapy have been identified, the fundamental mechanism remains not entirely recognized. Despite the lack of significant data, the SCs directly scavenge free radicals, secrete bioactive enzymes, and regulate mitochondrial activity, all of which contribute to an antioxidant stress role [57]. The SCs facilitate tissue regeneration primarily through the secretion of bioactive factors, secretome, or conditioned medium, resulting as a promising treatment option against skin aging [27].

7.2.1.4 Amniotic Cells

The two most prevalent varieties of pluripotent stem cells (PSCs) are iPSCs and embryonic stem cells (ESCs). Malignancy and inadequate transplanted cell viability rate are two major problems with ESC and iPSC technologies. PSCs have recently been identified from gestational tissues including placental membrane and amniotic fluid. Amniotic epithelial and mesenchymal cells are two types of human amnion-derived stem cells (hADSCs), a relatively novel source of SCs that have been shown to exhibit a number of beneficial traits. Furthermore, intravenous hADSCs showed no signs of hemolysis, allergic responses, and malignancies in recent investigation [58–60]. AmnMSC therapy gains popularity for the treatment of skin aging attributing to presence of cytokines and growth factors [61].

7.2.1.5 Placental Tissue

Both SCs and epithelial cells (ECs) that inhabit placental tissue have the ability to develop into wide range of tissues including cardiac, endothelial, pancreatic, pulmonary, neurological, myogenic, hepatogenic, osteogenic, and adipogenic. Placental cells can develop into all of these diverse types of tissues attributing to lineages arising from many areas of the placenta, such as the hematopoietic cells from the chorion, allantois, and yolk sac and the mesenchymal lineages from the chorion and the amnion [62]. MSC exosomes emphasize a number of skin-benefiting properties such as promoting angiogenesis in both acute and chronic wounds, enhancing collagen extracellular matrix development and deposition, stimulating fibroblast migration and proliferation, regulating inflammation and the immune response, controlling tissue reconfiguration to lessen unfavorable scarring. Additionally, MSC exosomes have the capacity to repair the dermal papilla cells inside the hair follicle *via* boosting the number of hair follicles in the anagen or growth phase [63–69].

7.3 Mechanisms of SC Action in Skin Regeneration

Stem cell action–based therapy has arisen as one of the most potential and powerful tools in improvement of skin repair and regeneration due to the ability of SCs to differentiate, autorenew, minimize immunogenicity, and produce an efficient secretome, among which the self-renewal and differentiation capacity of SCs into various cell types is of greater importance in biology of regeneration of injured tissue and physiologic tissue [70]. The method of cutaneous regeneration is essential for repairing skin defects and restoring the skin's lost integrity, barrier function, and tensile strength. Cutaneous wound healing is a complex process which includes inflammation, proliferation, and tissue remodeling [71]. Growing knowledge about SC biology has aided in the development of cell-based treatments that can influence several facets of skin repair [72]. Here in this review, we have summarized about few aspects of mechanism of skin regeneration (Figure 7.1) through SC action.

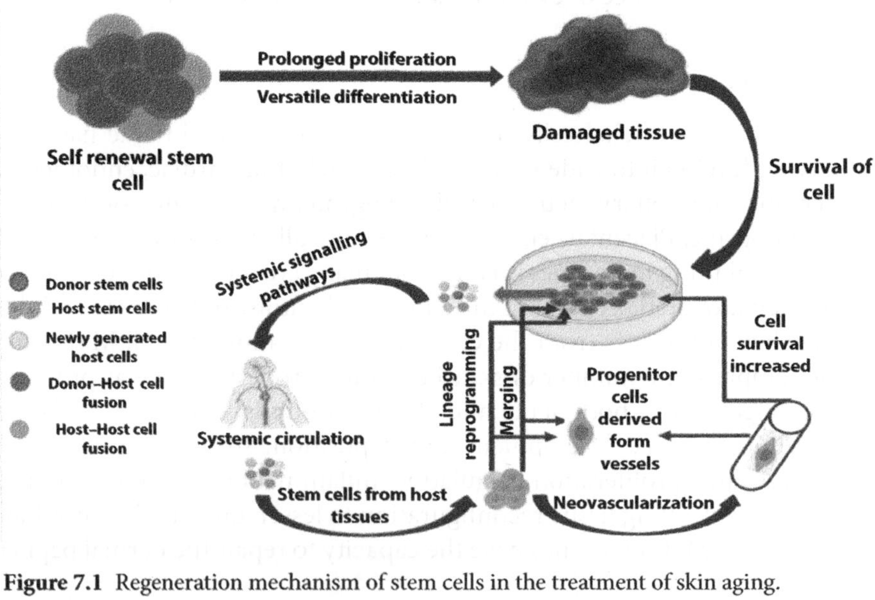

Figure 7.1 Regeneration mechanism of stem cells in the treatment of skin aging.

7.3.1 Epidermal SC Action in Skin Regeneration

Adult SCs known as epidermal stem cells (EpSCs) have three common traits: a greater capacity for self-renewal, a slower rate of cycling, and adherence to the epidermis' basement membrane. Stem cell "niches," i.e., microenvironment that are crucial for cutaneous repair and wound healing include both wingless/integrated (Wnt) and notch signaling systems [73]. Skin ECs' capacity to develop and/or preserve characteristics of versatile SCs has been related to Wnt signaling pathway. A multifaceted protein called β-catenin, which is stabilized when cells acquire a Wnt signal, is the key factor of the pathway. To activate members of the Lef/Tcf family of DNA-binding proteins, β-catenin is somehow necessary. When particular skin mesenchymal cells suppress a second signaling cascade, the bone morphogenetic protein (BMP) pathway, the multipotent ECs produce Lef1 and become committed to producing a hair follicle; thereby, the development of a hair follicle seems to be dependent on the Wnt and BMP signaling pathways [74, 75]. A highly prevalent signaling pathway, the Notch pathway, which comprises Notch receptor protein (Notch1-4), Notch ligand protein of adjacent cells (Jag1 to 2, and so forth), and DNA-binding protein, is found in signal transduction pathways that control cell fate across a range of organs. As a result of binding of Jag protein to the Notch receptor, the SCs retain the ability of multifunctional differentiation, whereas the inhibition of notch pathway may lead SC differentiation into terminally differentiated cell [76]. Furthermore, it has been reported that components of notch signaling cascade are widely spread in the ECs as well as the EpSCs and play a vital role in division and differentiation of epidermal cells. However, decreased expression of HES (hairy enhancer of split), a target gene involved in the Notch signaling pathway, results in restricted skin EpSC proliferation, thereby hindering epithelialization process, hence the disruption of cutaneous functional barrier and irregular hair growth occur [77, 78]. The Notch pathway also modulates SC adhesion, thereby affecting SC niche; as a consequence, various signals are transmitted to the SCs for self-renewal into different tissue cells [79].

Microribonucleic acids (micRNAs) are noncoding, small RNAs that govern posttranscriptional expression of genes by inhibiting the translation of messenger RNA (mRNA) or causing its destruction. Since micRNAs are major modulators of gene expression, they are essential for a variety of biological functions, including differentiation, homeostasis, and cell survival. However, abnormal expression of micRNA may result into various diseases [80]. Recently, a rise in interest in the functions of micRNAs in the maintenance of adult skin SC homeostasis and the formation of epidermal tissue has been reported. Studies demonstrated that, in various epidermal tissues of rodents, traces of numerous micRNA were detected, such as miR-141, miR-19b, miR-20, miR-200 (a, b, and c), and miR-429, and are found to be expressed in complete epidermal tissue layer [79]. It was also reported that the significance of the previously mentioned elevated micRNAs in the growth of epidermal tissue cells was confirmed by the upregulated expression of miR-203, miR-23b, miR-95, miR-210, miR 224, miR-26a, miR-200a, miR-27b, and miR-328 in differentiated keratinocytes and the decreased expression of miR-376a. By restricting p63 expression, miR-203 contributes to the retention of skin tissue structure and keratinocyte differentiation, resulting in the damage of EpSC traits and early differentiation [81]. A recent study demonstrated that EpSC-derived exosomes improved wound repair and scar restoration in experimental animals. EpSC-exosomes also increased the rate of skin tissue cell, nerves, and vessel regeneration, wherein these were achieved by impeding the activity of transforming growth factor β1 (TGF-β1) and its genes involved in the downstream pathway. EpSC-exosomes were rich in some particular micRNAs, such as, let-7a, miR-142-3p, miR-16 and miR-425-5p, as the study reported. MicroRNAs unique to epidermal stem cells-exosomes (EPSC), particularly miR-425-5p and miR-142-3p, were essential in preventing myofibroblast development by lowering dermal fibroblasts' production of TGF-β1 [82].

7.3.2 Mesenchymal SC Action in Skin Regeneration

MSCs are distributed over a wide area of tissues and are multipotent cells with the ability to differentiate and regenerate them. MSCs that are delivered endogenously or exogenously have the ability to travel and traffic to damaged tissue, aiding in the process of tissue healing. Based on their capacity to undergo differentiation and transdifferentiation into tissue-specific cells, fuse with native cells, and promote host cell viability and functional recovery by releasing a cascade of paracrine factors that influence the immune system's reaction and local microenvironment, MSCs may have therapeutic efficacy in cutaneous regeneration [72]. MSCs promote the development of vascular tissues and proliferation of cells during skin regeneration while reducing inflammation in skin damage lesion. MSCs facilitate wound repair in every stage [61].

Inflammatory phase, being one of the most crucial phases in skin regeneration process, promotes the activation of immune cells to destroy pathogen and facilitates wound healing. MSCs are reported to inhibit inflammatory responses in numerous mechanisms [61]. MSCs stimulate macrophages to polarize into an M2-like phenotype, which is a subset of macrophages that suppresses the immune system and lowers inflammation. Furthermore, MSC-induced M2-like phenotypic macrophage (MMSC) communications with natural killer (NK) cells suppress the expression of proteins linked to NK cell activation, including interferon-γ, NKp44 (natural killer cell activating receptor 44), CD25, and CD69. Moreover, MMSC can suppress T-cell division through stimulating regulated T-cell proliferation [83]. MSCs may transform into a variety of cutaneous cells, including endothelial cells, keratinocytes, fibroblasts, and others, depending on appropriate conditions. As a result, they can aid in wound re-epithelialization, blood vessel regeneration, and granulation tissue formation [84]. Re-epithelialization is the process by which ECs migrate and proliferate from the wound's surface toward the core in order to create a new layer of epithelium and replenish the skin's functional barrier [85]. Wnt/β-catenin signal transduction cascade involved in cell differentiation and proliferation plays a vital role in skin tissue regeneration. Recent studies have demonstrated that exosomes derived from HumUCMSCs have the ability to enhance the ability of re-epithelialization of burn wounds by raising the phosphorylation level of AKT and strengthening the downstream effects of AKT signaling, which in turn encourages the migration and proliferation of keratin-forming cells. Hence, AKT signal transduction pathway emerged as an important key factor in skin tissue regeneration [86]. Moreover, macrophages release TGF-α and epidermal growth factor (EGF), thereby leading to the modulation of keratinocyte proliferation by MSCs. Fibroblasts recruits fibronectin, keratinocyte growth factor, and EGF to promote keratinocyte migration and multiplication [87]. Taken into account, MSCs offer a variety of benefits when it comes to treating different types of wounds and can affect the healing process at several levels, such as inflammatory conditions, vascularity, and scarring, in the dermis and epidermis to enhance patients' standard of living as well as their survival probability [86].

7.4 Preclinical Applications of SC in Skin Aging

Stem cell therapies have shown significant effectiveness when other treatments have failed and are increasingly recommended for a variety of clinical disorders beyond typical sources to treat inherited blood illnesses [30, 31]. The *in vivo* SCs based therapeutic strategies (listed in Table 7.1) gain popularity to treat skin aging *via* regulating immune system with fostering an anti-inflammatory environment [19, 20].

Table 7.1 Preclinical study reports on SCTs in skin aging.

Stem cell therapies	Route of administration	Outcome	References
AdiSCs	Subcutaneous	Adipose-derived stem cell could aid in skin regeneration during aging of 6-week-old male and female mice (induced by subcutaneously administration of d-galactose once a day for 8 weeks) *via* trophic effects, antioxidation, and glycation inhibition.	[88]
		The stem cell treatment strategy became a potent approach to delay skin aging of young and aged mice *via* reducing expression of inflammation (interleukin 6 [IL-6], p65, and IL-1α) and aging (p21 and p53) inducing genes and growth factors (insulin-like growth factor 1, mammalian target of rapamycin) under nuclear receptor interacting protein 1 suppression.	[89]
Fullerenol-loaded AdiSCs		The potentiality of the strategy was observed in the treatment of d-galactose induced aging (*in vivo*) through the inhibition of protein overexpression of peroxisome proliferator-activated receptor-γ with the boosting of FoxO1 (forkhead box protein O1) protein expression.	[90]
Combination of stromal vascular fraction and AdiSCs		The improvement of collagen percentage in UVB-induced aged animal skin tissue with its regeneration ability confirmed the effectiveness of the treatment at a dose of 10^6 cells /100 μL/ injection.	[91]
Human AdiSCs-CM	—	The technique gained popularity in the treatment of UV-induced skin aging in *in vitro* human keratinocyte cell lines through the suppression of IL-6 and matrix metalloproteinase 1 (MMP-1) formation and amplification of the expression of heme oxygenase-1 (antioxidant gene) resulting in the inhibition of formation of reactive oxygen spices and mitigate photoaging.	[92]
Neural stem cell–conditioned medium	Topical	The stem cell and its released components (tissue inhibitor of metalloproteinase 1 and 2) were reported to be ameliorated UVB-induced skin photo *via* suppressing ROS generation and MMP expression with the activation of the NF-κB (nuclear factor κ-light-chain-enhancer of activated B cell) pathway *in vivo*.	[93]

(*Continued*)

Table 7.1 Preclinical study reports on SCTs in skin aging. (*Continued*)

Stem cell therapies	Route of administration	Outcome	References
BMSCs	Intradermal	Bone marrow–derived mesenchymal stem cell–loaded intradermal injections ameliorated the histopathological and ultramicroenvironmental aberrations of the 30 male, 200 ± 20 g weighed Sprague–Dawley rat's skin triggered by d-galactose *via* encouraging an antioxidant response and reducing oxidative stress carried on by aging.	[94]
	Intravenous	The treatment helped to improve aging and disorders associated with d-galactose–induced aging and successfully improved aging-related phenotypes *in vivo via* upregulating and downregulating the expression of inflammatory factor and the growth factor, respectively.	[95]
Umbilical cord MSCs	Topical	The conditioned serum-free media produced from umbilical cord–derived MSCs *in vivo* may offer protection against UV-induced photoaging *via* hindering the elevation of malondialdehyde and impeding the activities of glutathione peroxidase and superoxide dismutase.	[96]
	Subcutaneous	Dose-dependent antioxidant and proregenerative consequences were observed in aged nude mice model *via* altering the levels of both superoxide dismutase and malondialdehyde with the stimulation of type 1 collagen and vascular endothelial growth factor expressions resulting in the improvement of dermis width and skin configuration with the restoration of collagen.	[97]
Gelled secretome isolated from hWJ-MSCs (human Wharton jelly mesenchymal stem cells)	Topical	In the treatment of skin aging the isolated gelled secretome was found to be played a crucial role to restore damaged cells with enhanced collagen percentage *in vivo* through suppressing the expression of MMP-2, MMP-13, and P53.	[98]

7.5 Clinical Applications of SC in Skin Aging

To ameliorate human skin aging, a number of clinical trials have been carried out or are in progress to examine the safety, viability, and effectiveness of SC-based treatments.

According to a recent study conducted on 48 women between the ages of 41 and 60 years, AmnMSC-conditioned medium (AmnMSC-CM) demonstrated considerable reduction in pore scores, wrinkles, and spot polarized and spot UV as compared to the control group [99]. In order to improve the permeation of AmnMSC-CM, microneedling was used. Similarly, the use of AmnMSC-CM in conjunction with skin needling significantly enhanced the texture of the skin, promoted the formation of collagen and elastic fibers, and assisted in the treatment of face aging [100]. Evaluation of safety and efficacy of HumUCMSCs was conducted on 34 adults suffering from moderate to severe atopic dermatitis. A dose-dependent improvement was observed after a treatment with single dose of HumUCMSCs. Thereafter, more that 50% of the patients showed a significant improvement EASI (Eczema Area and Severity Index) score, as well as significant decline in SCORAD (SCORing Atopic Dermatitis) score and Investigator's Global Assessment score. Also, a decline in the concentrations of blood eosinophil level and serum immunoglobulin E level were seen after the treatment [101]. Randomized, prospective, split-face comparative study was conducted on 23 patients who had undergone ablative CO_2 fractional laser treatment on both cheeks. A cream and serum containing HumUCMSCs-conditioned media were administered to those 23 patients, which resulted in enhanced wound repair and improved post-therapeutic erythema, which effectively decreased time of curepost laser treatment [102]. Similar clinical trial was performed in 25 patients who underwent laser treatment, wherein moisturizers containing or not containing adipose-derived stem cell–conditioned medium (AdiSC-CM) plus 2% niacinamide was applied. The results demonstrated a significant improvement in wrinkles and other skin restoration parameters after the application of AdiSC-CM plus 2% niacinamide moisturizer. Additionally, the findings of the *in vitro* assays indicated that type I collagen was upregulated and that the production of proinflammatory cytokines, MMP-1, and MMP-2 was downregulated in a dose-dependent manner [103]. The combination treatment of human adipose tissue–derived SC–loaded exosome and microneedling became a great clinical promise for recovering skin aging of 28 individuals [104].

MSC-derived extracellular vesicles are of the following types such as apoptotic bodies (diameter: 500–800 nm), microvesicles (diameter: 150–500 nm), and exosomes (30–150 nm). MSCs-derived exosomes are not only involved in intercellular communication such as carrier in protein delivery, micRNA, and mRNA to target tissue cells, but are also involved in delivery of drugs, RNA, and enzymes wherein MSC-Exoms act as vehicle [105]. A clinical trial was conducted on 25 patients, wherein they were administered with AdiSC-exosomes along with ablative CO_2 fractional laser treatment. The AdiSC-Exoms-treated sides experienced less severe treatment-related erythema. It was observed that skin pore size considerably decreased compared to baseline on AdiSC-Exoms. Therefore, a combination of AdiSC-Exoms and laser therapy may have a synergistic effect on the effectiveness of therapies for atrophic acne scars [106]. An assessment of the possible therapeutic benefits of protein extracts (including IL-6, IL-8, and TGF-β) from AdiSCs in 30 female volunteers revealed that the intervention group significantly improved skin roughness, decreased wrinkle formation, enhanced skin brightness and elasticity, and reduced melanin index in comparison to the control group [107]. Typical drawbacks of these studies include their limited sample size and absence of long-term follow-up [108]. Another study reported the potential effect of human placental-derived MSC loaded exosomes against aging-induced stress [109]. Therefore, more research is needed to develop innovative methods for treating a variety of illnesses for which MSCs may be helpful.

7.6 Challenges and Considerations

SCTs related to skin aging is a quickly evolving area with considerable potential to transform dermatology and regenerative medicine. Nonetheless, the quest for SC-based treatments for skin aging encounters various scientific, ethical, regulatory, technical, and safety hurdles. Below are key considerations and challenges that need to be discussed [110].

7.6.1 Ethical and Regulatory Issues in SC Therapy

In clinical trials where SC therapies are used, obtaining informed consent from patients is critical. This involves clearly explaining the potential risks, benefits, and uncertainties linked to SC interventions for skin aging. Since SC therapies remain mostly experimental, it is important to grasp possible side effects and their long-term consequences [111]. With the emergence of SC-based treatments, there is a danger of uncontrolled commercialization of these therapies. Numerous commercial businesses are providing SC-based skin treatments without sufficient proof of safety or efficacy. This situation raises alarms about the exploitation of patients and the possible proliferation of pseudoscience within the beauty and skincare sectors [112]. Stem cell therapies aimed at addressing skin aging are required to pass stringent evaluations and receive authorization from regulatory bodies such as the US Food and Drug Administration and the

European Medicines Agency. The regulations should encompass not only clinical trials focusing on safety and effectiveness but also the ethical procurement of SCs. Nonetheless, the regulation of SC therapies can differ greatly from one country to another [113]. The heterogeneity of aging populations, variations in skin aging patterns, and the potential influence of placebo effects complicate the recruitment of volunteers for clinical trials focused on skin aging. Long-term clinical trials are crucial for accurately assessing the efficacy of SC treatments, as skin aging is a continuous process [114]. The International Society of Stem Cell Research (ISSCR) and pertinent agencies should contemplate prohibiting SC therapies that have not undergone clinical trial evaluation. Only authorized therapies should be provided to patients for a price, in accordance with the existing ISSCR guidelines [115]. Additional ethical and regulatory concerns with the collecting, storage, study, medical application, and disposal of SCs have been recognized. The ISSCR guidelines delineate significant issues and methods for their resolution. The guidelines regarding clinical studies do not adequately specify that SC therapy may be irreversible, potentially hindering participants from obtaining any advantages from the experiment or subsequent therapies [116].

7.6.2 Safety Concerns

Stem cell therapy for skin aging has promising potential for regenerative treatments; however, numerous safety problems require meticulous consideration [117]. The primary safety risks associated with SC treatment for skin aging are outlined below.

The potential for tumor formation is a significant concern associated with SCTs. PSCs, particularly iPSCs, can differentiate into various cell types; however, they also have the potential for uncontrolled or abnormal cell proliferation. Improper regulation or differentiation of SCs can lead to the formation of teratomas or other tumor types upon transplantation into skin tissue [118]. Cell reprogramming, such as the generation of iPSCs, can result in mutations contributing to genetic instability and elevate the risk of tumor genesis. Autologous SCs, derived from the patient, may also pose this risk if cultured for prolonged durations or subjected to adverse conditions during expansion [119]. Autologous SCs markedly diminish the likelihood of immunological rejection; conversely, allogeneic SCs may provoke an immune response, resulting in inflammation, graft rejection, or potentially severe immune-mediated damage to the skin or other organs [120]. SCs are cultivated under laboratory conditions prior to transplantation into the skin. If sterility is compromised during collecting, culture, or delivery, there is a danger of infection. This may lead to localized infections at the transplantation site or, in severe instances, systemic infection [121]. Utilizing SCs for wound healing in old skin carries an intrinsic risk of preexisting infection at the wound site, which complicates SC integration and heightens the probability of additional infection and protracted recovery [122]. An inflammatory reaction may result from the transplantation of SCs into aged or damaged skin tissue. The immunological milieu of aging skin is naturally changed, and the introduction of SCs may cause persistent low-grade inflammation. Fibrosis, an excessive accumulation of scar tissue that hinders skin regeneration and further damages the skin may result from this [123]. The ISSCR guidelines support for the recruitment of a varied array of trial participants, although they fail to consider the long-term risks and costs to the participants. A patient who consistently develops tumors following SC therapy may necessitate ongoing, lifelong surgical procedures for their excision. Patients may experience blindness as a result of SC therapy, necessitating perpetual aid for navigation and alterations to their living environments. In the absence of continued support for trial participants after a clinical study, disadvantaged individuals may incur more long-term expenses and face higher lifestyle challenges. Enhanced continuous support should be provided to trial participants instead of excluding the disadvantaged [124]. In numerous areas, SC therapies lack sufficient regulation, heightening the possibility of harmful procedures being presented to patients. The emergence of unlicensed SC clinics providing "miracle cures" for skin aging, lacking any scientific validation or oversight, presents considerable safety risks. Without proper guidelines and standards for SC preparation, culture, and application, the risk of utilizing substandard SCs in treatments is heightened. In the absence of quality control, patients may undergo treatments that are either inefficient or hazardous [125]. The ethical implications of SC source are essential for guaranteeing safe practices. The utilization of ESCs may provoke ethical dilemmas, and the exploitation of susceptible groups (e.g., women donating eggs for SC acquisition) is a significant concern within the overarching framework of SC research [126]. Although SCTs have promising opportunities for addressing skin aging, they are accompanied by several safety issues. Concerns such as tumorigenicity, immunological rejection, and the possibility of long-term negative impacts necessitate meticulous evaluation in both research and therapeutic settings. A deeper comprehension of SC dynamics coupled with stringent regulatory supervision is crucial to alleviate these dangers and guarantee the safe and successful application of SCs in dermatology [127].

7.6.3 Cost and Accessibility of SC Treatment

Although their high costs restrict their application to a limited percentage of the population, SCTs offer great potential to create revolution in healthcare [128]. Ensuring fair access to SC treatments and releasing their whole therapeutic potential depend on addressing these expenses. The high costs of SCTs range from the complexity of the technology to regulatory and manufacturing challenges. Intensive research and development require the successful development of SCTs which include costs for developing new treatments, conducting preclinical and clinical trials, and ensuring regulatory compliance [129]. Moreover, many SC treatments are still in the experimental or early clinical phases, so the return on investment for businesses creating these treatments is unknown, which can result in even more expensive products. Apart from research and development, complex and highly regulating manufacturing and quality control process also attribute the high cost of SCTs as strict quality control measures need to be taken in laboratory for ensuring safety and effectiveness of SCs [130]. Another difficulty that increases expenses is increasing manufacturing to satisfy patient demand; this is especially true for autologous treatments, which are tailored to unique patients. Apart from the aforementioned attributes, regulatory hurdles are the key factor that is responsible for the high costing of the SCTs due to the requirement of multiple phases of clinical trials to ensure its safety and efficacy by regulatory agencies such as US Food and Drug Administration [131]. Furthermore, postapproval, companies are required to persist in their investments in monitoring, compliance, and reporting to uphold regulatory approval, all of which add to elevated costs. One of the major features of SCTs is their ability to provide personalized treatments. These highly customized therapies make it challenging to attain economies of scale; hence, every treatment is intrinsically more expensive than mass-produced medications or treatments.

To address the above discussed challenges, some strategies need to be implemented to make it more affordable. Through the advancements in automation and manufacturing technology, the cost of SCTs can be decreased up to an extent. By automating critical stages in the production process, including cell culture and quality control, producers may diminish labor expenses and enhance productivity [132]. Moreover, advancements in biomanufacturing techniques, such as 3D bioprinting and large-scale bioreactors, may enable the production of SCs in larger volumes at reduced prices. One of the main factors influencing the cost of SCTs is regulatory approval; nevertheless, expediting the approval procedure may help save expenses. Promoting collaboration among regulatory bodies and researchers, together with the adoption of worldwide standards, may mitigate redundancy and expedite the approval process. Outcome-based pricing system could be adopted to develop more affordable SCTs, where the cost of the treatment is related to its effectiveness. If the therapy yields the anticipated outcomes, the whole payment is rendered; if not, the cost is diminished or reimbursed [133]. Another strategy entails the distribution of therapy expenses over time, with payments rendered in installments rather than in a lump sum. These payment options may alleviate the financial strain on patients and enhance the accessibility of high-cost medicines. By providing public funding and subsidies, government can help to make SCTs more cost-effective. By fostering collaboration among healthcare providers, biotech companies, and research institutions, it may be possible to reduce costs by utilizing shared infrastructure and resources. Establishing common biobanking facilities, for instance, might enable SC manufacture to be more affordable and assist to standardize it.

Despite holding significant potential for rejuvenating tissues and addressing aging, SCTs for aging are not yet widely approved by regulatory bodies; most treatments are available in private clinics and highly expensive. Accessibility is also limited due to geographical barriers, complex regulatory approval process, and some safety concerns, which have already been discussed. Through the implementation of above discussed strategies with ongoing research and technological advances, it is possible to develop more accessibility and affordable SCTs in the future [134].

7.7 Future Directions and Innovations

As the lifespan of individuals rises, the need for antiaging treatments escalates. Skin aging, as a direct representation of the aging process, is an important concern for both health and aesthetic purposes. The use of SCs' regenerative capabilities to restore and revitalize aging skin has made SCTs a viable treatment option. AdiSCs safeguard cells from oxidative injury by releasing growth factors (including HGF [hepatocyte growth factor] and vascular endothelial growth factor [VEGF]), cytokines such as IL-6 (by the enhancement of STAT3 [signal transducer and activator of transcription 3] and Nrf2 [nuclear factor erythroid 2-related factor 2] expression), and antioxidant enzymes (e.g., glutathione peroxidase [GPx], SOD [superoxide dismutase], and catalase). Protective factors including laminins, COL17A1 (collagen type XVII alpha 1 chain), Nrip1 (nuclear receptor interacting protein 1), and BAML1 (brain and muscle Arnt-like protein-1) contribute to skin aging by facilitating SC homeostasis. Consequently, these parameters may function as targets for antiaging interventions. Furthermore, the recent advancement of SC-based nanotechnologies has created a new era in SCTs. Therapeutic benefits, both physical and histological, were noted in nanofat grafting and SC-derived exosomes, with some results found in humans. Skin rejuvenation produces more pronounced results in a reduced timeframe. Nonetheless, some operations are invasive and may only be conducted at specific accredited centers. Moreover, obstacles related to healthcare regulations, inadequate survival rates of delivered cells, the potential for biological contamination, and high costing exist. In addition, a cell-free method, exemplified by the persistent release of adipose collagen fragments, has been proposed to mitigate the drawbacks of SC therapy. The observed modest problems necessitate the validation of the safety and feasibility of SC-based treatment in a larger cohort.

As research in this field progresses, SC-based treatments are projected to play a growing role in antiaging and dermatological therapy. Numerous advancements and possible future paths are emerging as a result of ongoing research, with an emphasis on honing SC treatments, boosting efficacy, and ensuring safety for long-term skin rejuvenation.

7.8 Conclusion

SCTs signify an enormous advance in regenerative medicine, especially in addressing the issues associated with aging. As our understanding of the basic mechanisms of aging advances, SCTs provide a novel method to revitalize old tissues, such as the skin, by facilitating regeneration and repair. SCTs possess the potential to restore diminished function, augment cellular turnover, and increase collagen formation, hence potentially reversing or substantially alleviating the visual and functional impacts of aging. The incorporation of emerging technologies, including personalized SCTs, iPSCs, exosome-based treatments, and sophisticated delivery methods such as nanotechnology, enhances the therapeutic efficacy of SCs in age-related disorders. The integration of SC treatments with other regenerative approaches, including growth factors and gene editing, is expected to improve therapeutic results and provide more effective, enduring remedies for aged skin. Nonetheless, despite the considerable promise of SCTs for aging, several hurdles persist. Concerns about safety, long-term implications, scalability, and ethical issues necessitate thorough study and meticulous regulation. It is essential to reconcile the potential of regenerative treatments with thorough clinical studies and evidence-based methodologies for confirming their extensive use in both clinical and cosmetic domains.

In conclusion, SC treatment is in the edge of addressing aging, with the potential to transform our approach to age-related ailments. Ongoing progress in research, technology, and clinical application will certainly provide more effective solutions, facilitating healthy aging and enhancing the quality of life for future generations.

Abbreviations

AdiSC-CM	Adipose-derived stem cell–conditioned medium
AdiSCs	Adipose-derived stem cells
AmnMSC-CM	Amniotic membrane stem cell–conditioned medium
BAML1	Brain and muscle Arnt-like protein-1
BMP	Bone morphogenetic protein
BMSCs	Bone marrow stem cells
CD	Cluster of differentiation
COL17A1	Collagen type XVII alpha 1 chain
DNA	Deoxyribonucleic acid
EASI	Eczema Area and Severity Index
ECs	Epithelial cells
EGF	Epidermal growth factor
EpSCs	Epidermal stem cells
ESCs	Embryonic stem cells
FoxO1 protein	Forkhead box protein O1
GPx	Glutathione peroxidase
hADSCs	Human amnion-derived stem cells
HES	Hairy enhancer of split
HGF	Hepatocyte growth factor
HSCs	Hematopoietic stem cells
HumUCMSCs	Human umbilical cord MSCs
hWJ-MSCs	Human Wharton jelly mesenchymal stem cells
IL	Interleukin
iPSCs	Induced pluripotent stem cells
ISSCR	International Society of Stem Cell Research
micRNAs	Microribonucleic acid
MMP	Matrix metalloproteinase
mRNA	Messenger RNA
MSCs	Mesenchymal stem cells
NF-κB	Nuclear factor kappa-light-chain-enhancer of activated B cells
NK cells	Natural killer cells

NKp44	Natural killer cell activating receptor 44
Nrf2	Nuclear factor erythroid 2–related factor 2
Nrip1	Nuclear receptor interacting protein 1
PSCs	Pluripotent stem cells
RNA	Ribonucleic acid
SCORAD	SCORing Atopic Dermatitis
SCs	Stem cells
SCTs	Stem cell therapies
SOD	Superoxide dismutase
STAT3	Signal transducer and activator of transcription 3
TGF	Transforming growth factor
VEGF	Vascular endothelial growth factor

AI disclosure statement: The author acknowledges the use of AI-based tools for minor language editing or grammatical correction. However, all intellectual contributions, analyses, and conclusions presented in this chapter are entirely the author's own.

References

1. Liu, H., Trends of population aging in China and the world as a whole. *Sci. Res. Aging*, 9, 12, 1–16, 2021.
2. Peng, D., The status quo of population aging and the development of the social security system in China. *Chin. Soc. Secur. Rev.*, 7, 31–47, 2023.
3. He, X., Wan, F., Su, W., Xie, W., Research progress on skin aging and active ingredients. *Molecules*, 28, 14, 5556, 2023.
4. Golden, J.S., Handfield, R.B., Daystar, J., Pires, S., An Economic Impact Analysis of the US Biobased Products Industry: 2023 Update, 2024.
5. Chaudhary, M., Khan, A., Gupta, M., Skin ageing: pathophysiology and current market treatment approaches. *Curr. Aging Sci.*, 13, 1, 22–30, 2020.
6. Curtis, R., Geesaman, B.J., DiStefano, P.S., Ageing and metabolism: drug discovery opportunities. *Nat. Rev. Drug Discov.*, 4, 7, 569–580, 2005.
7. Landau, M., Exogenous factors in skin aging. *Curr. Probl. Dermatol. Basel*, 35, 1, 2007.
8. Makrantonaki, E. and Zouboulis, C.C., Characteristics and pathomechanisms of endogenously aged skin. *Dermatology*, 214, 4, 352–360, 2007.
9. Varani, J., Spearman, D., Perone, P., Fligiel, S.E., Datta, S.C., Wang, Z.Q., Voorhees, J.J., Inhibition of type I procollagen synthesis by damaged collagen in photoaged skin and by collagenase-degraded collagen *in vitro*. *Am. J. Pathol.*, 158, 3, 931–942, 2001.

10. Zhang, S. and Duan, E., Fighting against skin aging: the way from bench to bedside. *Cell Transplant.*, 27, 5, 729–738, 2018.
11. Shin, J.W., Kwon, S.H., Choi, J.Y., Na, J.I., Huh, C.H., Choi, H.R., Park, K.C., Molecular mechanisms of dermal aging and antiaging approaches. *Int. J. Mol. Sci.*, 20, 9, 2126, 2019.
12. Ganceviciene, R., Liakou, A.I., Theodoridis, A., Makrantonaki, E., Zouboulis, C.C., Skin anti-aging strategies. *Dermato-endocrinology*, 4, 3, 308–319, 2012.
13. Khandpur, S., Gupta, S., Gunaabalaji, D.R., Stem cell therapy in dermatology. *Indian J. Dermatol. Venereol. Leprol.*, 87, 6, 753–767, 2021.
14. Aly, R.M., Current state of stem cell-based therapies: an overview. *Stem Cell Invest.*, 7, 8, 2020.
15. Schmauck-Medina, T., Molière, A., Lautrup, S., Zhang, J., Chlopicki, S., Madsen, H.B., Fang, E.F., New hallmarks of ageing: a 2022 Copenhagen ageing meeting summary. *Aging (Albany NY)*, 14, 16, 6829, 2022.
16. Panda, B., Sharma, Y., Gupta, S., Mohanty, S., Mesenchymal stem cell-derived exosomes as an emerging paradigm for regenerative therapy and nano-medicine: A comprehensive review. *Life*, 11, 8, 784, 2021.
17. Zhang, Y., Liu, Y., Liu, H., Tang, W.H., Exosomes: biogenesis, biologic function and clinical potential. *Cell Biosci.*, 9, 1–18, 2019.
18. Kandeel, M., Morsy, M.A., Alkhodair, K.M., Alhojaily, S., Mesenchymal stem cell-derived extracellular vesicles: An emerging diagnostic and therapeutic biomolecules for neurodegenerative disabilities. *Biomolecules*, 13, 8, 1250, 2023.
19. Karnas, E., Dudek, P., Zuba-Surma, E.K., Stem cell-derived extracellular vesicles as new tools in regenerative medicine-Immunomodulatory role and future perspectives. *Front. Immunol.*, 14, 1120175, 2023.
20. Gurunathan, S., Kang, M.H., Jeyaraj, M., Qasim, M., Kim, J.H., Review of the isolation, characterization, biological function, and multifarious therapeutic approaches of exosomes. *Cells*, 8, 4, 307, 2019.
21. Shi, C., Zhu, Y., Su, Y., Cheng, T., Stem cells and their applications in skin-cell therapy. *Trends Biotechnol.*, 24, 1, 48–52, 2006.
22. Chu, G.Y., Chen, Y.F., Chen, H.Y., Chan, M.H., Gau, C.S., Weng, S.M., Stem cell therapy on skin: Mechanisms, recent advances and drug reviewing issues. *J. Food Drug Anal.*, 26, 1, 14–20, 2018.
23. Kolios, G. and Moodley, Y., Introduction to stem cells and regenerative medicine. *Respiration*, 85, 1, 3–10, 2012.
24. Dekoninck, S. and Blanpain, C., Stem cell dynamics, migration and plasticity during wound healing. *Nat. Cell Biol.*, 21, 1, 18–24, 2019.
25. Lo, B. and Parham, L., Ethical issues in stem cell research. *Endocr. Rev.*, 30, 3, 204–213, 2009.
26. Fong, C.Y., Subramanian, A., Biswas, A., Gauthaman, K., Srikanth, P., Hande, M.P., Bongso, A., Derivation efficiency, cell proliferation, freeze–thaw survival, stem-cell properties and differentiation of human Wharton's jelly stem cells. *Reprod. Biomed. Online*, 21, 3, 391–401, 2010.

27. Li, X., Bai, J., Ji, X., Li, R., Xuan, Y., Wang, Y., Comprehensive characterization of four different populations of human mesenchymal stem cells as regards their immune properties, proliferation and differentiation. *Int. J. Mol. Med.*, 34, 3, 695–704, 2014.
28. Dakhore, S., Nayer, B., Hasegawa, K., Human pluripotent stem cell culture: current status, challenges, and advancement. *Stem Cells Int.*, 2018, 1, 7396905, 2018.
29. Liu, G., David, B.T., Trawczynski, M., Fessler, R.G., Advances in pluripotent stem cells: history, mechanisms, technologies, and applications. *Stem Cell Rev. Rep.*, 16, 1, 3–32, 2020.
30. Poliwoda, S., Noor, N., Downs, E., Schaaf, A., Cantwell, A., Ganti, L., Urits, I., Stem cells: a comprehensive review of origins and emerging clinical roles in medical practice. *Orthop. Rev.*, 14, 3, 37498, 2022.
31. Ringdén, O., Gustafsson, B., Sadeghi, B., Mesenchymal stromal cells in pediatric hematopoietic cell transplantation a review and a pilot study in children treated with decidua stromal cells for acute graft-versus-host disease. *Front. Immunol.*, 11, 567210, 2020.
32. Charles-de-Sá, L., Gontijo-de-Amorim, N.F., Rigotti, G., Sbarbati, A., Bernardi, P., Benati, D., Borojevic, R., Photoaged skin therapy with adipose-derived stem cells. *Plast. Reconstr. Surg.*, 145, 6, 1037e–1049e, 2020.
33. Seita, J. and Weissman, I.L., Hematopoietic stem cell: self-renewal versus differentiation. *Wiley Interdiscip. Rev. Syst. Biol. Med.*, 2, 6, 640–653, 2010.
34. Kaur-Bollinger, P., Gotze, K.S., Oostendorp, R.A., Role of secreted factors in the regulation of hematopoietic stem cells by the bone marrow microenvironment. *Front. Biosci.*, 17, 3, 876–891, 2012.
35. Comazzetto, S., Shen, B., Morrison, S.J., Niches that regulate stem cells and hematopoiesis in adult bone marrow. *Dev. Cell*, 56, 13, 1848–1860, 2021.
36. Adams, G.B. and Scadden, D.T., The hematopoietic stem cell in its place. *Nat. Immunol.*, 7, 4, 333–337, 2006.
37. Venkei, Z.G. and Yamashita, Y.M., Emerging mechanisms of asymmetric stem cell division. *J. Cell Biol.*, 217, 11, 3785–3795, 2018.
38. Siddiq, S., Pamphilon, D., Brunskill, S., Doree, C., Hyde, C., Stanworth, S., Bone marrow harvest versus peripheral stem cell collection for haemopoietic stem cell donation in healthy donors. *Cochrane Database Syst. Rev.*, 1, 1, 2009.
39. Cutler, C. and Antin, J.H., Peripheral blood stem cells for allogeneic transplantation: a review. *Stem Cells*, 19, 2, 108–117, 2001.
40. Singhal, S., Powles, R., Kulkarni, S., Treleaven, J., Sirohi, B., Millar, B., Mehta, J., Comparison of marrow and blood cell yields from the same donors in a double-blind, randomized study of allogeneic marrow vs blood stem cell transplantation. *Bone Marrow Transplant.*, 25, 5, 501–505, 2000.
41. Körbling, M. and Anderlini, P., Peripheral blood stem cell versus bone marrow allotransplantation: does the source of hematopoietic stem cells matter? *Blood J. Am. Soc. Hematol.*, 98, 10, 2900–2908, 2001.

42. Chamberlain, G., Fox, J., Ashton, B., Middleton, J., Concise review: mesenchymal stem cells: their phenotype, differentiation capacity, immunological features, and potential for homing. *Stem Cells*, 25, 11, 2739–2749, 2007.
43. Chen, J.S., Wong, V.W., Gurtner, G.C., Therapeutic potential of bone marrow-derived mesenchymal stem cells for cutaneous wound healing. *Front. Immunol.*, 3, 192, 2012.
44. Isakson, M., De Blacam, C., Whelan, D., McArdle, A., Clover, A.J.P., Mesenchymal stem cells and cutaneous wound healing: current evidence and future potential. *Stem Cells Int.*, 2015, 1, 831095, 2015.
45. Hocking, A.M. and Gibran, N.S., Mesenchymal stem cells: paracrine signaling and differentiation during cutaneous wound repair. *Exp. Cell. Res.*, 316, 14, 2213–2219, 2010.
46. Liu, H., McTaggart, S.J., Johnson, D.W., Gobe, G.C., Original article anti-oxidant pathways are stimulated by mesenchymal stromal cells in renal repair after ischemic injury. *Cytotherapy*, 14, 2, 162–172, 2012.
47. Wang, J., Liao, L., Tan, J., Mesenchymal-stem-cell-based experimental and clinical trials: current status and open questions. *Expert Opin. Biol. Ther.*, 11, 7, 893–909, 2011.
48. Johal, K.S., Lees, V.C., Reid, A.J., Adipose-derived stem cells: selecting for translational success. *Regener. Med.*, 10, 1, 923–940, 2015.
49. Ong, W.K. and Sugii, S., Adipose-derived stem cells: fatty potentials for therapy. *Int. J. Biochem. Cell Biol.*, 45, 6, 1083–1086, 2013.
50. Fraser, J.K., Wulur, I., Alfonso, Z., Hedrick, M.H., Fat tissue: an underappreciated source of stem cells for biotechnology. *Trends Biotechnol.*, 24, 4, 150–154, 2006.
51. Mazini, L., Rochette, L., Amine, M., Malka, G., Regenerative capacity of adipose derived stem cells (ADSCs), comparison with mesenchymal stem cells (MSCs). *Int. J. Mol. Sci.*, 20, 10, 2523, 2019.
52. Shukla, L., Morrison, W.A., Shayan, R., Adipose-derived stem cells in radiotherapy injury: a new frontier. *Front. Surg.*, 2, 1, 2015.
53. Strużyna, J. and Pojda, Z., Zastosowania komórek macierzystych z tkanki tłuszczowej w medycynie regeneracyjnej. *Plast. Surg. Burns/Chir. Plast. Oparzenia*, 3, 4, 2015.
54. Suh, A., Pham, A., Cress, M.J., Pincelli, T., TerKonda, S.P., Bruce, A.J., Shapiro, S.A., Adipose-derived cellular and cell-derived regenerative therapies in dermatology and aesthetic rejuvenation. *Ageing Res. Rev.*, 54, 100933, 2019.
55. Gimble, J.M., Katz, A.J., Bunnell, B.A., Adipose-derived stem cells for regenerative medicine. *Circ. Res.*, 100, 9, 1249–1260, 2007.
56. Gentile, P. and Garcovich, S., Adipose-derived mesenchymal stem cells (AD-MSCs) against ultraviolet (UV) radiation effects and the skin photoaging. *Biomedicines*, 9, 5, 532, 2021.
57. Ebrahimi, M.J., Aliaghaei, A., Boroujeni, M.E., Khodagholi, F., Meftahi, G., Abdollahifar, M.A., Sadeghi, Y., Human umbilical cord matrix stem cells

reverse oxidative stress-induced cell death and ameliorate motor function and striatal atrophy in rat model of Huntington disease. *Neurotoxic. Res.*, 34, 273–284, 2018.
58. Liu, Q.W., Huang, Q.M., Wu, H.Y., Zuo, G.S.L., Gu, H.C., Deng, K.Y., Xin, H.B., Characteristics and therapeutic potential of human amnion-derived stem cells. *Int. J. Mol. Sci.*, 22, 2, 970, 2021.
59. Miki, T., Stem cell characteristics and the therapeutic potential of amniotic epithelial cells. *Am. J. Reprod. Immunol.*, 80, 4, e13003, 2018, doi:10.1111/aji.13003.
60. Yang, P.J., Yuan, W.X., Liu, J., Li, J.Y., Tan, B., Qiu, C., Yu, L.Y., Biological characterization of human amniotic epithelial cells in a serum-free system and their safety evaluation. *Acta Pharmacol. Sin.*, 39, 8, 1305–1316, 2018.
61. Jo, H., Brito, S., Kwak, B.M., Park, S., Lee, M.G., Bin, B.H., Applications of mesenchymal stem cells in skin regeneration and rejuvenation. *Int. J. Mol. Sci.*, 22, 5, 2410, 2021.
62. Antoniadou, E. and David, A.L., Placental stem cells. *Best Pract. Res. Clin. Obstet. Gynaecol.*, 31, 13–29, 2016.
63. Medina, A. and Ghahary, A., Transdifferentiated circulating monocytes release exosomes containing 14-3-3 proteins with matrix metalloproteinase-1 stimulating effect for dermal fibroblasts. *Wound Repair Regener.*, 18, 2, 245–253, 2010.
64. Kotzerke, K., Mempel, M., Aung, T., Wulf, G.G., Urlaub, H., Wenzel, D., Braun, A., Immunostimulatory activity of murine keratinocyte-derived exosomes. *Exp. Dermatol.*, 22, 10, 650–655, 2013.
65. Hu, L.I., Wang, J., Zhou, X., Xiong, Z., Zhao, J., Yu, R., Chen, L., Exosomes derived from human adipose mensenchymal stem cells accelerates cutaneous wound healing *via* optimizing the characteristics of fibroblasts. *Sci. Rep.*, 6, 1, 32993, 2016.
66. Fang, S., Xu, C., Zhang, Y., Xue, C., Yang, C., Bi, H., Xing, X., Umbilical cord-derived mesenchymal stem cell-derived exosomal microRNAs suppress myofibroblast differentiation by inhibiting the transforming growth factor-β/SMAD2 pathway during wound healing. *Stem Cells Transl. Med.*, 5, 10, 1425–1439, 2016.
67. Bronckaers, A., Hilkens, P., Martens, W., Gervois, P., Ratajczak, J., Struys, T., Lambrichts, I., Mesenchymal stem/stromal cells as a pharmacological and therapeutic approach to accelerate angiogenesis. *Pharmacol. Ther.*, 143, 2, 181–196, 2014.
68. Liang, X., Zhang, L., Wang, S., Han, Q., Zhao, R.C., Exosomes secreted by mesenchymal stem cells promote endothelial cell angiogenesis by transferring miR-125a. *J. Cell Sci.*, 129, 11, 2182–2189, 2016.
69. Zhou, L., Wang, H., Jing, J., Yu, L., Wu, X., Lu, Z., Regulation of hair follicle development by exosomes derived from dermal papilla cells. *Biochem. Biophys. Res. Commun.*, 500, 2, 325–332, 2018.

70. Nourian, D.A., Mirahmadi, B.F., Chehelgerdi, M., RaeisiDehkordi, S., Skin tissue engineering: wound healing based on stem-cell-based therapeutic strategies. *Stem Cell Res. Ther.*, 10, 1–20, 2019.
71. Kanji, S. and Das, H., Advances of stem cell therapeutics in cutaneous wound healing and regeneration. *Mediators Inflamm.*, 2017, 1, 5217967, 2017.
72. Meiliana, A., Dewi, N.M., Wijaya, A., Stem cell therapy in wound healing and tissue regeneration. *Indones. Biomed. J.*, 8, 2, 61–70, 2016.
73. Kretzschmar, K. and Clevers, H., Wnt/β-catenin signaling in adult mammalian epithelial stem cells. *Dev. Biol.*, 428, 2, 273–282, 2017.
74. Alonso, L. and Fuchs, E., Stem cells of the skin epithelium. *Proc. Natl. Acad. Sci.*, 100, suppl_1, 11830–11835, 2003.
75. Beck, B. and Blanpain, C., Mechanisms regulating epidermal stem cells. *EMBO J.*, 31, 9, 2067–2075, 2012.
76. Shi, Y., Shu, B., Yang, R., Xu, Y., Xing, B., Liu, J., Xie, J., Wnt and Notch signaling pathway involved in wound healing by targeting c-Myc and Hes1 separately. *Stem Cell Res. Ther.*, 6, 1–13, 2015.
77. Aubin-Houzelstein, G., Notch signaling and the developing hair follicle, in: *Notch Signaling in Embryology and Cancer*, pp. 142–160, 2012.
78. Zhou, Z., Shu, B., Xu, Y., Liu, J., Wang, P., Chen, L., Xie, J., microRNA-203 modulates wound healing and scar formation *via* suppressing Hes1 expression in epidermal stem cells. *Cell. Physiol. Biochem.*, 49, 6, 2333–2347, 2018.
79. Yang, R., Wang, J., Chen, X., Shi, Y., Xie, J., Epidermal stem cells in wound healing and regeneration. *Stem Cells Int.*, 2020, 1, 9148310, 2020.
80. Ojeh, N., Pastar, I., Tomic-Canic, M., Stojadinovic, O., Stem cells in skin regeneration, wound healing, and their clinical applications. *Int. J. Mol. Sci.*, 16, 10, 25476–25501, 2015.
81. Hildebrand, J., Rütze, M., Walz, N., Gallinat, S., Wenck, H., Deppert, W., Knott, A., A comprehensive analysis of microRNA expression during human keratinocyte differentiation *in vitro* and *in vivo*. *J. Invest. Dermatol.*, 131, 1, 20–29, 2011.
82. Duan, M., Zhang, Y., Zhang, H., Meng, Y., Qian, M., Zhang, G., Epidermal stem cell-derived exosomes promote skin regeneration by downregulating transforming growth factor-β1 in wound healing. *Stem Cell Res. Ther.*, 11, 1–11, 2020.
83. Chiossone, L., Conte, R., Spaggiari, G.M., Serra, M., Romei, C., Bellora, F., Bottino, C., Mesenchymal stromal cells induce peculiar alternatively activated macrophages capable of dampening both innate and adaptive immune responses. *Stem Cells*, 34, 7, 1909–1921, 2016.
84. Solarte David, V.A., Güiza-Argüello, V.R., Arango-Rodríguez, M.L., Sossa, C.L., Becerra-Bayona, S.M., Decellularized tissues for wound healing: towards closing the gap between scaffold design and effective extracellular matrix remodeling. *Front. Bioeng. Biotechnol.*, 10, 821852, 2022.

85. Huang, S., Kuri, P., Aubert, Y., Brewster, M., Li, N., Farrelly, O., Rompolas, P., Lgr6 marks epidermal stem cells with a nerve-dependent role in wound re-epithelialization. *Cell Stem Cell*, 28, 9, 1582–1596, 2021.
86. Wu, S., Sun, S., Fu, W., Yang, Z., Yao, H., Zhang, Z., The role and prospects of mesenchymal stem cells in skin repair and regeneration. *Biomedicines*, 12, 4, 743, 2024.
87. Ellis, S., Lin, E.J., Tartar, D., Immunology of wound healing. *Curr. Dermatol. Rep.*, 7, 350–358, 2018.
88. Zhang, S., Dong, Z., Peng, Z., Lu, F., Anti-aging effect of adipose-derived stem cells in a mouse model of skin aging induced by D-galactose. *PLoS One*, 9, 5, e97573, 2014.
89. Hu, Y., Zhu, Y., Gerber, S.D., Osland, J.M., Chen, M., Rao, K.A., Yuan, R., Deletion of Nrip1 delays skin aging by reducing adipose-derived mesenchymal stem cells (ADMSCs) senescence, and maintaining ADMSCs quiescence. *Geroscience*, 43, 4, 1–19, 2021.
90. Li, W., Li, N., Sui, B., Yang, D., Anti-aging effect of fullerenol on skin aging through derived stem cells in a mouse model. *Exp. Ther. Med.*, 14, 5, 5045–5050, 2017.
91. Wang, J., Chen, Y., He, J., Li, G., Chen, X., Liu, H., Anti-aging effect of the stromal vascular fraction/adipose-derived stem cells in a mouse model of skin aging induced by UVB irradiation. *Front. Surg.*, 9, 950967, 2022.
92. Li, L., Ngo, H.T., Hwang, E., Wei, X., Liu, Y., Liu, J., Yi, T.H., Conditioned medium from human adipose-derived mesenchymal stem cell culture prevents UVB-induced skin aging in human keratinocytes and dermal fibroblasts. *Int. J. Mol. Sci.*, 21, 1, 49, 2019.
93. Hwang, I., Choi, K.A., Kim, M., Hong, S., Neural stem cells and the secreted proteins TIMPs ameliorate UVB-induced skin photodamage. *Biochem. Biophys. Res. Commun.*, 518, 2, 388–395, 2019.
94. Liu, Z., Hu, G.D., Luo, X.B., Yin, B., Shu, B., Guan, J.Z., Jia, C.Y., Potential of bone marrow mesenchymal stem cells in rejuvenation of the aged skin of rats. *Biomed. Rep.*, 6, 3, 279–284, 2017.
95. Liu, Q., Song, S., Song, L., Bi, Y., Zhu, K., Qiao, X., Ji, G., Mesenchymal stem cells alleviate aging *in vitro* and *in vivo*. *Ann. Transl. Med.*, 10, 20, 1092, 2022.
96. Liu, Q., Luo, Z., He, S., Peng, X., Xiong, S., Wang, Y., Gao, B.Z., Conditioned serum-free medium from umbilical cord mesenchymal stem cells has anti-photoaging properties. *Biotechnol. Lett.*, 35, 1707–1714, 2013.
97. Li, T., Zhou, L., Fan, M., Chen, Z., Yan, L., Lu, H., Shan, L., Human umbilical cord-derived mesenchymal stem cells ameliorate skin aging of nude mice through autophagy-mediated anti-senescent mechanism. *Stem Cell Rev. Rep.*, 18, 6, 2088–2103, 2022.
98. Widowati, W., Faried, A., Adam, A., Rahmat, D., Kusuma, H.S.W., Dewi, N.S.M., Vosough, M., Potential antiaging activity of secretome gel of human Wharton's jelly mesenchymal stem cells (hWJ-MSCs) in UV-induced mice models. *Iran. J. Basic Med. Sci.*, 27, 7, 868, 2024.

99. Prakoeswa, C.R.S., Pratiwi, F.D., Herwanto, N., Citrashanty, I., Indramaya, D.M., Murtiastutik, D., Rantam, F.A., The effects of amniotic membrane stem cell-conditioned medium on photoaging. *J. Dermatolog. Treat.*, 30, 5, 478–482, 2019.
100. El-Domyati, M., Moftah, N.H., Nasif, G.A., Ameen, S.W., Ibrahim, M.R., Ragaie, M.H., Facial rejuvenation using stem cell conditioned media combined with skin needling: a split-face comparative study. *J. Cosmet. Dermatol.*, 19, 9, 2404–2410, 2020.
101. Kim, H.S., Lee, J.H., Roh, K.H., Jun, H.J., Kang, K.S., Kim, T.Y., Clinical trial of human umbilical cord blood-derived stem cells for the treatment of moderate-to-severe atopic dermatitis: phase I/IIa studies. *Stem Cells*, 35, 1, 248–255, 2017.
102. Kim, J., Kim, B., Kim, S., Lee, Y.I., Kim, J., Lee, J.H., The effect of human umbilical cord blood–derived mesenchymal stem cell media containing serum on recovery after laser treatment: A double-blinded, randomized, split-face controlled study. *J. Cosmet. Dermatol.*, 19, 3, 651–656, 2020.
103. Lee, Y.I., Kim, S., Kim, J., Kim, J., Chung, K.B., Lee, J.H., Randomized controlled study for the anti-aging effect of human adipocyte-derived mesenchymal stem cell media combined with niacinamide after laser therapy. *J. Cosmet. Dermatol.*, 20, 6, 1774–1781, 2021.
104. Park, G.H., Kwon, H.H., Seok, J., Yang, S.H., Lee, J., Park, B.C., Park, K.Y., Efficacy of combined treatment with human adipose tissue stem cell-derived exosome-containing solution and microneedling for facial skin aging: a 12-week prospective, randomized, split-face study. *J. Cosmet. Dermatol.*, 22, 12, 3418–3426, 2023.
105. Lotfy, A., AboQuella, N.M., Wang, H., Mesenchymal stromal/stem cell (MSC)-derived exosomes in clinical trials. *Stem Cell Res. Ther.*, 14, 1, 66, 2023.
106. Kwon, H.H., Yang, S.H., Lee, J., Park, B.C., Park, K.Y., Jung, J.Y., Park, G.H., Combination Treatment with Human Adipose Tissue Stem Cell-derived Exosomes and Fractional CO_2 laser for acne scars: a 12-week prospective, double-blind, randomized, split-face study. *Acta dermato-venereologica*, 100, 18, 5913, 2020.
107. Wang, X., Shu, X., Huo, W., Zou, L., Li, L., Efficacy of protein extracts from medium of Adipose-derived stem cells *via* microneedles on Asian skin. *J. Cosmet. Laser Ther.*, 20, 4, 237–244, 2018.
108. Qian, H., Shan, Y., Gong, R., Lin, D., Zhang, M., Wang, C., Wang, L., Mechanism of action and therapeutic effects of oxidative stress and stem cell-based materials in skin aging: Current evidence and future perspectives. *Front. Bioeng. Biotechnol.*, 10, 1082403, 2023.
109. Matsuoka, T., Dan, K., Takanashi, K., Ogino, A., Early Effects of Porcine Placental Extracts and Stem Cell-Derived Exosomes on Aging Stress in Skin Cells. *J. Funct. Biomater.*, 15, 10, 306, 2024.

110. Marks, P.W., Witten, C.M., Califf, R.M., Clarifying stem-cell therapy's benefits and risks. *N. Engl. J. Med.*, 376, 11, 1007–1009, 2017.
111. Greely, H.T., Human brain surrogates research: the onrushing ethical dilemma. *Am. J. Bioethics*, 21, 1, 34–45, 2021.
112. Kiatpongsan, S. and Sipp, D., Monitoring and regulating offshore stem cell clinics. *Science*, 323, 5921, 1564–1565, 2009.
113. Chirba, M. and Noble, A., Our bodies, our cells: FDA regulation of autologous adult stem cell therapies. Bill of Health Petrie-Flom Center http://blogs.harvard.edu, 2013.
114. Dubinsky, P.M. and Henry, K.A., *The fundamentals of Clinical Research: A Universal Guide for Implementing Good Clinical Practice*, John Wiley & Sons, 2021.
115. Lovell-Badge, R., Anthony, E., Barker, R.A., Bubela, T., Brivanlou, A.H., Carpenter, M., Zhai, X., ISSCR guidelines for stem cell research and clinical translation: the 2021 update. *Stem Cell Rep.*, 16, 6, 1398–1408, 2021.
116. Harris, A.R., Walker, M.J., Gilbert, F., Ethical and regulatory issues of stem cell-derived 3-dimensional organoid and tissue therapy for personalised regenerative medicine. *BMC Med.*, 20, 1, 499, 2022.
117. Zhang, Y., Li, Q., Niu, Y., Wei, K., Wang, X., Niu, B., Zhang, L., Research progress on aging mechanism and drugs and the role of stem cells in anti-aging process. *Exp. Gerontol.*, 179, 112248, 2023.
118. Nussbaum, J., Minami, E., Laflamme, M.A., Virag, J.A., Ware, C.B., Masino, A., Murry, C.E., Transplantation of undifferentiated murine embryonic stem cells in the heart: teratoma formation and immune response. *FASEB J.*, 21, 7, 1345–1357, 2007.
119. Murry, C.E. and Keller, G., Differentiation of embryonic stem cells to clinically relevant populations: lessons from embryonic development. *Cell*, 132, 4, 661–680, 2008.
120. Glenn, J.D. and Whartenby, K.A., Mesenchymal stem cells: emerging mechanisms of immunomodulation and therapy. *World J. Stem Cells*, 6, 5, 526, 2014.
121. Kroon, E., Martinson, L.A., Kadoya, K., Bang, A.G., Kelly, O.G., Eliazer, S., Baetge, E.E., Pancreatic endoderm derived from human embryonic stem cells generates glucose-responsive insulin-secreting cells *in vivo*. *Nat. Biotechnol.*, 26, 4, 443–452, 2008.
122. Prokhorova, T.A., Harkness, L.M., Frandsen, U., Ditzel, N., Schrøder, H.D., Burns, J.S., Kassem, M., Teratoma formation by human embryonic stem cells is site dependent and enhanced by the presence of Matrigel. *Stem Cells Dev.*, 18, 1, 47–54, 2009.
123. Margiana, R., Markov, A., Zekiy, A.O., Hamza, M.U., Al-Dabbagh, K.A., Al-Zubaidi, S.H., Siahmansouri, H., Clinical application of mesenchymal stem cell in regenerative medicine: a narrative review. *Stem Cell Res. Ther.*, 13, 1, 366, 2022.

124. Volarevic, V., Markovic, B.S., Gazdic, M., Volarevic, A., Jovicic, N., Arsenijevic, N., Stojkovic, M., Ethical and safety issues of stem cell-based therapy. *Int. J. Med. Sci.*, 15, 1, 36, 2018.
125. Sipp, D., Caulfield, T., Kaye, J., Barfoot, J., Blackburn, C., Chan, S., Rasko, J.E., Marketing of unproven stem cell–based interventions: a call to action. *Sci. Transl. Med.*, 9, 397, eaag0426, 2017.
126. Petrof, G., Abdul-Wahab, A., McGrath, J.A., Cell therapy in dermatology. *Cold Spring Harbor Perspect. Med.*, 4, 6, a015156, 2014.
127. Hejrati, N., Wong, R., Khazaei, M., Fehlings, M.G., How can clinical safety and efficacy concerns in stem cell therapy for spinal cord injury be overcome? *Expert Opin. Biol. Ther.*, 23, 9, 883–899, 2023.
128. El Assaad, N., Chebly, A., Salame, R., Achkar, R., Atme, N.B., Akouch, K., Khalil, C., Anti-aging based on stem cell therapy: A scoping review. *World J. Exp. Med.*, 14, 3, 97233, 2024.
129. Dodson, B.P. and Levine, A.D., Challenges in the translation and commercialization of cell therapies. *BMC Biotech.*, 15, 1–15, 2015.
130. Trounson, A. and McDonald, C., Stem cell therapies in clinical trials: progress and challenges. *Cell Stem Cell*, 17, 1, 11–22, 2015.
131. Lysaght, T., Kerridge, I.H., Sipp, D., Porter, G., Capps, B.J., Ethical and regulatory challenges with autologous adult stem cells: a comparative review of international regulations. *J. Bioeth. Inq.*, 14, 261–273, 2017.
132. Sugarman, J., Barker, R.A., Kerridge, I., Lysaght, T., Pellegrini, G., Sipp, D., Tanner, C., Tackling ethical challenges of premature delivery of stem cell-based therapies: ISSCR 2018 annual meeting focus session report. *Stem Cell Rep.*, 11, 5, 1021–1025, 2018.
133. Sousa, A.C., Alvites, R., Lopes, B., Sousa, P., Moreira, A., Coelho, A., Maurício, A.C., 3D Printing/Bioprinting and Cellular Therapies for Regenerative Medicine: Current Advances, p. 52, 2024.
134. Diaz, Y., Addressing the High Costs of Stem Cell Therapies. *Stem Cell Res. Regener. Med.*, 7, 5, 255–256, 2024.

8

Future Window of Stem Cells in Skin Regeneration and Rejuvenation: Current Status and Future Aspect

Vivek Barik[1], Bramhajit Chatterjee[1], Snigdha Chatterjee[1], Swarupananda Mukherjee[2] and Piyali Khamkat[1]*

[1]Department of Pharmaceutical Technology, Brainware University, Barasat, Kolkata, India
[2]Department of Pharmaceutics, NSHM College of Pharmaceutical Technology, NSHM Knowledge Campus, Kolkata - Group of Institutions, Kolkata, West Bengal, India

Abstract

Stem cells (SCs), which are undifferentiated cells derived from embryonic and adult tissues, including skin, have the ability to create, preserve, and replace tissues as well as cells that have undergone terminal differentiation. In recent years, SC therapy has developed into a very innovative and exciting field of research. To regenerate damaged tissues, advanced tissue engineering techniques are especially useful for dermatological conditions such as wound healing, genodermatoses, and chronic autoinflammatory skin conditions. Mesenchymal SCs, induced pluripotent SCs, and embryonic SCs have shown potential impact in wound healing, skin elasticity, treating burns, and genetic skin disorders. SCs and their cytokines generate exosomes, which are essential in skin regeneration and rejuvenation. The complex procedure of regeneration is influenced by several factors, either separately or in combination, such as tissue remodeling, inflammation, and proliferation. Advancements in SC technology, including genome editing, three-dimensional printing, CRISPR, and nanotechnology, may overcome these challenges and improve treatment outcomes. Recent developments in preclinical and clinical research significantly improved SC therapy, providing a prospective treatment for several types of illnesses where existing conventional treatments are ineffective. This chapter addresses the way SC-based skin rejuvenation will contribute to skin therapies, antiaging treatments, wound and scar healing, and personalized healthcare.

**Corresponding author*: piyalikhamkat95@gmail.com

Sumel Ashique, Sanjay Dey, Biplab Debnath, Mohhammad Ramzan and Swarupananda Mukherjee (eds.) Stem Cell Therapy in Dermatological Disorders, (195–218) © 2025 Scrivener Publishing LLC

***Keywords*:** Stem cells, embryonic, exosomes, regeneration, rejuvenation, advancements, personalized

8.1 Introduction

As the largest organ in the body, the skin performs numerous activities, including immunological surveillance, vitamin D synthesis, sensory and thermoregulatory activities, and protection. It is made up of the dermis and epidermis, both of which are preserved and renewed by stem cells (SCs). Because they are largely inactive and undifferentiated, epidermal SCs (EpiSCs) can help heal wounds, self-renew tissue, and maintain homeostasis. Both internal and extrinsic factors contribute to facial aging, with genetics being the main internal contributor. Decreased epidermal thickness and keratinocyte atypia are the causes of extrinsic aging, commonly referred to as photoaging. UVB rays harm the stratum corneum of the epidermis, which results in collagen denature and water retention loss, which reduces skin suppleness and causes wrinkles [1]. Because skin SCs begin to decrease between the ages of 28 and 30 years, SC therapy can help counteract the damaging consequences of aging and wrinkles. Because skin SCs begin to decrease between the ages of 28 and 30 years, SC therapy can help counteract the damaging consequences of aging and wrinkles [2]. SCs have the ability to develop into multiple cell types. They are found in all bodily tissues and are crucial for tissue maintenance and injury repair. Numerous tissues, including the brain, heart, muscles, and bones, can be formed from SCs. The majority of SCs in the body simply contribute in tissue and organ maintenance and repair.

8.1.1 Concept of Skin Regeneration and Rejuvenation

Restoring the skin's young appearance and function, frequently through procedures such as collagen development and cell repair, is the ultimate objective of skin rejuvenation. Skin regeneration desires to fully replace damaged tissue with new and healthy tissues [1, 3]. When fibroblasts, the cells that generate collagen and keep the skin youthful, decline by an average of 2% annually after the ages of 28 to 30 years, skin aging occurs. Sagging, crow's feet, and wrinkles are some of the symptoms. Botox, fillers, and other skin treatments are available as treatment alternatives. Instead of treating the underlying damage, traditional treatments frequently concentrate on symptom management. Innovative methods for treating ailments such as burns, scars, chronic ulcers, chronic wounds, and aging skin are provided by regenerative medicine. Cutting-edge treatments such as tissue engineering, growth factors, and SC therapy have been developed to promote healing, reduce scarring, and preserve the integrity of the skin. By improving the fundamental causes and encouraging long-term regeneration, regenerative medicine also presents novel approaches to treating chronic skin disorders such as vitiligo, psoriasis, and eczema. The US Food and Drug Administration (FDA) has approved the nonsurgical, painless technique of skin rejuvenation with SCs in the United States, and it has been used for almost 20 years. The procedure consists of a punch biopsy, in which a 3- to 4-mm piece is removed behind the ear, and blood is collected; the patient's blood is sent to an SC laboratory, where the SCs are differentiated, and the desired quantity is achieved within 35 to 40 days. The procedure is easy and painless, with repeated applications at 1-month intervals. About 6 months after application, the skin SCs reach their 8- to 10-year-old level, allowing the patient to monitor the desired effect on their skin [2]. Chronic wounds that do not heal can result from a disruption in the complicated process of skin wound healing. Healing might be hampered by conditions such as trauma, advanced age, diabetes, renal disease, venous or arterial insufficiency, and local pressure effects. Regenerative therapies for skin diseases are generally considered safe, despite obstacles such as tissue origin, donor factors, isolation and culture procedures, potential negative effects, and ethical regulations. SCs and SC-derived products are used in clinical trials as more effective treatments. Reduced collagen and fibroblast synthesis, which throws off the fibroblasts' connection with other cells, is the main cause of skin aging. The characteristics of regenerative treatment include multidirectional differentiation, self-replication, immunomodulation, inflammatory management, angiogenesis, and hemostasis [1, 4].

8.1.2 Significance of SCs in Dermatology

Mesenchymal SCs (MSCs) are essential for wound healing because they influence tissue damage and the healing of injuries. MSCs are essential for wound healing because they influence tissue damage and the healing of injuries. Immunomodulation, antiapoptotic, proangiogenic, chemoattractive, and antifibrosis are some of their paracrine actions. Regenerative success is achieved when MSCs guide regeneration and restore regular function in the body. Hemostasis, inflammation, proliferation, and remodeling are the four stages of wound healing, considered a complex biological process involving several cellular, molecular, and signaling pathways.

Every stage of wound healing is mediated by MSCs, which also promote granulation and epithelialization and regulate inflammation. For severe wounds, they present an appealing alternative to current treatments [5]. These are found in dental pulp, bone marrow, adipose tissue, peripheral blood, placental tissue, umbilical cord, and synovial fluid. Because of their self-renewal and pluripotency, induced pluripotent SCs (iPSCs) are perfect for regenerative medicine and disease models. Because they can differentiate into all germ layers, they can be used to cure illnesses and wounds. The therapeutic potential of iPSCs is growing as a result of their excellent efficiency and safety in generating patient-specific cells [6]. The ability of embryonic SCs (ESCs) to proliferate or differentiate into any type of cell in the body makes them useful for the regeneration or restoration of damaged organs and tissue [7].

8.2 Types of SCs Used in Skin Regeneration and Rejuvenation

8.2.1 Mesenchymal SCs

The loss of function and structural deterioration that accompany aging skin can result in eczema, dermatitis, autoimmune disorders, and melanoma. Elastic fibers degenerate as the skin's collagen level declines and UV radiation exposure accelerates photoaging, or premature aging. The goal of rejuvenation treatments is to stop and reverse skin aging. Because of their effective re-epithelialization and release of growth factors required for skin regeneration, the mechanism of regeneration and rejuvenation by MSCs is indicated in Figure 8.1. Adipose-derived MSCs (AD-MSCs) have been receiving more interest in the field of skin antiaging therapy [8]. When photoaged skin is treated with AD-MSCs, the superficial layer of the dermis develops more elastic fibers, collagen and reticular fiber networks are altered, and solar elastosis is fully restored. Extracellular vesicles have also been employed as antiphotoaging agents and promote the proliferation of epidermal cells, lowering the generation of reactive oxygen species (ROS), and lessening wrinkles on the skin. The ability of human umbilical cord blood–derived MSCs is to modulate the immune system and proliferate quickly that has been investigated [9]. It has been discovered that amniotic membrane–derived MSCs (AM-MSCs) enhance photoaging because of their profusion, ease of acquisition, growth factors, and cytokines. In 2019, Prakoeswa *et al.* treated photoaged human patients with

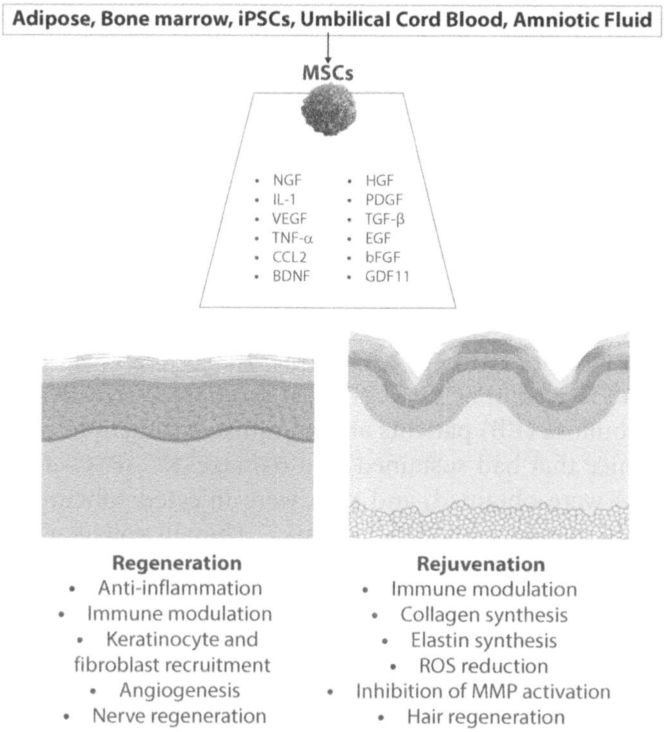

Figure 8.1 Regeneration and rejuvenation mechanism of MSCs (prepared by Biorender).

AM-MSC–conditioned media and demonstrated enhanced collagen synthesis, dermal fibroblast and epidermal keratinocyte migration and proliferation, and improved clinical photoaging [10]. Another study, conducted in 2017 by Liu *et al.*, on mouse models exposed to D-galactose–induced aging demonstrated the positive benefits of BM-MSCs on skin aging. By preventing lipid peroxidation and lowering malondialdehyde content, BM-MSC treatment decreased antioxidant activity, increased superoxide dismutase activity, and raised glutathione–peroxidase content [9]. By controlling the inflammatory response, encouraging cell division and neovascularization, and preventing scarring, SC exosomes are essential for wound healing. They enhance anti-inflammatory factors, suppress proinflammatory factors, and activate the PTEN/AKT signaling pathway. By encouraging the migration of fibroblasts and endothelial cells to the wound center, where they generate new granulation tissue and blood vessels, they also aid in the regeneration of skin tissue. Scar remodeling occurs as a result of collagen molecules cross-linking and fibroblast proliferation during the remodeling phase. By lowering inflammation and collagen deposition, SC exosomes can prevent the formation of scars. In general, wound healing and skin tissue regeneration depend on SC exosomes [11].

8.2.2 Induced Pluripotent SCs

Derived from skin or blood cells, iPSCs can be transformed into pluripotent states that resemble those of embryos, enabling the production of an infinite number of human cell sources for therapeutic purposes [12]. Research has demonstrated that iPSCs have the potential to rejuvenate and regenerate skin. In their 2018 study, Nakayama *et al.* differentiated keratinocyte-derived iPSCs (KC-iPSCs) from recessive dystrophic epidermolysis bullosa (EB) patients and administered them to immunocompromised mice that had sustained skin damage. KC-iPSC–derived MSCs (KC-iMSCs) were obtained, and they were injected subcutaneously and intravenously into immunodeficient mice. Human collagen VII was discovered at the dermal–epidermal interface following transplantation, signifying that the wound had healed successfully [13]. According to Kim *et al.* (2018), exosomes secreted by iMSCs (iMSCs-exo) promoted the growth of human dermal fibroblasts and keratinocytes [14]. In an immunodeficient mouse model, Veraitch *et al.* (2017) discovered that iMSCs enhanced the characteristics of dermal papilla cells and their hair-like structural shape [15]. Spitzhorn *et al.* (2019) discovered that regardless of donor age or MSC source, human iMSCs express genes linked to rejuvenation, such as CDKN1C, DNMT3B, GCNT2, INHBE, and POU5F1P1. This implies that iMSCs might have use in the medical field [16].

8.2.3 Embryonic SCs

The ability of ESCs to develop into all three germ layers was found by Evans *et al.* in 1981. Although ESCs can overcome the constraints of adult SCs, they are more likely to cause tumors and immunological rejection. One significant drawback is the ethical difficulty of embryo sacrifice [17]. ESC-derived MSCs (eMSCs) may be used in skin rejuvenation and regeneration treatments if successful. According to research by Yoon *et al.* (2018), eMSCs increase wound closure, vascular formation, and the production of collagen types I and III, fibronectin, and fibroblast-specific protein-1 in pressure ulcers [18].

8.2.4 Epidermal SCs

For wound healing, metabolism, and skin regeneration, skin SCs, mainly EpiSCs, are essential and are available in the basal layer of the epidermis. The "SC niches," or microenvironment, of SCs are a complex web of interconnected signaling pathways that regulate migration, proliferation, and differentiation. These niches are critical for wound healing and skin growth, and they include wingless/integrated (Wnt) and notch signaling pathways. These pathways are activated by damaged skin, which leads to the differentiation and growth of wound EpiSCs and helps in the development of wound healing [19, 20].

Since 1984, cultured epidermal cell sheets (CESs) have been used to treat burns, and it is generally agreed that giving patients CES can improve their chances of survival. EpiSCs can proliferate and differentiate, leading to the regeneration of epidermal tissue. EpiSCs grown in a carrier matrix supply the matrix elements required for wound healing in chronic wounds. Skin conditions such as EB, stable vitiligo, and limbal SC deficit may be treated using EpiSCs. They can be utilized as seed cells for wound healing and have demonstrated effectiveness comparable to that of conventional therapies. There are not many clinical studies on EpiSCs in wound healing, though. Tissue-engineered EpiSCs have been used as a possible treatment up to this point, but more research is required to determine their long-term effectiveness and associated clinical studies [19].

8.2.5 Adipose-Derived SCs

Adipose-derived SCs (ASCs) are available in adipose tissue and repair photoaging-related skin lesions by releasing cytokines and antioxidants that counteract the effects of ROS and UVB rays. They release growth factors to protect cells from oxidative cells, including vascular endothelial growth factor and hepatocyte growth factor. Photoaging is also manifested by hyperpigmentation, which mainly affects women due to exposure to UV radiation, hormonal changes, and genetic predisposition. Adipose-derived SCs secrete transforming growth factor $\beta 1$, which suppresses tyrosinase, an enzyme necessary for melanin synthesis. Recent studies report the presence of ASCs in subcutaneous tissue, suggesting their role in skin repair and wound healing [21].

8.2.6 Hair Follicle Epithelial SCs

Although hair follicle epithelial SCs are essential for wound healing, there is ongoing discussion on their requirement. Hair follicle cells speed up wound healing, although they are not required, according to studies. The anagen phase of hair follicle cycling is when wounds heal the fastest; however, other stages of the hair cycle also affect wound healing rates. Research has also demonstrated that mice lacking the Sept4/ARTS gene have more hair follicle SCs because they undergo less apoptosis. Although they are not required, hair follicle cells speed up the healing of wounds. According to a study by Ansell *et al.* in 2011, wound healing rates differ based on the stage of the hair cycle, with faster healing taking place during the *in vivo* anagen phase of hair follicle cycling [22, 23].

8.3 Emerging Technologies and Innovations

Skin tissue engineering is a complex procedure that includes choosing the right biomaterial, creating a suitable substrate, and choosing cells to mimic the structural and functional properties of the skin. Over the past years, the development of a proregenerative framework has been a primary emphasis in the research of wound repair and regeneration [24]. The primary objectives of wound management are to facilitate quick wound healing and skin regeneration by promoting vascularization in the reconstituted tissue. Over the past 10 years, numerous innovative techniques have developed prevascularized skin transplants using cellular and molecular biology [25].

8.3.1 Three-Dimensional Bioprinting for Skin Engineering

The use of three-dimensional (3D) bioprinting to heal and regenerate wounds has recently gained attention. 3D bioprinting may be tailored for skin morphology, with cells and other components meticulously arranged, facilitating the quick and dependable fabrication of bionic skin substitutes, thereby satisfying clinical and industrial demands. Moreover, it demonstrates exceptional performance characterized by high resolution, flexibility, repeatability, and elevated throughput, indicating significant potential for the production of tissue-engineered skin [26]. 3D bioprinting uses fast prototyping technology to sequentially print cells, development factors, and other biomaterials, therefore creating biomedical components that closely replicate the properties of natural tissues and organs [27]. Bioprinting is an extension of the quick prototyping or additive manufacturing (AM) method.

Additive manufacturing produces intricate 3D biocompatible structures through the automated deposition of biological materials onto a substrate, utilizing computer-aided design and computer-aided manufacturing technology. Three-dimensional bioprinting methods can be categorized into three main types depending on varying molding principles and printing materials: photocuring-based bioprinting, extrusion-based bioprinting, and droplet-based bioprinting [28].

8.3.2 CRISPR and Gene Editing Technologies in Skin Regeneration

CRISPR, an abbreviation for clustered regularly interspaced short palindromic repeats, is a revolutionary and potentially very powerful biomedical research tool due to its capacity to modify genetic material with unparalleled precision. CRISPR has much potential in skin regeneration to a multitude of dermatology diseases padding such as wound healing to tissue engineering improvement [29]. CRISPR regeneration is being explored as a therapeutic option for X-linked skin disorders. *Via* autonomous CRISPR detection mechanisms, scientists have used CRISPR to correct keratin gene defects responsible for EB simplex skin disorder that features fragile skin prone to blistering. By manipulating keratinocytes, the major cell type in the epidermis, scientists hope to restore protein normalcy and improve stiffness in the skin [30]. Chronic wounds such as diabetic ulcers represent a significant challenge to develop. Studies suggest that by knocking out the Ndrg2 gene using CRISPR/Cas9 in murine dendritic cells, regenerative capability is enhanced, which facilitates the healing of chronic wounds. The technique suggests that gene editing could modify the activity of immune cells to enhance tissue repair [31]. The successful therapy of this project relies on the successful introduction of CRISPR components into the dermal cells. Examples include dissolving microneedle patches, which have emerged as new techniques for delivering CRISPR/Cas9 into the dermis. Such methods present a promising avenue for disease treatment by delivering site-specific gene editing with low invasiveness, for example, psoriasis and atopic dermatitis [32].

8.3.3 Organ-on-a-Chip Models for Testing Skin Regeneration

Organ-on-a-chip technology is a significant development for biomedical engineering that provides dynamic microfluidic systems that mimic the functioning of human organs. Such technology models are specifically constructed to mimic the complex architecture and physiology of human skin, thus serving as important assessments for the testing of skin regeneration therapies. The skin-on-chip technologies include live skin cells integrated into microfluidic devices, thus making it possible to create a controlled set of conditions, replicating the structural and functional characteristics of human skin. These models usually have two or more layers, including the dermis being vascularized, to simulate blood flow. Dynamic perfusion in these devices improves food delivery and waste disposal, thereby enhancing tissue viability and physiological responses [33]. Skin-on-a-chip technologies let researchers observe the healing process in real time through controlled injury. This approach allows the changeable environment assessment of biological reactions and the effectiveness of putative treatments [34]. Under these models, an assessment of how drugs and cosmetics affect skin tissue can take place. They ensure execution assessments of efficacy and toxicity in a controlled environment closely resembling human physiology [35].

8.3.4 Nanotechnology in SC Therapy for Skin

At present, SC nanotechnology represents a new and exciting domain. Experimental research concerning SC interactions with nanostructures or nanomaterials has made some impressive advancements. The role of nanostructures, nanotechnology, and nanomaterials in SC therapy for degenerative diseases and injuries now seem to be on everyone's lips. The intersection of shape and properties of nanomaterials upon proliferation and differentiation of SCs is emerging as a new multidisciplinary area in the materials research and regenerative medicine arena [36]. At present, electrospun nanofibers of poly-ε-caprolactone (PCL) and polyvinyl alcohol have been characterized as safe, well-tolerated, and effective. They can serve as a smart skin cover dressing during the wound healing process, either independently or in conjunction with bioactive compounds. The integration of PCL nanofibers with embedded magnetic nanoparticles enhances the growth of MSCs *in vitro*. Furthermore, PCL nanofibers infused with natural extracts can safeguard skin MSCs from UV-induced aging, indicating a potential function in skin rejuvenation [37]. In the *in vitro* study, PCL nanofibers infused with silver and magnesium ions were shown to exhibit antimicrobial activity and a proangiogenic function toward wound repair when tested on vascular endothelial cells. Alternatively, the use of Ag-doped magnetite nanoparticles in the preparation of the PCL scaffolds improved cell adhesion and growth [38]. By increasing the silver concentration in the magnetite phase, parallel increases were observed in human melanocyte viability and antibacterial activity against *Escherichia coli* and

Staphylococcus aureus, along with a promotion of skin wound healing in rats. A three-layer nanofiber based on PCL containing chitosan and polyvinyl alcohol once combined with melatonin formed a new nanomaterial with improved hydration capacity and cell adhesion properties leading to reduced wound healing time, with total regeneration of the epithelial layer, reduced inflammatory cells, and collagen synthesis and remodeling of wounds [39]. Wound healing and skin rejuvenation are major challenges in regenerative medicine, yet the incorporation of nanotechnology leaps markedly forward in this field; newly discovered nanomaterials now facilitate innovative strategies in wound management, tissue regeneration, and thinning age-related morphogenetic changes.

8.4 Challenges in SC-Based Therapies

The major challenges of SC-based skin regeneration and rejuvenation need to be investigated such as regulating differentiation, restricted availability, guaranteeing cell survival, possible carcinogenesis, immunological rejection, and obstacles with large-scale production.

8.4.1 Immunogenicity and Rejection

In regenerative medicine, skin graft rejection is still a major problem. Although donor-derived dendritic cell–targeting treatments have increased longevity and tolerance, skin graft survival is not permanent. Bioengineered skin may be possible using SC-based treatments, but long-term restoration involves a series of procedures [40].

8.4.2 Safety Risks

Nonprofit groups including MEASURE, CoNCEPT, and CT-TRACS address safety issues and develop a clinical translation roadmap for SC therapy [41]. They draw attention to important elements, such as product attributes, medical records, and regulatory obligations. To create effective and safe SC products for clinical use, cutting-edge technologies and bioethical standards are required. High-quality SC products may result from cooperation between biologists, bioengineers, veterinarians, and physicians [42, 43].

8.4.3 Issues Related to Current Techniques and Technologies

Safety issues with SC treatment in regenerative medicine include immunological responses, self-renewal and differentiation, and a lack of standardized assessment standards. Its effectiveness depends on a robust regulatory structure, an ethical review procedure, and public education. Through interdisciplinary collaboration, technological development, and a solid ethical and legal foundation, SC therapy can enhance human health [44].

8.4.4 Contamination Issues

Preclinical testing, formal safety reviews, and consistent SC production techniques are essential to the therapy's success. Before being injected into the body, contamination issues need to be addressed, and cells need to be described. It can be difficult to maintain phenotypic and genetic behavior over an extended period of culture.

8.4.5 Difficulties in Developing Countries

Economic sector funding, international alliances, and exchange studies could help address the high cost of SC therapies, lack of infrastructure and skilled labor, and lack of government support that hamper many developing nations [45, 46].

8.5 Future Directions of SCs in Skin Regeneration and Rejuvenation

A reliable skin therapy method based on synthetic and bioengineered SCs would become possible because of their controlled differentiation and enhanced regenerative capabilities. SC-based therapies have shown great promise in skin rejuvenation, offering potential solutions for age-related deterioration, and wound healing. As research in this field progresses, future directions could enhance the efficacy and applicability of these therapies. Antiaging medicine will adopt personalized exosome therapies as a foundational treatment because these therapies consider specific genetic elements and environmental exposures in each patient [47, 48]. 3D bioprinting technicians could develop adapted skin tissues through the combination of SCs with biomaterials for skin grafting applications [49]. Artificial intelligence (AI) software systems should analyze patient information to generate predictions about the most suitable SC treatment plans for individual therapies. AI technologies used in drug screening enable scientists to discover compounds that improve both SC multiplication and skin tissue differentiation [50]. Bridging the gap between laboratory research and clinical application remains a major challenge. Standardized protocols for SCs preparation, quality control, and ethical considerations must be established. Regulatory approvals and long-term safety studies are necessary to ensure that SC-based skin rejuvenation therapies meet stringent medical standards. Public perception and awareness campaigns will be essential to promote responsible and ethical use of SC therapies [51].

8.5.1 Importance of AI

Modern clinics use ultramodern technology using AI for impulse control, allowing maximum effect in one pass, saving time and maintaining comfort, as confirmed by 85% of patients studied. Traditional microneedle radiofrequency is one of the facial rejuvenation methods used in aesthetic medicine clinics; it stimulates collagen, improves skin tone, and reduces wrinkles, but it can be painful and have long recovery times. The treatment increases the production of collagen and elastin, increasing skin thickness, firmness, and elasticity. It provides a lifting effect without surgery, natural appearance, quick results, and improved skin tone. It is more comfortable than traditional fractional radiofrequency, safe for all skin types, and has been proven in 14 clinical trials [52]. By evaluating patient data, AI can predict appropriate SC therapy regimens, and drug screening tools can find substances that promote SC proliferation and skin tissue differentiation.

8.5.2 Personalized Medicine

Personalized medicine with SCs has an impact as an innovative treatment approach for various skin diseases. This method customizes therapies according to each person's genetic and phenotypic profiles boosting effectiveness and reducing side effects [53, 54]. Cancer SCs act as biomarkers to personalize treatment and group patients based on their tumor profiles [55]. iPSCs help to discover drugs and screen for toxicity allowing doctors to test drugs for specific patients [56, 57]. Due to genetic and metabolic variations, the medical paradigm has difficulties in understanding the wide range of patient responses to medications. By generating tissues from each patient's SCs, SCs provide a way to overcome this issue by producing an organ "mini-me" in the laboratory. By exposing these customized tissue models to different medications, scientists can watch how they respond in actual situations. This strategy could guarantee that patients receive therapies that are most appropriate for their particular biological composition and lessen undesirable medication reactions. In both disease research and regenerative medicine, iPSCs represent a revolutionary development. Reprogrammed from a patient's somatic cells, these cells can develop into nearly any kind of human cell. Researchers can replicate and analyze the condition in a dish once the cells have been stimulated to develop into disease-specific cell types. This offers a dynamic platform for analyzing the subtleties of diseases in ways that were previously unimaginable [58].

8.5.3 Genome Editing

By targeting aging-related genes and altering aging-related pathways, CRISPR-Cas genome editing technology may be able to revitalize old SCs. This strategy may prolong the functional lifespan of SCs and increase the effectiveness of SC-based treatments for age-related illnesses by delaying the aging process of SCs and boosting their capacity for regeneration. Research on aging and regenerative medicine will be significantly impacted by this technology. By replacing or repairing damaged tissue, regenerative SCs can improve quality of life, reverse disease symptoms, and constrain its progression. They are also being investigated for the treatment of chronic tissue inflammation and age-related macular degeneration. The effectiveness and specificity of these treatments are further improved by the use of cutting-edge gene editing technologies such as CRISPR, providing new opportunities for aging-related personalized medicine. The technique of epigenetic rejuvenation aims to repair the aging-induced reduction in SC activity that is impacted by alterations in epigenetic markers. Tools based on CRISPR, such as dCas9, can deliver epigenetic enzymes that change the chromatin state and target particular gene loci. By doing this, SC populations are successfully revitalized, regaining their functionality and increasing their potential for treatment. This method could increase the longevity of SCs and enhance treatments for age-related illnesses that rely on SCs. It is an effective regenerative medicine technology that aims to improve healthier aging outcomes and restore youthful traits [59, 60].

8.6 Current Clinical Trials Updates

Due to promising preclinical results, SCs have been used for skin rejuvenation and regeneration. ASC injections reduced wrinkles and enhanced skin thickness and collagen content in rats with UVB-induced wrinkles.

Table 8.1 Clinical trial updates for skin rejuvenation with the use of SCs.

Details of the study	Number of patients	Observations	References
Following liposuction, either expanded MSCs or autologous fat and SVF were injected into the ears under local anesthesia.	1 Male and 5 Female patients	According to histologic analysis, both treatment groups were more hydrated and had less reticular cutaneous elastosis than the baseline.	[65]
Half of the face was randomly assigned to receive therapy with microneedling + stem cell–conditioned media every 4 weeks for three sessions, while the other half received microneedling alone.	15 Females	By comparing the treatment side to the baseline and placebo, there was a notable improvement in dermal collagen, fine wrinkles, patient satisfaction scores, and the doctor's overall assessment 4 weeks into the course of therapy.	[61]
Using a fanning technique, platelet-rich plasma) and nanofat suspensions are injected into the face once.	48 Male and 2 female patients	All patients reported improvements in texture, elasticity, shine, firmness, small wrinkles, and skin hydration.	[66]
Had three treatments of microneedling to the face every 2 weeks, along with normal saline or a medium conditioned by amniotic membrane SCs.	48 Females	Comparing clinical assessments of pores, wrinkles, spot-polarized, and spot-UV features at 8 weeks to a placebo, significant improvements were observed.	[10]

The combination of SCs and microneedling promoted epidermal penetration. SC-conditioned medium improved patient satisfaction by decreasing wrinkles, erythema pigmentation, and pore size in two randomized, controlled, and blinded split-facial trials compared with a placebo [61]. Patient satisfaction and all objective clinical assessment scores were enhanced when laser resurfacing and SC therapy were combined [62]. Following liposuction, autologous fat transfer techniques improve skin regeneration by minimizing wrinkles, pores, and pigmentation [63]. There are three types of fat transfer: macrofat, microfat, and nanofat. Macrofat is used to lipofill significant flaws, while nanofat is used to regenerate skin and microfat has applications for delicate regions [45, 64]. Table 8.1 represents recent clinical trial updates for skin rejuvenation with the use of SCs. SCs, are essential for healing severe wounds such as burns, injuries, and diabetes. Through the release of growth factors and the promotion of anti-inflammatory, antioxidative, and antiapoptotic qualities, they indirectly contribute in wound healing [45].

8.7 Regulatory Challenges for SC-Based Therapies

Clinical and commercial SC-based products (SCBPs) are particularly vulnerable to safety, effectiveness, and quality-related regulatory concerns. Assays for prospective contamination, karyotype testing, and advancement for the necessary cell population are all essential components of safety testing. After safety has been scientifically proven, the product needs to succeed in clinical efficiency by *in vitro* functional assays, which need to be confirmed to satisfy regulatory standards. To ensure excellent quality, the product has to meet certain requirements. Animal models have intrinsic drawbacks such as the use of immunocompromised small animals for disorders such as orthopedics. Issues such as biological traits, *in vitro* manipulation, cell persistence, injection route, and intended clinical usage should be taken seriously while choosing the best model for tumorigenicity investigations. With less preclinical data, clinical trials should be carried out with the utmost care to ensure safety and ethical considerations. Guidance on these clinical issues can be found in the rules and regulations governing SCBPs in the United States, India, and the European Union (EU) [67, 68].

The three directives that form the foundation of European cell therapy law are Directive 2003/63/EC, Directive 2001/20/EC, and Directive 2004/23/EC. These guidelines specify particular conditions and classify cell therapy products as clinical products. Opinions regarding the effectiveness, safety, and quality of advanced therapy medicinal products are prepared by the European Medicines Agency. However, the EU has not yet given SC-based medical products a marketing authorization. EU guidelines should be followed in clinical development programs, and clinical trials should show that they are safe and effective [69]. During clinical trials, the use of biomarkers and long-term efficacy monitoring should be taken into considerations. Cell therapy products, comprising 361 and 351 products, are covered by the US Code of Federal Regulations. The FDA has released guidance documents on biologic, drugs, and device regulations and monitors these products under the Good Tissue Practice. SC-based treatments are categorized according to the criteria they are intended to treat, and federal funding is the only restriction. ApliGraf®, Cardiel®, and Epicel® have received FDA approval; minimally modified therapies do not need FDA approval. In addition to minimally modified SC products, doctors can prescribe off-label. The "Ethical Guidelines for Biomedical Research on Human Subjects" published by the Indian Council of Medical Research (ICMR) in 2006 classified SC research into three categories: allowed, restricted, and prohibited. Clinical trial with clinical-grade SCs is one of the acceptable categories; nonetheless, it must adhere to GCP and ICMR principles. Collaborations with international corporations involving foreign-imported SC products are among the restricted categories. The guidelines emphasize the importance of reviewing and keeping updated on human SC research and therapy. Because there are no particular laws for SC therapies, the Indian sector is currently at a crossroads in deciding how to commercially license these products [70].

There are currently not any regulations developed by the International Conference on Harmonization of Technical Requirements for Registration of Pharmaceuticals for Human Use that are specific to SCBPs [69]. Still, international organizations such as the Hinxton Group and the International Society for Stem Cell Research have released nonbinding standards and codes of practice. Requiring quality-controlled cells, proven safety and efficacy in animal models, SC expertise in peer review, and voluntarily informed permission, these criteria highlight the clinical translation of SCs [71, 72].

8.8 Conclusion

SC treatments have the potential to have a big influence on humanity when paired with biomedical engineering and cell rejuvenation. Recent studies on MSCs show their potential in skin therapy, improving immune function, antioxidant activity, cell proliferation, and overall skin morphology.

However, challenges such as low tissue frequency and limited proliferative potential remain. MSCs, a promising target for skin therapy research, have shown great potential due to their high self-renewal capacity and differentiation ability. Further studies are needed to ensure safe implementation of these methods. This review addresses the promise of SC treatments for wounds that do not heal and skin conditions. Choosing the appropriate SC type is crucial for full skin regeneration. A combination strategy with smart matrices including nanoparticles may accelerate wound healing and replicate the characteristics of original tissue. To investigate long-term impacts, more clinical trials are required.

AI disclosure statement: The author acknowledges the use of AI-based tools for minor language editing or grammatical correction. However, all intellectual contributions, analyses, and conclusions presented in this chapter are entirely the author's own.

References

1. Tumbar, T., Guasch, G., Greco, V., Blanpain, C., Lowry, W.E., Rendl, M., *et al.*, Defining the Epithelial Stem Cell Niche in Skin. *Science (1979)*, 303, 5656, 359–63, 2004 Jan 16.
2. Skin Rejuvenation Treatment with Stem Cells, GenCell [Internet], [cited 2025 Mar 17]. Available from: https://www.gencell.com.ua/en/skin-rejuvenation-with-stem-cells.
3. Khamkat, P., Barik, V., Mohapatra, S., Karati, D., Mukherjee, S., Current Approaches on Transfersomal Patch: A Noninvasive Innovative Booster for Improved Transdermal Drug Delivery. *Curr. Pharm. Biotechnol.*, 25, 12–20, 2024 Aug 23.
4. Ojeh, N., Pastar, I., Tomic-Canic, M., Stojadinovic, O., Stem Cells in Skin Regeneration, Wound Healing, and Their Clinical Applications. *Int. J. Mol. Sci.*, 16, 10, 25476–501, 2015 Oct 23.
5. Ogliari, K.S., Marinowic, D., Brum, D.E., Loth, F., Stem cells in dermatology. *An. Bras. Dermatol.*, 89, 2, 286–91, 2014 Apr.
6. Aboul-Soud, M.A.M., Alzahrani, A.J., Mahmoud, A., Induced Pluripotent Stem Cells (iPSCs)—Roles in Regenerative Therapies, Disease Modelling and Drug Screening. *Cells*, 10, 9, 2319, 2021 Sep 5.
7. Lovell-Badge, R., Anthony, E., Barker, R.A., Bubela, T., Brivanlou, A.H., Carpenter, M., *et al.*, ISSCR Guidelines for Stem Cell Research and Clinical Translation: The 2021 update. *Stem Cell Rep.*, 16, 6, 1398–408, 2021 Jun 8.

8. Jo, H., Brito, S., Kwak, B.M., Park, S., Lee, M.G., Bin, B.H., Applications of Mesenchymal Stem Cells in Skin Regeneration and Rejuvenation. *Int. J. Mol. Sci.*, 22, 5, 2410, 2021 Feb 27.
9. Liu, Z., Hu, G.D., Luo, X.B., Yin, B., Shu, B., Guan, J.Z., *et al.*, Potential of bone marrow mesenchymal stem cells in rejuvenation of the aged skin of rats. *Biomed. Rep.*, 6, 3, 279–84, 2017 Mar.
10. Prakoeswa, C.R.S., Pratiwi, F.D., Herwanto, N., Citrashanty, I., Indramaya, D.M., Murtiastutik, D., *et al.*, The effects of amniotic membrane stem cell-conditioned medium on photoaging. *J. Dermatolog. Treat.*, 30, 5, 478–82, 2019 Jul 4.
11. Jin, Y., Li, S., Yu, Q., Chen, T., Liu, D., Application of stem cells in regeneration medicine. *MedComm (2020)*, 4, 4, e291, 2023 Aug 17.
12. Induced pluripotent stem cells, UCLA BSCRC [Internet], [cited 2025 Mar 17]. Available from: https://stemcell.ucla.edu/glossary/induced-pluripotent-stem-cells.
13. Nakayama, C., Fujita, Y., Matsumura, W., Ujiie, I., Takashima, S., Shinkuma, S., *et al.*, The development of induced pluripotent stem cell-derived mesenchymal stem/stromal cells from normal human and RDEB epidermal keratinocytes. *J. Dermatol. Sci.*, 91, 3, 301–10, 2018 Sep.
14. Kim, S., Lee, S., Kim, H., Kim, T., Exosomes Secreted from Induced Pluripotent Stem Cell-Derived Mesenchymal Stem Cells Accelerate Skin Cell Proliferation. *Int. J. Mol. Sci.*, 19, 10, 3119, 2018 Oct 11.
15. Veraitch, O., Mabuchi, Y., Matsuzaki, Y., Sasaki, T., Okuno, H., Tsukashima, A., *et al.*, Induction of hair follicle dermal papilla cell properties in human induced pluripotent stem cell-derived multipotent LNGFR(+)THY-1(+) mesenchymal cells. *Sci. Rep.*, 7, 1, 42777, 2017 Feb 21.
16. Spitzhorn, L.S., Megges, M., Wruck, W., Rahman, M.S., Otte, J., Degistirici, O., *et al.*, Human iPSC-derived MSCs (iMSCs) from aged individuals acquire a rejuvenation signature. *Stem Cell Res. Ther.*, 10, 1, 100, 2019 Dec 18.
17. Volarevic, V., Markovic, B.S., Gazdic, M., Volarevic, A., Jovicic, N., Arsenijevic, N., *et al.*, Ethical and Safety Issues of Stem Cell-Based Therapy. *Int. J. Med. Sci.*, 15, 1, 36–45, 2018.
18. Yoon, D., Yoon, D., Sim, H., Hwang, I., Lee, J.S., Chun, W., Accelerated Wound Healing by Fibroblasts Differentiated from Human Embryonic Stem Cell-Derived Mesenchymal Stem Cells in a Pressure Ulcer Animal Model. *Stem Cells Int.*, 2018, 1–12, 2018 Dec 30.
19. Yang, R., Wang, J., Chen, X., Shi, Y., Xie, J., Epidermal Stem Cells in Wound Healing and Regeneration. *Stem Cells Int.*, 2020, 1–11, 2020 Jan 29.
20. Carsin, H., Ainaud, P., Le Bever, H., Rives, J.M., Lakhel, A., Stephanazzi, J., *et al.*, Cultured epithelial autografts in extensive burn coverage of severely traumatized patients: a five year single-center experience with 30 patients. *Burns*, 26, 4, 379–87, 2000 Jun.
21. Surowiecka, A. and Strużyna, J., Adipose-Derived Stem Cells for Facial Rejuvenation. *J. Pers. Med.*, 12, 1, 117, 2022 Jan 16.

22. Takeo, M., Lee, W., Ito, M., Wound Healing and Skin Regeneration. *Cold Spring Harbor Perspect. Med.*, 5, 1, a023267–a023267, 2015 Jan 1.
23. Ansell, D.M., Kloepper, J.E., Thomason, H.A., Paus, R., Hardman, M.J., Exploring the hair growth-wound healing connection: Anagen phase promotes wound re-Epithelialization. *J. Invest. Dermatol.* [Internet], 131, 2, 518–28, 2011 Jan 1 [cited 2025 Mar 21]. Available from: https://www.jidonline.org/action/showFullText?pii=S0022202X15351654.
24. Bhardwaj, N., Chouhan, D., Mandal, B.B., 3D functional scaffolds for skin tissue engineering, in: *Functional 3D Tissue Engineering Scaffolds*, pp. 345–65, Elsevier, Woodhead Publishing, 2018.
25. Chouhan, D., Dey, N., Bhardwaj, N., Mandal, B.B., Emerging and innovative approaches for wound healing and skin regeneration: Current status and advances. *Biomaterials*, 216, 119267, 2019 Sep.
26. Weng, T., Zhang, W., Xia, Y., Wu, P., Yang, M., Jin, R., *et al.*, 3D bioprinting for skin tissue engineering: Current status and perspectives. *J. Tissue Eng.*, 12, 20417314211028574, 2021 Jan 13.
27. Gu, B.K., Choi, D.J., Park, S.J., Kim, Y.J., Kim, C.H., 3D Bioprinting Technologies for Tissue Engineering Applications. *Adv. Exp. Med. Biol.*, 1078, 15–28, 2018.
28. Gudapati, H., Dey, M., Ozbolat, I., A comprehensive review on droplet-based bioprinting: Past, present and future. *Biomaterials*, 102, 20–42, 2016 Sep.
29. Jiang, F. and Doudna, J.A., CRISPR–Cas9 Structures and Mechanisms. *Annu. Rev. Biophys.*, 46, 1, 505–29, 2017 May 22.
30. Baker, C. and Hayden, M.S., Gene editing in dermatology: Harnessing CRISPR for the treatment of cutaneous disease. *F1000Res*, 9, 281, 2020 Oct 7.
31. Henn, D., Zhao, D., Sivaraj, D., Trotsyuk, A., Bonham, C.A., Fischer, K.S., *et al.*, Cas9-mediated knockout of Ndrg2 enhances the regenerative potential of dendritic cells for wound healing. *Nat. Commun.*, 14, 1, 4729, 2023 Aug 7.
32. News: CRISPR Delivery To The Skin With Microneedles, CRISPR Medicine [Internet], [cited 2025 Mar 14]. Available from: https://crisprmedicinenews.com/news/crispr-delivery-to-the-skin-with-microneedles/.
33. Risueño, I., Valencia, L., Jorcano, J.L., Velasco, D., Skin-on-a-chip models: General overview and future perspectives. *APL Bioeng.*, 5, 3, 030901, 2021 Sep 1.
34. Huh, D., Leslie, D.C., Matthews, B.D., Fraser, J.P., Jurek, S., Hamilton, G.A., *et al.*, A Human Disease Model of Drug Toxicity–Induced Pulmonary Edema in a Lung-on-a-Chip Microdevice. *Sci. Transl. Med.*, 4, 159, 159ra147, 2012 Nov 7.
35. Groeber, F., Holeiter, M., Hampel, M., Hinderer, S., Schenke-Layland, K., Skin tissue engineering — *In vivo* and *in vitro* applications. *Adv. Drug Deliv. Rev.*, 63, 4–5, 352–66, 2011 Apr.
36. Dong, Y., Wu, X., Chen, X., Zhou, P., Xu, F., Liang, W., Nanotechnology shaping stem cell therapy: Recent advances, application, challenges, and future outlook. *Biomed. Pharmacother.*, 137, 111236, 2021 May.

37. Bellu, E., Garroni, G., Cruciani, S., Balzano, F., Serra, D., Satta, R., *et al.*, Smart Nanofibers with Natural Extracts Prevent Senescence Patterning in a Dynamic Cell Culture Model of Human Skin. *Cells*, 9, 12, 2530, 2020 Nov 24.
38. Ahmed, M.K., Zayed, M.A., El-dek, S.I., Hady, M.A., El Sherbiny, D.H., Uskoković, V., Nanofibrous ε-polycaprolactone scaffolds containing Ag-doped magnetite nanoparticles: Physicochemical characterization and biological testing for wound dressing applications *in vitro* and *in vivo. Bioact. Mater.*, 6, 7, 2070–88, 2021 Jul.
39. Daňková, J., Buzgo, M., Vejpravová, J., Kubíčková, S., Sovková, V., Vysloužilová, L., *et al.*, Highly efficient mesenchymal stem cell proliferation on poly-ε-caprolactone nanofibers with embedded magnetic nanoparticles. *Int. J. Nanomed.*, 10, 7307–17, 2015 Dec.
40. Kim, J.Y., Kang, B.M., Lee, J.S., Park, H.J., Wi, H.J., Yoon, J.S., *et al.*, UVB-induced depletion of donor-derived dendritic cells prevents allograft rejection of immune-privileged hair follicles in humanized mice. *Am. J. Transplant.*, 19, 5, 1344–55, 2019 May 1.
41. Cell Therapy - TRAcking, Circulation, & Safety (CT-TRACS) - HESI - Health and Environmental Sciences Institute [Internet], [cited 2025 Mar 18]. Available from: https://hesiglobal.org/cell-therapy-tracking-circulation-safety-ct-tracs/.
42. Frati, P., Scopetti, M., Santurro, A., Gatto, V., Fineschi, V., Stem Cell Research and Clinical Translation: A Roadmap about Good Clinical Practice and Patient Care. *Stem Cells Int.*, 2017, 1–8, 2017.
43. Zaman, W.S.W.K., Karman, S.B., Ramlan, E.I., Tukimin, S.N.B., Ahmad, M.Y.B., Machine Learning in Stem Cells Research: Application for Biosafety and Bioefficacy Assessment. *IEEE Access* [Internet], 9, 25926–45, 2021 Feb 2 [cited 2025 Mar 18]. Available from: https://pure.ulster.ac.uk/en/publications/machine-learning-in-stem-cells-research-application-for-biosafety.
44. Baldari, S., Di Rocco, G., Piccoli, M., Pozzobon, M., Muraca, M., Toietta, G., Challenges and Strategies for Improving the Regenerative Effects of Mesenchymal Stromal Cell-Based Therapies. *Int. J. Mol. Sci.*, 18, 10, 2087, 2017 Oct 2.
45. Chakraborty, D., De, A., Chakraborty, D., De, A., Stem-cell therapy in dermatology – Challenges and opportunities. *Indian J. Skin Allergy* [Internet], 3, 2, 93–105, 2024 Oct 17 [cited 2025 Mar 18]. Available from: https://skinallergyjournal.com/stem-cell-therapy-in-dermatology-challenges-and-opportunities/.
46. Ikehara, S., Grand challenges in stem cell treatments. *Front. Cell Dev. Biol.*, 1, 2, 2013 Oct 10.
47. Prodinger, C.M., Reichelt, J., Bauer, J.W., Laimer, M., Current and Future Perspectives of Stem Cell Therapy in Dermatology. *Ann. Dermatol.*, 29, 6, 667, 2017.
48. Chu, G.Y., Chen, Y.F., Chen, H.Y., Chan, M.H., Gau, C.S., Weng, S.M., Stem cell therapy on skin: Mechanisms, recent advances and drug reviewing issues. *J. Food Drug Anal.*, 26, 1, 14–20, 2018 Jan.

49. Davis, L.E., Shalin, S.C., Tackett, A.J., Current state of melanoma diagnosis and treatment. *Cancer Biol. Ther.*, 20, 11, 1366–79, 2019 Nov 2.
50. The Surgeon General's Call to Action to Prevent Skin Cancer - PubMed [Internet], [cited 2025 Mar 19]. Available from: https://pubmed.ncbi.nlm.nih.gov/25320835/.
51. Ciążyńska, M., Kamińska-Winciorek, G., Lange, D., Lewandowski, B., Reich, A., Sławińska, M., *et al.*, The incidence and clinical analysis of non-melanoma skin cancer. *Sci. Rep.*, 11, 1, 4337, 2021 Feb 22.
52. AI skin rejuvenation: artificial intelligence in medicine [Internet], [cited 2025 Mar 20]. Available from: https://wellclinic.pl/en/ai-skin-rejuvenation/.
53. Farabi, B., Roster, K., Hirani, R., Tepper, K., Atak, M.F., Safai, B., The Efficacy of Stem Cells in Wound Healing: A Systematic Review. *Int. J. Mol. Sci.*, 25, 5, 3006, 2024 Mar 5.
54. Hasanzadeh, A., Ebadati, A., Dastanpour, L., Aref, A.R., Sahandi Zangabad, P., Kalbasi, A., *et al.*, Applications of Innovation Technologies for Personalized Cancer Medicine: Stem Cells and Gene-Editing Tools. *ACS Pharmacol. Transl. Sci.*, 6, 12, 1758–79, 2023 Dec 8.
55. Kaur, T., Khan, N., Pasha, Z., Bhat, R., Virupakshappa, D., Bharisharanesha, R., *et al.*, Stem Cells: Innovations, Applications, and Future Directions. *J. Pharm. Bioallied Sci.*, 16, Suppl 4, S3041–3, 2024 Dec.
56. Abdulmalek, O.A.A.Y., Husain, K.H., AlKhalifa, H.K.A.A., Alturani, M.M.A.B., Butler, A.E., Moin, A.S.M., Therapeutic Applications of Stem Cell-Derived Exosomes. *Int. J. Mol. Sci.*, 25, 6, 3562, 2024 Mar 21.
57. Petrus-Reurer, S., Romano, M., Howlett, S., Jones, J.L., Lombardi, G., Saeb-Parsy, K., Immunological considerations and challenges for regenerative cellular therapies. *Commun. Biol.*, 4, 1, 798, 2021 Jun 25.
58. Applications of Stem Cells in Personalised Medicine [Internet], [cited 2025 Mar 20]. Available from: https://oxfordglobal.com/nextgen-biomed/resources/applications-of-stem-cells-in-personalised-medicine.
59. CRISPR-Cas Genome Editing For Rejuvenation Of Aging Stem Cells [Internet], [cited 2025 Mar 20]. Available from: https://www.cellandgene.com/doc/crispr-cas-genome-editing-for-rejuvenation-of-aging-stem-cells-0001.
60. Ji, S., Xiong, M., Chen, H., Liu, Y., Zhou, L., Hong, Y., *et al.*, Cellular rejuvenation: molecular mechanisms and potential therapeutic interventions for diseases. *Signal Transduct. Target. Ther.*, 8, 1, 116, 2023 Mar 14.
61. Seo, K.Y., Kim, D.H., Lee, S.E., Yoon, M.S., Lee, H.J., RETRACTED ARTICLE: Skin rejuvenation by microneedle fractional radiofrequency and a human stem cell conditioned medium in Asian skin: a randomized controlled investigator blinded split-face study. *J. Cosmet. Laser Ther.*, 15, 1, 25–33, 2013 Feb 1.
62. Zhou, B.-R., Zhang, T., Bin Jameel, A.A., Xu, Y., Xu, Y., Guo, S.-L., *et al.*, The efficacy of conditioned media of adipose-derived stem cells combined with

ablative carbon dioxide fractional resurfacing for atrophic acne scars and skin rejuvenation. *J. Cosmet. Laser Ther.*, 18, 3, 138–48, 2016 Apr 2.

63. Coleman, S.R. and Katzel, E.B., Fat Grafting for Facial Filling and Regeneration. *Clin. Plast. Surg.*, 42, 3, 289–300, 2015 Jul.
64. Tonnard, P., Verpaele, A., Peeters, G., Hamdi, M., Cornelissen, M., Declercq, H., Nanofat Grafting. *Plast. Reconstr. Surg.*, 132, 4, 1017–26, 2013 Oct.
65. Charles-de-Sá, L., Gontijo-de-Amorim, N.F., Maeda Takiya, C., Borojevic, R., Benati, D., Bernardi, P., *et al.*, Antiaging Treatment of the Facial Skin by Fat Graft and Adipose-Derived Stem Cells. *Plast. Reconstr. Surg.*, 135, 4, 999–1009, 2015 Apr.
66. Taub, A., Bucay, V., Keller, G., Williams, J., Mehregan, D., Multi-Center, Double-Blind, Vehicle-Controlled Clinical Trial of an Alpha and Beta Defensin-Containing Anti-Aging Skin Care Regimen With Clinical, Histopathologic, Immunohistochemical, Photographic, and Ultrasound Evaluation. *J. Drugs Dermatol.* [Internet], 17, 4, 426–41, 2018 [cited 2025 Mar 18]. Available from: https://pubmed.ncbi.nlm.nih.gov/29601620/.
67. Rayment, E.A. and Williams, D.J., Concise Review: Mind the Gap: Challenges in Characterizing and Quantifying Cell- and Tissue-Based Therapies for Clinical Translation. *Stem Cells*, 28, 5, 996–1004, 2010 May 1.
68. Liras, A., Future research and therapeutic applications of human stem cells: general, regulatory, and bioethical aspects. *J. Transl. Med.*, 8, 1, 131, 2010 Dec 10.
69. George, B., Regulations and guidelines governing stem cell based products: Clinical considerations. *Perspect. Clin. Res.*, 2, 3, 94, 2011.
70. Halme, D.G. and Kessler, D.A., FDA Regulation of Stem-Cell–Based Therapies. *N. Engl. J. Med.*, 355, 16, 1730–5, 2006 Oct 19.
71. An international consortium on stem cells, ethics and law. *Rev. Derecho Genoma Hum.* [Internet], 24, 251–5, 2006 Jan [cited 2025 Mar 18]. Available from: https://pubmed.ncbi.nlm.nih.gov/17124979/.
72. Martell, K., Trounson, A., Baum, E., Stem Cell Therapies in Clinical Trials: Workshop on Best Practices and the Need for Harmonization. *Cell Stem Cell*, 7, 4, 451–4, 2010 Oct.

9

Stem Cell Therapy in Dermatology: Challenges and Opportunities

Garima Chandak[1], Sumel Ashique[1], Shubham Kumar[1], Narendra Kumar Pandey[1], Dileep Singh Baghel[1], Bimlesh Kumar[1*], Saurabh Singh[1] and Omji Porwal[2]

[1]*School of Pharmaceutical Sciences, Lovely Professional University Phagwara, Punjab, India*

[2]*Qaiwan International University, Department of Pharmacy, Kurdistan, Sulaymaniyah, Iraq*

Abstract

Utilizing stem cells' properties as an immunomodulator and regenerative approach in the treatment of autoimmune skin diseases, such as systemic lupus erythematosus, systemic sclerosis, pemphigus vulgaris, psoriasis, etc.; skin pigmentation; chronic ulcers; scarring; and even aesthetic medicine has varying degrees of success. Because of their capacity to regulate inflammatory responses and differentiate into skin-specific lineages, mesenchymal stem cells, epidermal stem cells, and induced pluripotent stem cells are the most studied multipotent cell types for regenerative skin treatment. Even though stem cell therapy has a lot of potential, there are a lot of obstacles in the way of its practical application in dermatology. These include hazards such as immunological rejection, tumor formation, high production costs, scaling up challenges, and ethical concerns. Furthermore, the absence of defined techniques for cell isolation, culture, and transplantation casts doubt on the efficacy and reproducibility of treatments. Nonetheless, chances to improve the safety, effectiveness, and applicability of stem cell–based therapies are presented by developments in more recent technologies like three-dimensional bioprinting, gene editing, and biomateriality. This chapter discusses the latest advancements, difficulties, and potential for sustainable and focused therapies in the field of stem cell therapy for dermatological disorders.

**Corresponding author*: bimlesh1Pharm@gmail.com; bimlesh_11@rediffmail.com; ORCID: 0000-0001-8072-5172

Sumel Ashique, Sanjay Dey, Biplab Debnath, Mohhammad Ramzan and Swarupananda Mukherjee (eds.) Stem Cell Therapy in Dermatological Disorders, (219–270) © 2025 Scrivener Publishing LLC

Keywords: Stem cell differentiation, cellular therapy, dermatological treatment, regenerative medicine, skin rejuvenation

9.1 Introduction

9.1.1 Overview of Stem Cell Therapy

Stem cells are undifferentiated, self-replicating cells present in many tissues, capable of giving rise to many different types of cells [1]. This peculiar characteristic of plasticity holds therapeutic potential and hence emerged as a promising new field that offers unique ways in getting rid of certain diseases and skin conditions too. The specialized microenvironment in which stem cells are present called "stem cell niche" works as a unit in maintenance of homeostasis and repair damaged tissues during an injury [2]. They show three main features:

9.1.1.1 Self-Renewal

This refers to their ability to undergo multiple cycles of asymmetrical cell divisions allowing stem cells to create differentiated cell types as well as identical versions of their parent cells preserving the original undifferentiated cell group.

9.1.1.2 Differentiation

The stem cells have the potential to transform into specialized cell types within their native stem cell location, such as muscle cells, neurons, or skin cells.

9.1.1.3 Plasticity

It is defined as the ability of adult stem cells to cross the cellular lineage barriers and differentiate into cell types that differ from their original tissue [3].

Stem cells are further classified based on their differentiating capacity [20]. The capacity of a unique way of differentiation makes them different, which are given below:

- **Totipotent/omnipotent cell:** Ability to differentiate into any possible cell type including placental cells and is described by the initial stage of embryonic cell divisions known as morula cells

- **Pluripotent cell:** Capability to change into every mature cell existing within the body across the three germ layers, excluding placental tissue. Pluripotent cells include embryonic stem cells (ESCs) isolated from an early-stage embryo (blastocyst)
- **Multipotent cell:** Has a specialized potential to develop into multiple distinct mature body cell type from a specific limited connected cell type group, e.g., hematopoietic (adult) stem cells belong to this category because they transform into red and white blood cells or platelets and mesenchymal stem cells (MSCs) (bone, cartilage, connective tissue, and adipose tissue)
- **Oligopotent cell:** Ability to differentiate into a limited number of cell types, e.g., myeloid stem cells or adult lymphoid
- **Unipotent cell:** These cells specialize in producing a single cell type, such as muscle stem cells [4, 20]

Over the past decades, the field of stem cell therapy has seen revolutionary evolution [4]. In the 1960s, bone marrow transplantation received a breakthrough after scientists found hematopoietic stem cells (HSCs) [5]. In 1998, human ESC (HESC) isolation created major medical progress, which enabled additional potential uses in regenerative medicine [6]. The advancement of stem cell science occurred in 2006 through the discovery of induced pluripotent stem cells (iPSCs) by Shinya Yamanaka removed ethical concerns associated with ESCs providing highly effective stem cells [7].

9.1.2 Previous Perspective on Dermatological Applications

Traditional dermatological treatments relied only on topical medications, conducting surgical procedures, and using systemic drug therapies to manage symptoms without addressing skin disease origins or damage triggers. Earlier, burn injury treatment utilized skin grafts; however, donor site limitations and graft rejection remained significant challenges. The application of corticosteroids along with immunosuppressant drugs was used for psoriasis and vitiligo, but these medications frequently caused side effects and recurrence [8]. In early 2000s, discovery of adult stem cells in the skin tissue opened the opportunities to regenerative medical methods. Skin-derived precursors along with MSCs accelerate wound healing while providing better scar outcomes and controlling immune responses within autoimmune skin conditions. Hence, the emergence of stem cell research has undergone fundamental change emphasizing cell regenerative therapies and therapeutic advances beyond standard method of disease control. Persistent wounds such as diabetic ulcers and pressure sores patients became the first to receive advantages from stem cell therapy. Experimental studies revealed that MSCs boost angiogenesis, accelerate skin cell growth, and minimize inflammation, creating a foundation for future clinical studies. The field of stem cells revealed promising results for combating hair loss problems including alopecia. Laboratory-produced growth factors derived from stem cells activate dormant hair follicles, leading to new hair production and healthier head skin [9].

9.1.3 Scope and Significance in Dermatology

The wide-ranging potential of stem cell therapy not only addresses medical but also cosmetic concerns. The therapeutic approach provides promising solutions to established treatment-resistant dermatological conditions while displaying skin tissue regeneration abilities. The following are the key aspects:

9.1.3.1 Management of Autoimmune Skin Disorders

In conditions such as psoriasis, vitiligo, and systemic sclerosis, the immune system shows dysregulation followed by persistent inflammation. Immunomodulatory behavior of MSCs and iPSCs alongside reduces inflammatory cytokines while reinforcing tissue recovery processes. These therapeutic approaches offer a new treatment strategy for the control of these hard-to-treat disorders [10].

9.1.3.2 Treatment of Chronic Ulcers and Wounds

The healthcare field faces substantial challenge in treating nonhealing wounds such as diabetic ulcers, venous leg ulcers, and pressure sores. Stem cells, specifically MSCs, show their ability to control wound inflammation while speeding up healing by enhancing both epithelialization and angiogenesis processes [11].

9.1.3.3 *Burns, Scar Reduction, and Rejuvenation*

The consequences of burn injuries are severe scarring and long-term impairment of functions. Regenerative approaches based on stem cells work to reconstruct damaged dermis along with damaged epidermis while minimizing scar development. Autologous skin-derived stem cells demonstrate effective outcomes according to preclinical and clinical trials. Healthcare practitioners are using stem cell therapy to meet rising demand for antiaging treatments by enhancing skin elasticity, tackling wrinkles, and providing hydration to aging skin. Recent research shows that Adipose-derived stem cells (ADSCs) release factors including growth factors and cytokines, widely studied for their ability to revitalize aging skin structures [12].

9.1.3.4 *Hair Regeneration*

Millions of individuals worldwide experience hair loss disorders, primarily including androgenetic alopecia. The potential to regenerate hair follicles and improve hair density comes from stem cells of ADSCs as well as dermal papilla-derived cells. The cosmetic dermatologic community has started to acknowledge this solution [12].

9.1.3.5 *Treatment of Pigmentation Disorders*

Stem cell–derived components restore melanocytes and control pigment production to give a uniform skin tone. Therefore, it can address disorders such as melasma and postinflammatory hyperpigmentation [13].

9.1.3.6 *Development of Bioengineered Skin Substitutes*

Stem cells are used in creating functional skin replacements through bioengineered skin substitutes for treating extensive burns and skin defect. These substitutes replicate natural skin structures and functions, resulting in better outcomes than conventional grafting approaches [13].

9.2 Types of Stem Cells in Dermatology

9.2.1 Embryonic Stem Cells

Pluripotent cells residing in the blastocyst inner cell mass have the capability to develop all body organs apart from the placenta [14].

Characteristics: Pluripotency, self-renewal
Applications: Wound healing, skin grafting, disease modeling such as epidermolysis bullosa (EB)
Challenges: Ethical concerns, risk of teratoma formation due to uncontrolled differentiation

9.2.2 Mesenchymal Stem Cells

They function as stromal cells exhibit multilineage differentiation. Found in many tissues including skin, bone marrow, umbilical cord, adipose tissue, endometrial polyps, etc. [54].

Characteristics: Multipotency, immunomodulatory properties
Applications: Wound healing, scars treatment, skin rejuvenation
Challenges: Standardized procedures for cell isolation, culture, and differentiation, allogeneic MSCs can trigger immune system rejection

9.2.3 Induced Pluripotent Stem Cells

Embryonic-like cells produced by converting somatic cells into pluripotent state by genetic engineering. Induced stem cells show the capability to develop into multiple forms of specialized cells [15].

Characteristics: Pluripotent, patient-specific
Applications: Skin regeneration, disease modeling such as psoriasis and vitiligo, drug testing
Challenges: Genomic instability, risk of tumorigenicity, high cost, and complex

9.2.4 Hematopoietic Stem Cells

Immature cells that develop into erythrocytes, lymphocytes, and platelets. They are generally found in bone marrow and peripheral blood [16].

Characteristics: Regeneration, hematopoiesis
Applications: Wound healing, treatment of autoimmune skin disease (systemic lupus erythematosus [SLE], scleroderma, and cutaneous vasculitis) and chronic skin conditions (eczema)
Challenges: Limited availability and Invasive procedure, risk of causing graft-versus-host disease in allogenic transplants

9.2.5 Skin-Derived Precursors

Adult stem cell presents within the dermis and mostly isolated from dermal papilla and hair follicle bulges [17].

Characteristics: Multipotency, accessibility
Applications: Tissue engineering, hair regeneration, and wound healing
Challenges: Lack of regulatory protocols and understanding of their niche, difficult to achieve reliable consistency and scalability for clinical applications

9.3 Mechanisms of Action

9.3.1 Stem Cell Differentiation in Skin Repair

Stem cells have the ability to give rise to specialized cell lineages and are capable of self-renewal [18]. Traditionally, there are two types of stem cells: ESCs and somatic stem cells. Due to the pluripotent nature of ESCs, they can generate descendants that belong to all three germ layers comprising ectoderm, mesoderm, and endoderm [6]. Unlike ESCs, somatic stem cells are multipotent or unipotent cells having restricted plasticity, found in many tissues including skin and bone marrow; i.e., HSCs produce mature blood cell lineages, whereas neural stem cells produce neuronal and glial cells. Professor Yamanaka's laboratory accomplished immense success in Kyoto by reprogramming differentiated, adult somatic cells to generate iPSCs. The process of reprogramming is done by exogenous addition of four transcription factors (Oct-3/4, Sox2, c-Myc, and Klf4) delivered using retroviral transduction. Modern molecular tools are used for identifying multiple varieties of skin stem cells [19, 20].

The skin harbors various stem cell populations, each residing in distinct niches and characterized by specific markers. Epidermal stem cells are primarily located in the basal layer of the epidermis and are identified by markers such as p63, $\beta 1^{high}$/melanoma chondroitin sulfate proteoglycan þ (MCSPþ), and a6^{high}/CD71^{dim} [21]. MSCs are found in the dermis (divided into mesodermal derivatives and some neural cells), expressing markers such as CD70, CD90, and CD105 [22]. Follicular stem cells, residing in the follicle bulge region, play a crucial role in hair follicle regeneration and are marked by K15, CD34, Lgr5, Sox9, Lhx2, NFATC1, K15, NFIB, PHLDA1,

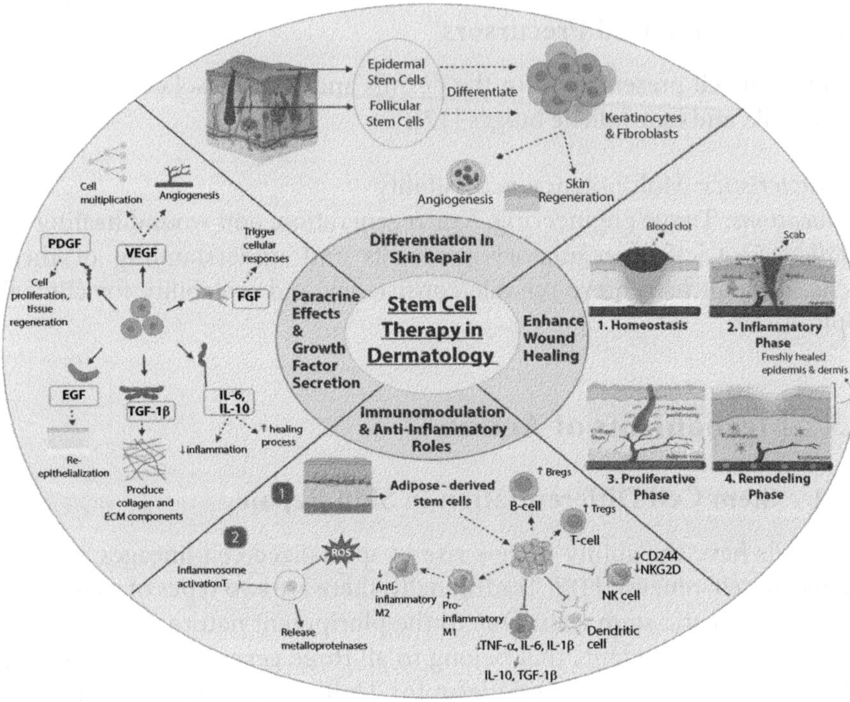

Figure 9.1 Mechanism of action of stem cell therapy in dermatology. 1) Differentiation in skin repair: epidermal and follicular stem cells differentiate to form specific skin cell types, such as keratinocytes and fibroblasts that promote angiogenesis and skin regeneration. 2) Paracrine effects and growth factor secretion: stem cells produce various growth factors such as vascular endothelial growth factor (VEGF), transforming growth factor-β (TGF-β), platelet-derived growth factor (PDGF), epidermal growth factor (EGF), fibroblast growth factors (FGFs), and interleukins (e.g., IL-6, IL-10). All of them show different property to promote tissue repair and regeneration. 3) Immunomodulation and anti-inflammatory roles: MSCs, most commonly adipose-derived stem cells, are proliferated to modulate the action of immune cells, such as B cells and T cells by increasing Bregs and Tregs in the body, inhibit natural killer (NK) cell (by blocking CD244 and NKG2D pathway), dendritic cell, and proinflammatory cytokines (TNF-α, IL-6, and IL-1β). Moreover, they release anti-inflammatory cytokines (IL-10, TGF-β) and induce conversion of m1 to m2 phenotype. 4) Enhance wound healing: this mechanism involves four process; homeostasis, inflammatory, proliferative, and remodeling phase. (Source: Biorender).

CD200, K19, and bromodeoxyuridine dye retention [23]. Melanocyte stem cells, responsible for pigment production, are localized in the follicle bulge region and hair germ, expressing specific cell markers such as Oct, Sox, and Pax3 [24]. Sebaceous gland stem cells, which contribute in maintaining homeostasis, are found within the sebaceous glands and the infundibulum, marked by Blimp1 [25]. Additionally, neural progenitor cells present in the follicle dermal papillae, divided into neural and glial cell lineages, express markers such as vimentin, nestin, S100B protein, Pax6, and GFAP [26]. Lastly, HSCs also located in the follicle dermal papillae (divided into erythroid and myeloid lineages) and can be identified by markers including CD34, CD45, CD38, endothelial protein C receptor, and aldehyde dehydrogenase [27]. These diverse stem cell populations contribute to the maintenance, regeneration, and function of the skin and its appendages.

The scientific evidence reported that epidermal stem cells exist in low numbers, divide infrequently, and produce short-lived, rapidly dividing cell networks, enabling epidermal regeneration. They are also responsible for repairing skin injuries. Most epidermal stem cells are located in the basal layer of the epidermis, and a few might be found in the bulge region of the hair follicle and the base of the sebaceous glands [28]. In skin injury, both epidermal and follicular stem cells are involved in the re-epithelialization of wounds [29, 30]. In the case of a full-thickness wound, both the cells from the hair follicle move toward the wound area. Epidermal stem cells are reported and reactivated in response to skin injury and contribute to skin regeneration on the cellular level (Figure 9.1) [31].

9.3.2 Paracrine Effects and Growth Factor Secretion

The main therapeutic advantages of MSC therapy derived through the secretion of various growth factors and cytokines acting on nearby cells through a paracrine phenomenon. The repair and regeneration of injured skin depend on this mechanism to regulate different cellular and molecular procedures for successful restorative processes. Through indirect mechanism, this process mediates wound healing by influencing cell migration patterns, growth, differentiation, and inflammatory bodily responses. The constituent of secretome's diverse factors are vascular endothelial growth factor (VEGF), HGF, IGF-1, SDF1α, TGF-β, and TSG-6 [32].

VEGFs secreted by stem cells bind to receptors present on endothelial cells, resulting in cell multiplication and migration to form new blood vessels, thereby promoting angiogenesis and vascularization within the wound tissue [33]. Transforming growth factor β (TGF-β) induces fibroblasts to produce collagen and other extracellular matrix (ECM) components, maintaining structural integrity of the newly formed tissue. It also transforms the wound from inflammatory to proliferative phase through regulation of immune cell activity [34]. Platelet-derived growth factor (PDGF) binds to target cells, enhancing cell proliferation and contributing significantly to tissue regeneration. Epidermal growth factor (EGF) plays a key role in re-epithelialization by binding to receptors on keratinocytes, activating signaling pathways that drive cell growth and migration across the wound. Similarly, fibroblast growth factors bind with receptor tyrosine kinases on target cells, triggering cellular responses necessary for tissue repair, including cell proliferation, differentiation, and angiogenesis. Additionally, interleukins (IL-6 and IL-10) exhibit a dual role in wound healing. IL-6, in the early stages, induces an inflammatory response to combat pathogens, whereas IL-10 subsequently activates anti-inflammatory pathways, stimulating tissue repair and balancing the healing process. These growth factors collectively coordinate various phases of wound healing, ensuring effective tissue regeneration and restoration of normal function [35].

9.3.3 Immunomodulation and Anti-Inflammatory Roles

Skin conditions such as psoriasis, atopic dermatitis, SLE, and chronic wounds are frequently caused by dysregulated immune responses and uncontrolled inflammation. Current treatments seek to reduce the immune responses but have certain limitations such as systemic side effects and partially effective. Therefore, stem cells, particularly MSCs, directly interact with immune cells to reduce skin inflammation by blocking proinflammatory factors and releasing anti-inflammatory cytokines, thus improving immune responses at injury or disease sites [36]. ADSCs are a type of MSCs obtained from adipose tissue. These cells derive from both skin cells and the mesoderm while maintaining a close relationship to epidermal cells. Compared to bone marrow MSCs (BM-MSCs), umbilical cord MSCs, etc., ADSCs demonstrate better immunomodulatory effects. Under the effects of anesthesia, ADSCs can be extracted in large quantities from subcutaneous tissue cells through repeated liposuction procedures [37]. Hence, ADSCs are the ideal choice for stem cell therapy used in immune-related skin conditions as they regulate immune responses and support skin regeneration by facilitating production of keratinocytes and the secretome [38].

Paracrine mechanism involves direct and indirect interactions of stem cells during immunomodulation. In direct cell-to-cell interactions, stem cells control T cells, macrophages, B cells, natural killer (NK) cells, and dendritic cells to promote immune tolerance. They also inhibit overactivated autoimmune responses leading T immune-related skin diseases. The cells exercise their immunomodulatory properties through selective control of immune cell phenotypes and secretory responses in addition to establishing either immunological stagnation or active anti-inflammatory conditions. However, indirect mechanism involves release of many bioactive molecules such as cytokines, growth factors, and extracellular vesicles. It involves inhibition of proinflammatory cytokines (tumor necrosis factor α [TNF-α], IL-6, and IL-1β), release of anti-inflammatory cytokines (IL-10, TGF-β), and inducing macrophage polarization by converting proinflammatory M1 phenotype into anti-inflammatory M2 phenotype [39]. Further, stem cells control the complement cascade, thus protecting damaged tissue during inflammatory skin problems.

9.3.3.1 Anti-Inflammatory Mechanisms

The overproduction of ROS causes skin inflammation and tissue damage. When stem cells function, they produce antioxidant enzymes such as superoxide dismutase and catalase to break down ROS. The skin inflammatory disorders include NLRP3 inflammasomes, which play a role in their progression. Stem cells inhibit inflammasome activation by reducing the production of both IL-1β and IL-18. Stem cells control the production and breakdown of ECM in fibrosis and chronic wounds by releasing metalloproteinases, promoting tissue repair and reducing scarring [40].

9.3.4 Enhancing Wound Healing

Wounds are the physical injuries developed due to internal and external skin breakage leading to opening or rupture of the skin. Proper and instant wound healing serves as a vital process for restoring functional tissues and repairing damaged structural disorders [41]. Wound healing is an active metabolic process that requires abundant nutrients and oxygen supply. The sequential phases of wound healing are inflammatory phase, proliferative phase, and remodeling phase [42].

During the inflammatory phase, blood clotting and homeostasis are two vital mechanisms that occur after an injury to control bleeding by platelet clumping at the injury site. This process lasts about 2 days. The inflammatory stage emerges after homeostasis reaches equilibrium.

It also involves modulation of inflammatory response and suppression of ROS [43]. Proliferative phase starts on the third day of injury and lasts for about 2 weeks. The development of this phase enables TGF-β released by macrophages, T lymphocytes, and platelets to become essential controllers of fibroblast cellular function. TGF-β reduces protease fabrication and activates the production of tissue inhibitor of metalloproteinase. This phase completes with angiogenesis, fibroblast migration, and epithelialization [42]. Remodeling phase is the last phase of wound healing that spans approximately 1 to 2 years or more. This phase maintains equilibrium between the formation and degradation of ECM components. Regular epithelial growth and wound tissue formation signify the progression. As the process of wound healing progresses, the number of macrophages and fibroblasts decreased due to apoptosis. Following this phenomenon, metabolic activity at the wound area diminished due to stop of the blood flow to the injury site; hence, wound completely heals [44].

9.4 Current Applications of Stem Cell Therapy in Dermatology

9.4.1 Chronic Wounds

Stem cell therapy is considered safe, effective, and viable for the treatment of chronic wounds and ulcers as per numerous clinical and preclinical studies. They promote wound healing by directing migration of fibroblast cells to the injury site so they can localize in the wound bed. After localization, fibroblasts aid in recovery of wound by producing collagen and other essential ECM components. They also downregulate matrix metalloproteinase-1 (MMP-1) activity, a primary collagenase responsible for destroying ECM structure and ICAM1, a protein used in inflammation. Hence, stem cells function as a defense mechanism to preserve the ECM and maintain dermal structural integrity. Wounds treated with MSCs show higher elastin levels that confer tissue resiliency not usually observed in normal wound healing. They also support re-epithelialization process by facilitating cell growth, differentiation, and migration to create a multilayered epidermis. The application of SCs to chronic wounds is promoting angiogenesis through upregulation of VEGF and Ang-1 and increased microvessel density across the entire wound bed, so that nutrients and oxygen can be provided to developing cells to increase their longevity [45].

9.4.2 Ulcers

Chronic ulcers such as diabetic foot ulcers, arterial ulcers, venous leg ulcers, and pressure sores, characterized by prolonged healing and substantial morbidity. Earlier, diabetic foot was treated with medication and surgery to normalize blood flow. However, the issue of foot ischemia was not resolved by medical treatment, causing the risk of amputation to the patients. Diabetic foot was also associated with cardiovascular and cerebrovascular diseases as patients were unable to tolerate the discomfort of distal bypass surgery. Hence, stem cell transplantation attracted attention in treating chronic ulcers as it aims to improve blood supply and alleviate limb ischemia [46]. The activation of p38 mitogen-activated protein kinase signaling pathway allows MSCs to produce VEGF and HGF proteins. The injured area contains TNF-α, which functions as a necessary factor for secretion of VEGF and HGF production. Growth factors and cytokines manage cellular responses while affecting immune cell activities. These molecules serve vital roles during cell-to-cell interactions among fibroblasts, keratin-forming cells, immune cells, and endothelial cells. Different types of stem cells can improve chronic ulcers by enhancing the expression of VEGF, promoting proliferation, formation of new blood vessels, and collagen deposition; reduce excessive inflammation; and release various growth factors and cytokines essential for wound healing [11]. In most clinical trials, bone marrow–derived MSCs (BM-MSCs) are used for cutaneous wound healing, whereas in diabetic ulcer, ADSCs and umbilical cord–derived cells have shown its potential in clinical studies [47].

9.4.3 Psoriasis

Chronic inflammatory skin conditions caused by rapid growth of keratinocyte cell and immune dysfunction result in scaly, erythematous plaques. HSC transplantation procedure replaces faulty immune cells with health ones, thereby providing resetting the immune system. They have shown potential in treating psoriasis patients by decreasing both cutaneous symptoms and systemic inflammatory markers. Another widely researched stem cell for psoriasis is MSCs. They act by modulating hyperactive immune response by balancing T_H1, T_H17, and Treg cell activity. Also, they repair and restore damaged skin through differentiation of MSCs into fibroblasts and keratinocytes [48].

9.4.4 Vitiligo

Vitiligo is an autoimmune disorder where melanocytes are being destroyed by immune system causing depigmentation on the skin. Epidermal stem cells can restore pigmentation by regenerating melanocytes and repopulating areas that lost pigmentation. Similarly, iPSCs taken from the patient's somatic cell to differentiate into melanocytes provide uniform skin tone. MSCs play a vital role in controlling immune response by preventing T-cell–mediated destruction of melanocytes and secretion of growth factors to improve skin health [49].

9.4.5 Scleroderma

The disease is characterized by excessive accumulation of collagen in the tissue causing severe skin thickening and tissue fibrosis. MSCs derived from adipose tissue and bone marrow decrease tissue scarring by decreasing fibroblast activity and collagen production. It also enhances blood circulation toward infected tissue by promoting angiogenesis. Autologous HSC transplantation stops disease progression by destroying immune cells that attack the body and establishes new immune cells from healthy sources. iPSCs have shown the potential to regenerate damaged tissue and treat fibrosis [50].

9.4.6 Pemphigus Vulgaris

Pemphigus vulgaris is an autoimmune blistering disorder that develops when autoantibodies attack desmosomal proteins and destroy cell adhesive connections, leading to blister formation. MSCs act by suppressing the B-cell functions responsible for autoantibody production. These cells facilitate tissue restoration along with wound healing actions in blistered skin regions. The procedure of hematopoietic stem cell transplantation (HSCT) provides immune system reconstruction for refractory patients. Allogenic HSCT allows replacement of ill-functioning B cells with new healthy immune cells [51].

9.4.7 Systemic Lupus Erythematosus

Systemic lupus erythematosus is a systemic autoimmune disease causing rashes, photosensitivity, and chronic inflammation to the skin. MSCs restrain T-cell and B-cell activation processes while supporting regulatory T cells.

HSCs reset the immune system by introducing healthy cells that replace damaged immune cells. iPSCs repair the damaged tissue. Therefore, stem cells reduce inflammation and promote better overall skin condition [52].

9.4.8 Stem Cell–Based Approaches for Skin Aging

In a recent study, Wang *et al* reported that MSCs extracted from healthy mice show antiaging effects after intravenous delivery into older mice [53]. MSCs, specifically adipose stem cells, have shown its ability in improving aged skin by enhancing production of angiogenesis growth factors [54]. The antiaging mechanisms involve multiple therapeutic activities including protection against chronic inflammation, promotion of tissue renewal, and controlling cellular senescence [55]. Stem cell therapy can prevent skin aging through various mechanism. Stem cells produce exosomes and secretomes that activate vital signaling pathways (NF-κB, AP-1, MAPK, P-AKT, NRF2, SIRT-1), which help in skin regeneration. Regulation of oxidative stress is a key factor of skin aging that causes cellular damage by producing free radicals. Hence, the upregulation of antioxidant enzymes such as HO-1 and NQO1 helps scavenge reactive oxygen species (ROS), while exosomes further enhance antioxidant defenses by increasing the levels of SOD1, CAT, FRAP, and GP, thereby maintaining redox balance in skin cells. Protection against extracellular matrix (ECM) degradation is achieved through enhanced synthesis of ECM components like elastin, collagen, and hyaluronic acid, along with the inhibition of matrix metalloproteinases (MMPs). Additionally, fibroblast activity is stimulated to support tissue regeneration. It also regulates cell activity; for example, the aging skin shows signs of senescence together with apoptosis and decreased proliferation. Therefore, stem cell therapy works by reducing cell senescence (SA-β-gal), apoptosis, and cell cycle arrest (P53, P12, P16) and enhancing autophagy, cell migration, and cell proliferation (Ki67). It helps to regulate inflammatory pathway where exosomes decrease inflammation by downregulating IL-6, IL-1, TNF-α, and increasing TGF-β [56].

9.4.9 Applications in Hair Regeneration

Millions of people worldwide experience alopecia, caused by various factors such as genetic, hormonal imbalance, environmental exposure, and autoimmune conditions. Over-the-counter drugs such as minoxidil and propecia are taken on a continuous basis to show the result. Topical application of minoxidil to scalp skin increases blood circulation through vasodilation, thus delivery abundant nutrients and oxygen to hair follicles.

Finasteride, inhibitor of 5-α reductase enzyme, results in decreased DHT levels near follicles [57]. This drug is limited to male patients as it is contraindicated in pregnant women due to teratogenic effect. Stem cell therapy is a novel method used to manage hair loss by addressing fundamental causes of deteriorating hair follicles and encouraging natural hair growth process. Scientists introduced "stem cell educator therapy," a process involving autologous blood circulation through a closed-loop system with multipotent stem cells (cord blood stem cells). The treated blood was returned back to the patient after the "education" process. This therapy produced a ring of TGF-β, which created protection for the hair follicles against an attack from the immune system. Hence, this research shows stem cells potential to regulate hair growth patterns in cases of alopecia [58].

9.4.10 Burn and Scar Treatment

The application of stem cells in burn wound healing presents promising results by promoting rapid healing, superior scar results, and increased regeneration and reformation of skin tissues and their appendages along with reduced inflammatory responses and minimizing scarring, fibrosis, and infection [59]. Recent medical research demonstrated that rather than postengraftment proliferation and differentiation, the therapeutic efficacy of stem cell comes from their ability to generate paracrine signaling molecules. Proteomic analysis of BM-MSCs, adipose-derived MSCs (AD-MSCs), and umbilical cord–derived MSCs (UC-MSCs) found that chemokines, cytokines, and various growth factors are secreted to boost fibroblast, keratinocytes, and endothelial cell proliferation and migration. The secretome profile of MSCs enables wound healing and thus making stem cell products such as exosomes and conditioned media suitable for burn wound treatments [60].

As per research, stem cell therapy minimizes systemic inflammation by enhancing anti-inflammatory cytokines and decreasing proinflammatory cytokines. They also regulate cellular functions of T cells, B cells, macrophages, and NK cells. Moreover, they also promote bacterial clearance from the body. Additionally, stem cells contribute to wound healing by differentiating into various cell types such as myofibroblasts, dermal fibroblasts, antigen-presenting cells (APCs), and lymphoid cells, while also activating resident stem cells to support tissue repair and regeneration. Hence, they play a direct role in wound repair by embracing wound tissue for healing and restoration [61].

Stem cells have shown their therapeutic potential not only in acute care but also in scarring. Studies show that keloid implantation animal models when treated with AD-MSC–derived conditioned media have reduced inflammation and fibrotic tissue formation [62]. Similarly, use of BM-MSC derived conditioned media showed the same results. Hence, stem cells are better than traditional treatment approaches used for burn care [63].

9.5 Advances in Stem Cell Delivery Systems

9.5.1 Three-Dimensional Bioprinting and Skin Substitutes

Three-dimensional (3D) bioprinting refers to the 3D deposition of bioink with cells using computer-controlled printers to create functional tissue equivalents. The primary characteristics of skin bioprinting over traditional cell therapy systems are technical adaptability and reproducibility, alongside enabling production of patient-specific constructs. However, its applications are limited due to less availability of printable polymers and solvents [64].

In recent years, development of 2D skin substitutes took place based on various processes such as electrospinning, casting, freeze drying, and 3D bioprinting. 3D skin substitutes, despite being thin and flat, gained more popularity than 2D skin substitutes because of superior cellular viability and tissue healing properties. The 3D-printed structures show potential to repair, restore, and regenerate skin tissue and accelerate wound healing and skin-related pharmacology, toxicology, and drug development [65]. The first and foremost research on skin bioprinting was published in 2009, in which primary human dermal fibroblasts were incorporated into a collagen hydrogel [66]. Bioinks play a critical role in 3D bioprinting to restate the physiological and pathological environment. They are designed as hydrogel biomaterials containing specific cell types. Hydrogels are the 3D network of hydrophilic polymer that exhibits an excellent capacity to absorb water up to 10 times of their initial weight. They are classified into two main groups, namely, natural polymers (e.g., collagen, gelatin, alginate) and synthetic polymers (e.g., PCL [polycaprolactone], PLA [polylactic acid], ABS [acrylonitrile butadiene styrene]). They have shown promising results on wound healing because of moist environment and a cell carrier. Most common hydrogel combinations used for coculture of keratinocytes/fibroblasts involves double-nozzle printing systems made with alginate/gelatin [67]. Moreover, by controlling various printing parameters such as temperature, extrusion rates, and geometry, along with physical and chemical characteristics of bioink characteristics (e.g., cell density, crosslinking, and viscosity), the working of the bioprinting construct can be optimized in terms of cell viability, wound healing, and vascularization. The core use of skin bioprinting technology is wound healing; however, it has also been widely used in diagnostic applications such as pharmaceutical, toxicity, and cosmetic testing. However, reconstructive surgery and scarless wound healing are the primary clinical benefits of skin bioprinting technology showing great potential through research approaches [68].

9.5.2 Scaffold-Based Delivery

Scaffold-based delivery systems utilize biocompatible materials to create a 3D network that mimics the skin's ECM, facilitating the transport and delivery of stem cells, growth factors, and therapeutic agents to injury or disease sites. These tissue scaffolds play a crucial role in skin healing by supporting fundamental cellular processes such as adherence, movement, and differentiation. Additionally, they function as reservoirs for the gradual release of bioactive molecules such as VEGF, PDGF, and anti-inflammatory cytokines, which promote angiogenesis, reduce inflammation, and accelerate tissue regeneration. Various types of scaffolds are used in dermatological applications [69]. Hydrogel-based scaffolds, which form hydrated polymer networks, provide a conducive environment for cell survival and proliferation. They are extensively utilized to transport MSCs for chronic wound treatment and scar minimization while also aiding in the controlled release of growth factors to enhance angiogenesis [70]. Electrospun nanofibrous scaffolds, produced through electrospinning, generate nanoscale fibers that mimic tissue structures and support the adhesion and differentiation of stem cells [71]. Decellularized dermal scaffolds, derived from human or animal tissues, retain the ECM while removing cellular components, making them valuable in regenerative applications. Composite scaffolds, designed by combining two or more materials, including natural and synthetic polymers, enhance functionality. For instance, wound infection healing benefits from the integration of MSCs with antimicrobial agents, whereas chronic wounds can be effectively treated through the sustained delivery of bioactive molecules [72].

9.5.3 Injectable Stem Cell Therapies

A revolutionary approach in regenerative dermatology is injectable stem cell therapy. It is defined as direct injection of stem cells into damaged or diseased skin tissues to promote repair and functioning. This minimally invasive technique delivers stem cells to the target site by using biocompatible media such as hydrogels, platelet-rich plasma, and saline solutions. This specific delivery method enables local functioning of stem cells, which increase effectiveness while reducing systemic side effects [73]. They can also manage to create a controlled microenvironment that supports survival, multiplication, and functioning of stem cells. Stem cell–derived bioactive molecules produce growth factors including VEGF, EGF, and TGF-β and help in angiogenesis, re-epithelialization, and activating ECM remodeling. Multiple dermatological complications such as chronic wounds, burn injuries, and scars have shown therapeutic results from injectable stem cell therapies by restoring the skin components and texture. Injectable therapy of stem cells represents a promising frontier in regenerative medicine, promoting rapid wound healing, minimizing scarring while they rebuild skin resilience and initial smoothness, and thus providing patients with an efficient therapeutic option for serious dermatological wounds [74].

9.5.4 Nanotechnology in Stem Cell Delivery

The term "nano" in nanotechnology generally represents dimensions spanning from 1 to 100 nm. This specified range is significant because it bridges the gap between atomic or molecular structures and macroscopic systems. Nanomaterials exhibit unique properties that differ from bulk counterparts due to quantum phenomena and greater surface area-to-volume ratios. These features include elevated chemical reactivity, electrical conductivity, mechanical strength, and optical properties [75]. Nanotechnology fused with stem cell sciences provides fascinating probabilities for understanding prevention of human disease, theranostics, and importance of molecular equipment in the regulation of stem cell differentiation. Interaction of stem cell with nanotechnology improves treatment outcomes in many ways. Stem cells reach precise disease locations through nanomaterials, which provide directed movement and controlled release mechanisms of stem cells [76].

Nanoparticles play a significant role in dermatological applications by facilitating targeted delivery, improving stability, and enhancing therapeutic effects. Polymeric nanoparticles, such as those made from poly(lactic-co-glycolic acid) (PLGA), PLA, and polyethylene glycol, enable the controlled release of bioactive molecules. PLGA nanoparticles loaded with MSC-derived exosomes promote wound healing by enhancing angiogenesis and ECM remodeling [77]. Lipid-based nanoparticles, including liposomes, solid lipid nanoparticles (SLNs), and nanostructured lipid carriers, are used for the transport of stem cell–derived cytokines and growth factors, aiding in scar reduction and skin rejuvenation. Liposomes also improve the penetration of therapeutic agents across the skin, whereas SLNs enhance drug stability. Metallic nanoparticles, such as gold, silver, and titanium dioxide nanoparticles, exhibit diverse applications. Gold nanoparticles are used for skin rejuvenation and wound healing due to their antioxidant and anti-inflammatory properties, whereas silver nanoparticles possess antimicrobial activity for infected wound treatment. Titanium dioxide nanoparticles, widely used in sunscreens, provide protection against UV radiation [78]. Magnetic nanoparticles enable targeted stem cell delivery through external magnetic fields, ensuring higher cell concentrations at injury sites to accelerate healing in burns and chronic wounds. Silica-based nanoparticles serve as carriers for stem cell–secreted factors, facilitating controlled release and skin regeneration while also aiding in the treatment of vitiligo by transporting melanocyte-precursor stem cells [79]. Protein-based nanoparticles, composed of silk fibroin, collagen, or gelatin, mimic ECM components, supporting stem cell adhesion, growth, and differentiation into keratinocytes and fibroblasts. Additionally, nanomaterials are utilized for the topical delivery of retinoids, antioxidants, botulinum toxin, and growth factors for skin rejuvenation. Quantum dots are specialized nanoparticles used in diagnosing cutaneous tumors and sentinel lymph nodes. Carbon-based nanoparticles, particularly carbon nanotubes, are linked to nucleic acids and antibodies, enabling their binding to receptors. Their electrical conductivity varies based on binding interactions, making them useful in diagnosing cutaneous infections and potential malignancies [80].

9.6 Challenges in Stem Cell Therapy for Dermatology

9.6.1 Regulatory and Ethical Concerns

The implementation of stem cell therapy in dermatology faces numerous barriers for its widespread adoption. They require solutions to ethical issues, technical hurdles, and economic barriers to become safe, effective, and widely accepted for its clinical use. The research field of stem cells faces specific legitimate dilemmas, which directly affect both science and medical practice. ESCs possess high-differentiation power and self-renewal but often raise moral concerns as obtained from human embryos [81]. Three major challenges arise during a frozen embryo donation, involving voluntary agreements/informed consent of both women and couple donating embryo, consent from gamete donors involved in creation of embryo, and confidentiality of donor details [82]. Due to ethical issues, numerous nations have tightened restrictions that limit access to research and clinical practices.

Somatic cell nuclear transfer (SCNT) offers potential to develop pluripotent stem cells from patient-specific DNA, which could help in tailored therapies and disease modeling. However, this technique triggers substantial moral values. Critics argue against creating embryos specifically for research due to respect for human life, while supporters establish that SCNT-derived entities remain different from traditional embryos [83]. The use of SCNT for human reproduction faces ethical controversies due to high risk of congenital defects and some belief that it demeans human dignity by treating clones as manufactured goods rather than blessings. Additionally, proposals that use animal oocytes for hybrid embryo creation raise further ethical concerns, which fears about chimeric entities. Overall, the moral along with scientific domains intersects with cultural factors in all discussions related to SCNT [84]. Medically necessary abortions serve as the source to obtain pluripotent stem cells. However, the use of fetal tissue for research purpose creates ethical dilemmas because it is associated with abortion, which multiple people strongly oppose. The federal government allows fetal tissue research when the tissue donation process starts only after women complete their pregnancy termination decisions. Federal regulations stipulate such a requirement because it reduces the risk that a woman would act based on tissue donation in making abortion choices. IPSCs overcome ethical concerns associated with ESC research, as they are obtained from somatic cells without using embryos or oocytes. Regulatory procedures delay the approval process of stem cell therapies because they need to satisfy both safety protocols and demonstrate sufficient therapeutic value [82].

9.6.2 Risk of Tumorigenesis

HESCs have the ability to self-renew and proliferate while maintaining its cellular identity and pluripotency. Reprogramming somatic cells into human iPSCs (HiPSCs) requires transcription factors (OCT4, SOX2, MYC, and Krupple-like factor 4 [KLF4]) that are highly expressed in case of cancer. HiPSCs are commonly derived using integrating vectors, thus causing a risk for genetic alterations and reprogramming factor reactivation at later stages. They are more prone to chromosomal aberrations as compared to HESCs [85]. Owing to genetic and epigenetic factors, HiPSCs exhibit greater tumorigenic properties than HESCs and cause a risk for the development of teratocarcinomas and somatic tumors [86]. The tumorigenic risk of MSCs remains lower than ESCs and iPSCs, yet safety concerns do remain. The cellular properties of MSCs consist of multipotency where they can transform into specifically limited number of cell types and are less likely to induce teratoma formation, although, in some situations, MSCs can contribute to tumor enlargement due to genetic mutations or abnormalities occur in chromosomes during MSC expansion in culture. Further, studies suggest that MSCs might convert into tumor-promoting cancer-associated fibroblasts, which promote tumor growth [87].

9.6.3 Immune Rejection and Compatibility Issues

Immune rejection occurs after recipient's immune system identifies stem cells as foreign substances and destroys them. This response makes the therapy ineffective, leading to transplant failure. Immune systems are divided into innate and adaptive immunity, playing a pivotal role in rejecting foreign cells. The innate immune system serves as the body's first line of defense, detecting potential threats through nonspecific mechanisms. Although MSCs exhibit immune-privileged properties, they can still be identified by innate immune components [88]. Pattern recognition receptors (PRRs), such as Toll-like receptors on macrophages, dendritic cells, and neutrophils, detect danger signals known as damage-associated molecular patterns or pathogen-associated molecular patterns from stressed or dying transplanted stem cells. Activation of PRRs triggers a cytokine cascade, leading to the release of proinflammatory cytokines such as IL-1β, IL-6, and TNF-α, which contribute to immune rejection [89]. Additionally, the highly polymorphic major histocompatibility complex (MHC) molecules vary between individuals, triggering a strong T-cell–mediated immune response against transplanted allogeneic cells. MHC molecules are classified into two groups: MHC class I, expressed on all nucleated cells, and MHC class II, primarily found on APCs. $CD8^+$ cytotoxic T cells require MHC class I molecules for activation, while MHC class II molecules are necessary to activate $CD4^+$ helper T cells, which further stimulate immune cells, including cytotoxic T cells. NK cells also play a role in immune rejection by targeting transplanted stem cells that fail to express MHC class I molecules, similar to their action against cancer cells [90].

Surface antigens present on transplanted stem cells can activate the complement cascade, resulting in the formation of the membrane attack complex, which destroys the cells through lysis. Additionally, macrophages mediate the rejection process by recognizing complement protein-marked stem cells as nonself and eliminating them *via* opsonization [91]. The adaptive immune system is primarily responsible for chronic or delayed rejection, providing long-lasting immune protection against foreign cells.

T lymphocytes initiate stem cell destruction through perforin–granzyme pathways or Fas–FasL interactions, whereas B-lymphocytes contribute to rejection by producing antibodies against the transplanted cells. Several factors further influence immune rejection in dermatological stem cell therapies. Furthermore, minor histocompatibility antigens contribute to immune rejection even when there is an identical HLA match between donor and recipient, as subtle antigenic differences between cells can still provoke an immune response [92].

9.6.4 Limitations in Stem Cell Sourcing and Culturing

Despite stem cells showing great potential in treating varieties of skin conditions, they face barrier to implementing this into clinical practice due to problems in sourcing and culturing of stem cells. As already discussed, use of stem cells in dermatology is limited due to ethical and regulatory concerns. It is difficult to harvest epidermal stem cells and used in skin repair and rejuvenation, in sufficient numbers without affecting the neighboring tissue [93]. Cultured stem cells are prone to epigenetic alterations, uncontrolled differentiation, and tumorigenicity due to prolonged culture. Moreover, they also have a short shelf life and might lose their viability or regenerative property during cryopreservation or clinical delivery of skin stem cells [94].

9.6.5 Cost and Accessibility

The widespread implementation of stem cell therapy encounters major obstacles related to cost and accessibility. One of the primary concerns lies in high expenses linked to stem cell acquisition. Transforming experimental stem cell therapies from laboratories to medical applications requires thorough preclinical research and clinical trials, which are both time-consuming and financially intensive. Regulatory approvals create additional hurdles because they need extensive quality assurance assessment together with safety evaluations, which push up the total expenses [95], in addition to creating regional stem cell banks, reforming insurance policies, and campaign for public awareness about stem cell donation along with private sector financial incentives. The combined effect of such coordinated interventions will make stem cell therapy in dermatology affordable, accessible, and globally beneficial to patients worldwide [96].

9.7 Opportunities and Emerging Trends

More effective dermatological treatments are being made possible by the developments of stem cell therapy, which promises futures for conditions such as burns, scars, aging, and hereditary skin problems. Among the most significant new developments and prospects are a few below:

9.7.1 Advances in Gene Editing (CRISPR and Beyond)

Gene editing technologies have revolutionized biological research, and it may be greatly helped in the study and treatment of inherited and acquired human disorders [97]. The CRISPR (Clustered Regularly Interspaced Short Palindromic Repeats) system uses programmable RNA-guided CRISPR-associated (Cas) nucleases to add, modify, or delete genetic material from certain regions of the genome. CRISPR-Cas nucleases are easier to design and execute than other site-specific nucleases, such as transcription activator-like effector nucleases, zinc finger, and meganucleases, primarily due to the easy modification of a guide RNA sequence [98]. Gene editing technologies, particularly CRISPR-Cas9, have now become possible in order to make accurate modifications in the genome of living things. The technique promises dermatology as an area of treatment for enhancing tissue regeneration, repairing genetic anomalies, and revitalizing skin cells [99].

9.7.1.1 Gene Editing in Dermatology

9.7.1.1.1 Gene therapy for Genetic Skin Disorders

Somatic gene editing has been concentrated on conditions such as EB, characterized by exquisitely fragile skin. Skin stem cells from the patient themselves may be edited directly for their mutations with the aid of CRISPR-Cas9 [100]. These genetically modified cells are reinfused into the patient. A study demonstrated the successful correction of mutations in human epidermolysis bullosa (EB) skin stem cells using CRISPR technology, paving the way for the development of potentially curative therapies.

9.7.1.1.2 Aging and Regeneration

One of the most interesting applications of CRISPR is reversing skin aging. To regenerate old skin, scientists are trying to modify some genes responsible for aging, like those responsible for the formation of collagen. Researchers have shown that CRISPR can be applied to alter stem cells in a manner that promotes renewal of skin cells and reduces the appearance of aging indicators [101].

9.7.1.1.3 Reversing UV Damage

The most important reason DNA damage causes aging and mutations due to the effect of UV radiations on the skin is its effects on skin cells. CRISPR could reverse the UV-induced mutations in skin cells, which will increase the repairing and regenerative abilities of skin cells toward damages tissues. Scientists are still exploring whether CRISPR is capable of fixing UV-induced mutations and inhibiting the development of skin cancer like basal cell carcinoma [102].

9.7.1.1.4 Beyond CRISPR: Other Techniques

Although CRISPR-Cas9 remains the most prevalent gene editing, it is in development to offer more precision and less toxicity. These are expected to further advance skin repair and regeneration in dermatology [103].

9.7.2 Personalized Medicine Approaches

Personalized medicines adapt treatment plans based on each patient's own genetic and molecular profile. More and more, stem cell–based treatments are being tailored to maximize healing and regeneration, especially for diseases such as psoriasis, eczema, and skin cancer [104].

9.7.2.1 Personalized Medicine in Dermatology

9.7.2.1.1 Autologous Stem Cell Grafts

The application of personalized medicine in dermatology is based on the use of autologous stem cells [105]. These are cells taken from the patient's body, and they may be used to create personalized skin grafts, which will be less likely to be rejected by the immune system. For instance, the iPSCs are derived from the patient's blood or skin cells and are reprogrammed to become pluripotent cells that may develop into functional skin cells [106].

9.7.2.1.2 Tailored Skin Cancer Therapies

Personalized therapy is being explored for skin malignancies such as melanoma. CSCs can be isolated, genetically modified, and assayed *in vitro* from patient's tumor to identify the most effective immunotherapy or chemotherapy regimens [107]. This is particularly relevant for advanced melanoma, as certain tumor markers are being targeted by stem cell–based therapies or personalized vaccines [108].

9.7.2.1.3 Psoriasis and Eczema

Other chronic inflammatory skin conditions, such as psoriasis, are also being researched using stem cell–based personalized therapy strategies. The use of the patient's own cells is expected to increase skin regeneration, lessen side effects, and reduce inflammation through more individualized and efficient therapies [125]. It affects more than 125 million people worldwide, or 2% to 3% of the world's population. While psoriasis can lead to psoriatic arthritis, it is most common in adults between the age of 15 and 25 years. It occurs more commonly at higher latitudes and is more common among Caucasians than in other ethnic groups. Although it is classified as an autoimmune disease, psoriasis has a familial predisposition, but there is no association of autoantigens with the disease [109].

9.7.3 Integration with Artificial Intelligence and Machine Learning

Artificial intelligence (AI) and machine learning (ML) are the new tools changing the dermatological world in an unprecedented way with improved diagnostic, treatments planning accuracy, and speed, including creating stem cell treatments. AI established great expectations in this respect and evolved into a synergistic strategy that is enhancing human knowledge [110, 111]. AI refers to the ability of a computer to understand and learn from external information and apply the gained insight to predict and self-correct in similar or unfamiliar situations [112]. AI uses automated algorithms to solve problems and analyze large datasets, identify patterns, and predict outcomes that are impossible with human intelligence. AI-based solutions have the potential to enhance precision medicine by supporting clinical decision support in real time [113].

9.7.3.1 Applications of AI and ML in Dermatology

9.7.3.1.1 Predicting Treatment Outcomes

Analyses of large amounts of data from these therapies by stem cells can help the AI model predict how a particular patient's skin will respond to a given treatment with stem cells [114]. For instance, AI will help predict whether a patient's stem cells can repair tissue effectively or if other alternatives of therapy must be pursued through the evaluation of genetic data [115].

9.7.3.1.2 Automating Skin Diagnosis

AI systems can recognize skin conditions from the images. For example, deep learning algorithms have been used to train the networks to realize early detection of melanoma and other skin malignancies. Patients who may be suitable for stem cell–based regenerative treatments can be selected with the help of this [116].

9.7.3.1.3 AI-Optimized Stem Cell Therapy Developments

AI may be used to design new treatments involving stem cells that are tailored to a particular patient's history, given the data produced by trails in the clinic and current research [117]. For example, algorithms might predict which type of stem cells—such as epidermal or MSCs—would do best for a given type of wound healing [118].

9.7.4 Novel Biomaterials for Skin Tissue Engineering in Stem Cell Therapy

The efficacy of stem cell–based skin regeneration treatments depends on the developments of biomaterials. Biomaterials work as scaffolds, providing the necessary space for cell proliferation and differentiation, mimicking the natural environment of the skin. Hydrogels are widely used in tissue engineering due to their ability to mimic the physical properties of skin and provide stem cells with a hydrated environment. Hydrogel-based scaffolds for skin regeneration can be loaded with bioactive molecules such as cytokines or growth factors that promote cell differentiation and proliferation [119]. The most recent innovative material is nanofibrous scaffolds produced through methods such as electrospinning. These scaffolds allow for a high surface area for the adherence of cells and stimulate cellular infiltration because they are similar in structure to the ECM [120]. Nanofibers are particularly useful in applications such as burn wounds and chronic ulcers since they create an environment that is conducive to the development of skin stem cells into keratinocytes and fibroblasts, which are necessary for regeneration of the skin [121]. The advent of 3D bioprinting has led to the emergence of skin tissues engineering. Its application can exactly deposit bioinks, which include collagen, fibrin, and stem cells; hence making it possible for the construction of multilayered skin structures that could correspond to the shape of epidermis, dermis, and subcutaneous layers in healthy skin. Bioactive glasses and ceramics are gaining importance in accelerating angiogenesis and epithelialization during the healing process of wounds. These substances can be added to skin scaffolds to promote tissue healing, aid stem cell activity, and hasten healing—especially in burn and chronic wounds [122]. Finally, collagen and fibrin-based scaffolds are critical because they are biocompatible natural polymers and possess the ability to promote tissue integration and the development of stem cells. The preparation of biodegradable matrices for skin regeneration using such scaffolds often facilitates tissue healing in most injuries. By creating a favorable milieu for skin tissue regeneration and repair, these biomaterials—which range from hydrogels to bioactive glasses—collectively are improving the results of stem cell–based treatments in dermatology [123].

9.8 Preclinical and Clinical Studies

9.8.1 Preclinical Findings

A respected number of animal-based studies have been conducted to analyze the therapeutic possibilities of MSCs for different skin wounds, such as thermal burns, diabetic wounds, radiation burns, acute skin wounds, and pressure sores. These studies exhibit considerable variations, including cell source, animal model, and stem cell delivery methods. Several studies discovered that MSCs can differentiate into skin-specific cells at the wound regions. The transplantation of MSCs increases levels of pro-healing factors at the affected sites along with repairing tissue strength. Systemic cell delivery methods are easy to operate, yet the efficiency of cell engraftment at wounds remains low. Alternative methods have been used to enhance cell retention at the wounds through local suspension injection or scaffold-based cell grafting procedures. Through these transplantation methods, MSC cells produced enhanced wound healing results, which incorporated reduced inflammation coupled with improved angiogenesis, better granulation, tissue growth, and faster wound closure [124]. Rodent models represent a favored system in MSC-based wound healing research because of their easy availability and low expenses. The development of animal models that precisely replicate human wound healing processes should serve as the priority for future research since models based on porcine skin wounds have proven effective [124].

A study was conducted using *in vivo* mouse showed that immature adipocytes start to activate during the proliferative phase jointly with mature adipocytes and fibroblast movement. The wounds of lipoatrophic mice heal differently than normal mice because it appears that adipocytes serve as fundamental components of wound repair communications between cells to drive both fibroblast movement and their operational capacity. Multiple research findings have established that AD-MSCs accelerate the process of re-epithelialization while decreasing wound sizes during the healing process (Table 9.1) [125].

9.8.2 Status of Ongoing Clinical Trials

Table 9.1 Clinical trials related to stem cell therapy in treatment of various skin conditions.

Title	Study design	Phase	Description	Status	Outcome	Reference
ASCs to Treat Chronic Wounds (NCT05797688)	Open-label	Phase 1	Ten patients undergo standard wound debridement with concurrent fat harvest subcutaneously into the wound site. Patients were monitored weekly for 24 weeks, but after 3 months if wound size does not decrease in size by at least 40%, a maximum of 2 cycles of fat harvest and cell injection may be given.	Withdrawn	It determines the use of ASCs in treating chronic wounds.	[126]
Healing Chronic Venous Stasis Wounds With Autologous Cell Therapy (or SVF) (NCT02961699)	Prospective, randomized, single site	NA	Subjects were classified into two groups: treatment subjects (24 patients) and control subjects (12 patients). The treatment group underwent a small liposuction procedure, whereas the control group received debridement and normal dressing changes as per the standard of care.	Unknown status	Assessed the efficacy of SVF therapy in improving tissue healing process	[127]

(*Continued*)

Table 9.1 Clinical trials related to stem cell therapy in treatment of various skin conditions. (*Continued*)

Title	Study design	Phase	Description	Status	Outcome	Reference
Adipose Derived Regenerative Cellular Therapy of Chronic Wounds (NCT02092870)	Open-label	Phase 2	Patient received a single treatment with ASCs through multiple cellular injections given within and around the wound. The volume for each injection remained below 250 μL.	Completed	Evaluated the safety of ASC therapy in healing wound by measuring the percent change in wound size	[128]
Rising Tide—Amniotic Tissue(s) Treatments for Chronic Diabetic Foot Ulcers (NCT06681428)	Randomized controlled clinical trial	NA	Placental allograft is applied topically on to the open surface of chronic diabetic foot ulcer, followed by dressing as per standard of care. Outcome of the treatment was measured by percentage of index ulcers healed at 12 weeks.	Ongoing	Evaluated the efficacy of amnion/chorion allograft in treatment of chronic diabetic foot ulcers	[129]

(*Continued*)

Table 9.1 Clinical trials related to stem cell therapy in treatment of various skin conditions. *(Continued)*

Title	Study design	Phase	Description	Status	Outcome	Reference
Stem Cell Transplant to Treat Patients With Systemic Sclerosis (NCT00058578)	Open-label	Phase 1	Before the transplant, the patient received daily injections of G-CSF (Neupogen) for 5–6 days to stimulate the production of WBCs. After collection of WBCs, special agents called monoclonal antibodies are used to select $CD34^+$ stem cells from collected WBCs. Finally, the collected $CD34^+$ cells will then be infused into the patient. After transplant, patients were monitored closely once a year for 2 years.	Completed	Elucidated the efficacy and safety of autologous T-cell depleted HSC transplantation in patients with systemic sclerosis	[130]

(Continued)

Table 9.1 Clinical trials related to stem cell therapy in treatment of various skin conditions. (*Continued*)

Title	Study design	Phase	Description	Status	Outcome	Reference
Allogeneic Hematopoietic Cell Transplantation for Severe Systemic Sclerosis (NCT00622895)	Open-label	Phases 1 and 2	Patients were given chemotherapy and low-dose radiation regimen, including cyclophosphamide given on first and second day intravenously, and fludarabine given intravenously for 5 days. After completing this, patients exposed to a single low dose of total body irradiation. The next day, patients received the allogenic bone marrow transplant, followed by high dose of cyclophosphamide on third and fourth days to prevent graft rejection and graft-versus-host disease. On the fifth day, patients received tacrolimus for 6 months, mycophenolic acid for about 5 weeks, and daily injections of a growth factor called G-CSF to increase the count of WBCs until normal levels are reached.	Completed	It determined potential safety of reduced-intensity conditioning followed by allogenic hematopoietic cell transplantation for patients with severe systemic sclerosis.	[131]

(*Continued*)

Table 9.1 Clinical trials related to stem cell therapy in treatment of various skin conditions. (*Continued*)

Title	Study design	Phase	Description	Status	Outcome	Reference
Pilot Trial of Mesenchymal Stem Cells for Systemic Lupus Erythematosus (NCT03171194)	Open-label	Phase 1	All the participants received infusion, usually in arm or hand, followed by receiving standard of care and their safety will be monitored throughout the study.	Completed	It provided the valuable safety data of using umbilical cord–derived MSCs in SLE patients.	[132]
Stem Cell Therapy to Improve Burn Wound Healing (NCT02104713)	Open-label	Phase 1	The first group containing five subjects will receive the minimum dose of allogenic MSCs, whereas the second group of five participants will receive a higher dose, and this will be repeated for the third and fourth receiving higher dose.	Completed	It determined the safety of allogeneic stem cell therapy from healthy donors.	[133]

(*Continued*)

Table 9.1 Clinical trials related to stem cell therapy in treatment of various skin conditions. (*Continued*)

Title	Study design	Phase	Description	Status	Outcome	Reference
Clinical Application of Stem Cell Educator Therapy in Alopecia Areata (NCT04011748)	Randomized, parallel, and double-label trial	Phase 2	The therapeutic potential of stem cell educator therapy for the treatment of AA by using topical minoxidil (5%) as control. Primary endpoint of the treatment was % change in scalp hair growth, which is measured with Severity of Alopecia Tool score.	Unknown status	Using topical minoxidil as control, this trial assessed the potential of stem cell therapy for the treatment of alopecia.	[134]
Safety Study of Bone Marrow Derived Stem Cells on Patients With Cutaneous Photoaging (NCT01771679)	Open-label	Phases 1 and 2	This study determines whether a one-time injection of allogeneic mesenchymal bone marrow cells is safe and beneficial in the treatment of photodamage on the face.	Unknown status	It determined potential use of allogeneic mesenchymal bone marrow cells for treating photodamaged skin.	[135]

9.8.3 Case Studies and Success Stories

A 19-year-old male college student reported a 5-year history of severe plaque psoriasis that appeared demarcated, round, hyperpigmented lesions covered by thick white scales distributed throughout his entire body. The patient also experienced itching and declared the pain intolerable. Due to this, patient also suffered from depression and had suicidal thoughts. The patient had the medical history of receiving several systemic treatments including methotrexate, acitretin, cyclosporine, and etanercept without beneficial results. Therefore, allogenic MSC-based therapy was offered as an option based on earlier reports of success. The procedure for isolation and expansion of MSCs followed the methodology outlined by Mitrano *et al* [136]. A biopsy of gingival connective tissue was used from a healthy adult donor. The tissue was minced for explant culture. The resulting cells were cultured to increase the numbers for 3- to 4-week *in vitro* conditions, followed by cryopreservation. During treatment days, MSCs were thawed and made a suspension in 5 mL normal saline solution and given as intravenous bolus. The patient received two successive weekly administrations of MSCs (3×10^6/kg per infusion). The patient experienced no adverse reactions along with improvement of plaques. After 5 weeks, three more MSC infusions were given to the patient weekly. Complete remission of psoriatic lesions became visible exactly 1 week following the final medication administration. Three years have now passed since the patient successfully recovered from psoriasis attacks [137].

A study was conducted by Hong *et al* to evaluate the therapeutic effects of culture pure melanocytes in 102 patients suffering from vitiligo. Of 102 patients, 12 had generalized vitiligo, and 90 had localized vitiligo. Vitiliginous lesions treated were located on the face (n = 26), neck (n = 16), neck (n = 16), trunk (n = 29), arm and leg (n = 12), and hand and foot (n = 23). Before 2 to 4 hours of transplantation, lidocaine cream was topically applied, followed by removing the epidermis of the affected area by cleaning with 70% alcohol and treated with ultrapulse CO_2 laser (pulse rate = 30-50 Hz at an energy level of 225 mJ per pulse). Lastly, melanocyte suspension was applied to the laser denuded area at a density of 600 to 1000 cells/mm^2. Vaseline gauze was used to cover the transplant area, followed by soaking the gauze with F12 medium and finally securing with surgical tape. After 10 days, dressings were removed. The results showed 86% of the patients had a satisfactory result with repigmentation of 50% or more [138].

A 38-year-old male patient was suffering from psoriasis vulgaris for 2 years. Physical examination showed numerous erythematous plaques with numerous silvery scales present all over the scalp along with behind the ears. With the help of Psoriasis Scalp Severity Index, severity of the disease was assessed, and it was found to be 28. Adipose tissue was collected by a procedure called lipoaspiration under aseptic conditions from a healthy volunteer. MSCs were isolated from adipose tissue by standard enzymatic digestion method with 0.1% collagenase type I. The concentrated MSC-CM was aliquoted and stored at −20°C until use. MSC-CM was topically applied on the affected areas once daily for over a period of 1 month. It was observed that the number of scales was reduced after 2 weeks of topical application of MSC-CM, ultimately improving the quality of life [139].

A 45-year-old female patient reported a 5-month history of alopecia areata. The preliminary examination of scalp showed presence of small bare nonscarring patch in the vertex area. Using the Severity of Alopecia Tool (SALT) score, severity of the disease was assessed, and it was found to be 6%. The patient complained of mild anxiety and nervousness due to persistent disease. Adipose tissue (100 mL) was collected from the waist area of the patient by lipoaspiration under aseptic environment. MSCs were isolated from adipose tissue by mechanical agitation including vertexing. The 30% MSC-CM in cream base was topically applied on the affected area of scalp daily for a period of 2 months. During this, patient did not receive any other treatment for alopecia. The patient was continuously monitored at 1 and 3 months after first application of MSC-CM for any adverse events, and no any events were reported. The result was that after 3 months, there was visible increase in hair count and density in the affected scalp region along with marked reduction in circumference of hairless patch. The SALT score was also reduced to approximately 0% with continued improvement. After hair growth, patient was out of alopecia-related anxiety alongside boost confidence level and improved quality of life [140].

A patient aged 67 years complained of chronic ischemic foot ulcer on the right toe second metatarsal for approximately 8 to 10 years. She was not able to walk and sleep due to intense pain. The patient had a history of smoking and hypertension that contributed as a risk factor for vascular disease, whereas she reported no past trauma to her second toe. The arteriogram scan showed that none of the right foot viable arteries could support for surgical or percutaneous revascularization procedures. Therefore, patient chose to receive own stem cells to regenerate ischemic tissues. The medical team performed bone marrow aspiration from the patient's iliac crest under general anesthesia to retrieve 500 mL of the sample. The mononuclear cells were isolated by running density gradient centrifugation followed by concentrating into 30 mL. The cell suspension was administered into gastrocnemius muscle of the affected leg skeletal muscle at 41 specific points. The patient was kept on continuous 12-week observations, which included clinical check-ups, arteriograms, and noninvasive vascular studies. The patient experienced significant improvements such as reduction in pain, enhanced mobility, healing of ulcer, and improved vascular indicators. Posttreatment arteriograms revealed enhanced circulation along with doubled ankle-brachial index from baseline (0.4 to 0.79). Furthermore, toe pressures and transcutaneous oxygen pressure measurement also improved, leading to better tissue perfusion [141].

9.9 Regulatory and Ethical Considerations

The regulatory framework for stem cell therapy is continuously growing to fulfill emerging requirements for standardized practices and patient safety concerns. Global regulatory bodies such as US Food and Drug Administration (FDA), European Medicines Agency (EMA), Japan's Pharmaceuticals and Medical Devices Agency, and Central Drug Standard Control Organization (CDSCO) of India look after the development and marketing authorization of stem cell–based treatments. The processing of stem cells derived from cord blood for unrelated allogeneic purposes needs approval from the FDA through licensing for their distribution. The FDA demands manufacturers to present their production methods through an evaluation procedure enabling the agency to verify manufacturing steps that guarantee purity and potency standards [142]. Regulatory guidelines for stem cell therapy are different. For example in the United States, the FDA mandates the Regenerative Medicine Advance Therapy designation under the 21st Century Cures Act, requiring biologics licensing for allogeneic stem cell products. In the European Union, the EMA requires approval for Advanced Therapy Medicinal Product, such as gene/cell therapies under regulation (EC) no. 1394/2007 [143]. In Japan, conditional approvals for regenerative therapies are governed under the PMD (Pharmaceuticals and Medical Devices) Act, whereas ASRM (The Act on the Safety of Regenerative Medicine) Act governs clinical trials for minimally manipulated cells, and in India, the CDSCO requires an IND application for clinical trials as per the New Drugs and Clinical Trials Rules [144]. The ethical dilemmas related to stem cell use are complex with multiple facets, reflecting deep-seated fundamental moral values, scientific and technological advancements, and societal implications. However, countries such as Germany prohibit research of HESCs, whereas others such as the United States and the United Kingdom allow the research under strict oversight, thus creating global disparities in therapeutic development [145].

9.10 Future Directions

The use of stem cell therapy promises to transform the treatment of many skin disorders and is positioned on the eve of radical transformation for the field of dermatology [146]. Patient-derived cells could be reprogrammed by means of advanced techniques of cell engineering, such as iPSCs, which limit the immunological rejection and moral dilemma but do not provide individualized therapy [147]. Advancements, such as 3D bioprinting, allow the creation of detailed skin grafts that closely mirror the architecture of real skin and improve the results of wound healing [148]. Gene-edited stem cells, using CRISPR/Cas9 techniques, offer the possibility of curing hereditary skin problems, but targeted delivery systems using nanotechnology can enhance the precision of stem cell therapies [149]. Even with all this, much work remains to be done. Standardization of the procedures will need to ensure consistent treatment results in clinical applications. In miscellaneous circumstances, the determination of biomarkers for patient classification could improve therapy accuracy further [50]. Top priority needs to be given to well-planned phase III studies for ascertaining the safety and efficacy of MSC-based treatments to be utilized in burns and chronic wounds [150]. The proper direction of research efforts requires knowledge gaps; for instance, women's health and diabetic retinopathy research focuses on the need for focused investigations to answer pressing clinical problems [151]. As the field progresses, it will be essential that developments in stem cell research yield affordable treatments satisfying the requirements of many patient groups, ultimately altering the face of contemporary medicine and closing critical gaps in healthcare.

Better integration of research findings into routine practice could be facilitated through using technologies such as telemedicine, improving training programs for medical personnel, and implementing evidence-based methods into practice. Through these interlinked areas, the healthcare system can improve delivery and responsiveness of patient care toward changing requirements in society [152]. For example, laboratory techniques frequently need to be modified to properly match clinical situations, even if they may show effectiveness in controlled conditions [153].

Ultimately, developing a learning healthcare system that prioritizes both practice and research would make it possible to translate scientific discoveries into real health advantages in a more effective way for patients [154].

9.11 Conclusion

Patient protection and social justice are prioritized when bioethical issues are shifted from embryo status to all stem cell applications. Public benefits, a fundamental ethical tenet of stem cell research, may be restricted by overcommercialization and restrictive patents. Standardized testing, ethical regulatory frameworks, and better donor acquisition are necessary to advance stem cell–based dermatological treatments. New technologies such as 3D bioprinting and gene editing will increase the accessibility and uptake of treatments. Global dermatological care will be transformed by collaborative research initiatives that will propel patient-specific, minimally invasive medicines.

AI disclosure statement: The author confirms that no artificial intelligence (AI)-generated content or AI-assisted writing tools were used in the preparation of this chapter (including summaries and the abstract). All text, analysis, and conclusions are the result of the author's original work.

References

1. Ramalho-Santos, M. and Willenbring, H., On the origin of the term "stem cell." *Cell Stem Cell [Internet]*, 1, 1, 35–8, 2007 Jun 7 [cited 2025 Jan 29]. Available from: https://pubmed.ncbi.nlm.nih.gov/18371332/.
2. Schofield, R., The relationship between the spleen colony-forming cell and the haemopoietic stem cell. *Blood Cells*, 1–2, 7–25, 1978, europepmc.org [Internet]. [cited 2025 Jan 29]; Available from: https://europepmc.org/article/med/747780.
3. Herzog, E.L., Chai, L., Krause, D.S., Plasticity of marrow-derived stem cells. *Blood*, 102, 10, 3483–93, 2003 Nov 15.
4. Monti, M., Perotti, C., del Fante, C., Cervio, M., Redi, C.A., Stem cells: sources and therapies. *Biol. Res. [Internet]*, 45, 3, 207–14, 2012 [cited 2025 Jan 29]. Available from: https://pubmed.ncbi.nlm.nih.gov/23283430/.
5. Thomas, E.D., Lochte, H.L., Lu, W.C., Ferrebee, J.W., Intravenous infusion of bone marrow in patients receiving radiation and chemotherapy. *N. Engl. J. Med. [Internet]*, 257, 11, 491–6, 1957 Sep 12 [cited 2025 Jan 29]. Available from: https://pubmed.ncbi.nlm.nih.gov/13464965/.

6. T, J.A., IE, J., S, S.S., W, M.A., S, J.J., M, V.S., *et al.*, Embryonic stem cell lines derived from human blastocysts. *Science [Internet]*, 282, 5391, 1145–7, 1998 Nov 6 [cited 2025 Jan 29]. Available from: https://pubmed.ncbi.nlm.nih.gov/9804556/.
7. Takahashi, K. and Yamanaka, S., Induction of pluripotent stem cells from mouse embryonic and adult fibroblast cultures by defined factors. *Cell [Internet]*, 126, 4, 663–76, 2006 Aug 25 [cited 2025 Jan 29]. Available from: https://pubmed.ncbi.nlm.nih.gov/16904174/.
8. Williams, H.C., Strengths and limitations of evidence-based dermatology. *Indian J. Dermatol.*, 59, 2, 127–33, 2014 Mar.
9. Chen, M., Przyborowski, M., Berthiaume, F., Stem Cells for Skin Tissue Engineering and Wound Healing. *Crit. Rev. Biomed. Eng.*, 37, 4–5, 399–421, 2009.
10. Chang, W.L., Lee, W.R., Kuo, Y.C., Huang, Y.H., Vitiligo: An Autoimmune Skin Disease and its Immunomodulatory Therapeutic Intervention. *Front. Cell Dev. Biol.*, 9, 797026, 2021 Dec 14.
11. Yu, Q., Qiao, G., Wang, M., Yu, L., Sun, Y., Shi, H., *et al.*, Stem Cell-Based Therapy for Diabetic Foot Ulcers. *Front. Cell Dev. Biol.*, 10, 812262, 2022 Feb 1.
12. Gasteratos, K., Kouzounis, K., Goverman, J., Autologous Stem Cell-derived Therapies for Androgenetic Alopecia: A Systematic Review of Randomized Control Trials on Efficacy, Safety, and Outcomes. *Plast. Reconstr. Surg. Glob. Open [Internet]*, 12, 2, e5606, 2024 Feb 13 [cited 2025 Feb 20]. Available from: https://pmc.ncbi.nlm.nih.gov/articles/PMC10863936/.
13. Dutra Alves, N.S., Reigado, G.R., Santos, M., Caldeira, I.D.S., Hernandes H dos, S., Freitas-Marchi, B.L., *et al.*, Advances in regenerative medicine-based approaches for skin regeneration and rejuvenation. *Front. Bioeng. Biotechnol.*, 13, 1527854, 2025 Feb 12.
14. Rippon, H.J. and Bishop, A.E., Embryonic stem cells. *Cell Proliferation [Internet]*, 37, 1, 23–34, 2004 Feb 1 [cited 2025 Jan 29]. Available from: https://onlinelibrary.wiley.com/doi/full/10.1111/j.1365-2184.2004.00298.x.
15. Okita, K. and Yamanaka, S., Induced pluripotent stem cells: opportunities and challenges. *Philos. Trans. R. Soc., B [Internet]*, 366, 1575, 2198–207, 2011 Jul 4 [cited 2025 Jan 29]. Available from: https://royalsocietypublishing.org/doi/10.1098/rstb.2011.0016.
16. Eaves, C.J., Hematopoietic stem cells: concepts, definitions, and the new reality. *Blood [Internet]*, 125, 17, 2605–13, 2015 Apr 23 [cited 2025 Jan 29]. Available from: https://dx.doi.org/10.1182/blood-2014-12-570200.
17. Toma, J.G., McKenzie, I.A., Bagli, D., Miller, F.D., Isolation and Characterization of Multipotent Skin-Derived Precursors from Human Skin. *Stem Cells* [Internet], 23, 6, 727–37, 2005 Jun 1 [cited 2025 Jan 29]. Available from: https://dx.doi.org/10.1634/stemcells.2004-0134.

18. Arias, A.M., Drosophila melanogaster and the development of biology in the 20th century. *Methods Mol. Biol.* [Internet], 420, 1–25, 2008 [cited 2025 Jan 29]. Available from: https://pubmed.ncbi.nlm.nih.gov/18641938/.
19. Takahashi, K., Tanabe, K., Ohnuki, M., Narita, M., Ichisaka, T., Tomoda, K., *et al.*, Induction of pluripotent stem cells from adult human fibroblasts by defined factors. *Cell* [Internet], 131, 5, 861–72, 2007 Nov 30 [cited 2025 Jan 29]. Available from: https://pubmed.ncbi.nlm.nih.gov/18035408/.
20. Chu, G.Y., Chen, Y.F., Chen, H.Y., Chan, M.H., Gau, C.S., Weng, S.M., Stem cell therapy on skin: Mechanisms, recent advances and drug reviewing issues. *J. Food Drug Anal.* [Internet], 26, 1, 14–20, 2018 Jan 1 [cited 2025 Jan 29]. Available from: https://pubmed.ncbi.nlm.nih.gov/29389549/.
21. Senoo, M., Pinto, F., Crum, C.P., McKeon, F., p63 Is essential for the proliferative potential of stem cells in stratified epithelia. *Cell* [Internet], 129, 3, 523–36, 2007 May 4 [cited 2025 Jan 29]. Available from: https://pubmed.ncbi.nlm.nih.gov/17482546/.
22. Garzón, I., Miyake, J., González-Andrades, M., Carmona, R., Carda, C., Del, M., *et al.*, Wharton's Jelly Stem Cells: A Novel Cell Source for Oral Mucosa and Skin Epithelia Regeneration. *Stem Cells Transl. Med.* [Internet], 2, 8, 625, 2013 Aug 1 [cited 2025 Jan 29]. Available from: https://pmc.ncbi.nlm.nih.gov/articles/PMC3726142/.
23. Nowak, J.A., Polak, L., Pasolli, H.A., Fuchs, E., Hair follicle stem cells are specified and function in early skin morphogenesis. *Cell Stem Cell* [Internet], 3, 1, 33–43, 2008 Jul 3 [cited 2025 Jan 29]. Available from: https://pubmed.ncbi.nlm.nih.gov/18593557/.
24. Nishimura, E.K., Jordan, S.A., Oshima, H., Yoshida, H., Osawa, M., Moriyama, M., *et al.*, Dominant role of the niche in melanocyte stem-cell fate determination. *Nature* [Internet], 416, 6883, 854–60, 2002 Apr 25 [cited 2025 Jan 29]. Available from: https://pubmed.ncbi.nlm.nih.gov/11976685/.
25. Horsley, V., O'Carroll, D., Tooze, R., Ohinata, Y., Saitou, M., Obukhanych, T., *et al.*, Blimp1 defines a progenitor population that governs cellular input to the sebaceous gland. *Cell* [Internet], 126, 3, 597–609, 2006 Aug 4 [cited 2025 Jan 29]. Available from: https://pubmed.ncbi.nlm.nih.gov/16901790/.
26. Klassen, H., Schwartz, M.R., Bailey, A.H., Young, M.J., Surface markers expressed by multipotent human and mouse neural progenitor cells include tetraspanins and non-protein epitopes. *Neurosci. Lett.*, 312, 3, 180–2, 2001 Oct 26.
27. Taussig, D.C., Pearce, D.J., Simpson, C., Rohatiner, A.Z., Lister, T.A., Kelly, G., *et al.*, Hematopoietic stem cells express multiple myeloid markers: implications for the origin and targeted therapy of acute myeloid leukemia. *Blood* [Internet], 106, 13, 4086–92, 2005 Dec 15 [cited 2025 Jan 29]. Available from: https://dx.doi.org/10.1182/blood-2005-03-1072.
28. Fuchs, E., Skin stem cells: rising to the surface. *J. Cell Biol.* [Internet], 180, 2, 273–84, 2008 Jan 28 [cited 2025 Jan 29]. Available from: https://pubmed.ncbi.nlm.nih.gov/18209104/.

29. Ito, M., Liu, Y., Yang, Z., Nguyen, J., Liang, F., Morris, R.J., *et al.*, Stem cells in the hair follicle bulge contribute to wound repair but not to homeostasis of the epidermis. *Nat. Med.* [Internet], 11, 12, 1351–4, 2005 Dec [cited 2025 Jan 29]. Available from: https://pubmed.ncbi.nlm.nih.gov/16288281/.
30. Taylor, G., Lehrer, M.S., Jensen, P.J., Sun, T.T., Lavker, R.M., Involvement of follicular stem cells in forming not only the follicle but also the epidermis. *Cell* [Internet], 102, 4, 451–61, 2000 Aug 18 [cited 2025 Jan 29]. Available from: https://pubmed.ncbi.nlm.nih.gov/10966107/.
31. Langton, A.K., Herrick, S.E., Headon, D.J., An extended epidermal response heals cutaneous wounds in the absence of a hair follicle stem cell contribution. *J. Invest. Dermatol.* [Internet], 128, 5, 1311–8, 2008 [cited 2025 Jan 29]. Available from: https://pubmed.ncbi.nlm.nih.gov/18037901/.
32. Sid-Otmane, C., Perrault, L.P., Ly, H.Q., Mesenchymal stem cell mediates cardiac repair through autocrine, paracrine and endocrine axes. *J. Transl. Med.*, 18, 1, 336, 2020 Dec 1.
33. Bao, P., Kodra, A., Tomic-Canic, M., Golinko, M.S., Ehrlich, H.P., Brem, H., The Role of Vascular Endothelial Growth Factor in Wound Healing. *J. Surg. Res.*, 153, 2, 347–58, 2009 May.
34. Valluru, M., Staton, C.A., Reed, M.W.R., Brown, N.J., Transforming Growth Factor-? and Endoglin Signaling Orchestrate Wound Healing. *Front. Physiol.*, 2, 89, 2011.
35. Liu, Z., Yang, J., Chen, Y., Chen, C., Wang, J., Lee, Y.M., *et al.*, P311 Facilitates the Angiogenesis and Wound Healing Function of MSCs by Increasing VEGF Production. *Front. Immunol.*, 13, 821932, 2022 Jan 28.
36. Ribeiro, A., Laranjeira, P., Mendes, S., Velada, I., Leite, C., Andrade, P., *et al.*, Mesenchymal stem cells from umbilical cord matrix, adipose tissue and bone marrow exhibit different capability to suppress peripheral blood B, natural killer and T cells. *Stem Cell Res. Ther.*, 4, 5, 125, 2013 Oct 15.
37. Strioga, M., Viswanathan, S., Darinskas, A., Slaby, O., Michalek, J., Same or Not the Same? Comparison of Adipose Tissue-Derived Versus Bone Marrow-Derived Mesenchymal Stem and Stromal Cells. *Stem Cells Dev.*, 21, 14, 2724–52, 2012 Sep 20.
38. Hoang, D.M., Pham, P.T., Bach, T.Q., Ngo, A.T.L., Nguyen, Q.T., Phan, T.T.K., *et al.*, Stem cell-based therapy for human diseases. *Signal Transduction Targeted Ther.*, 7, 1, 272, 2022 Aug 6.
39. Al-Ghadban, S. and Bunnell, B.A., Adipose tissue-derived stem cells: Immunomodulatory effects and therapeutic potential. *Physiology* [Internet], 35, 2, 125–33, 2020 [cited 2025 Jan 30]. Available from: https://journals.physiology.org/doi/10.1152/physiol.00021.2019.
40. Kim, D.J., Mustoe, T., Clark, R.A., Cutaneous wound healing in aging small mammals: a systematic review. *Wound Repair Regener.* [Internet], 23, 3, 318–39, 2015 May 1 [cited 2025 Jan 30]. Available from: https://onlinelibrary.wiley.com/doi/full/10.1111/wrr.12290.

41. Murthy, S., Gautam, M.K., Goel, S., Purohit, V., Sharma, H., Goel, R.K., Evaluation of *in vivo* wound healing activity of Bacopa monniera on different wound model in rats. *Biomed. Res. Int.* [Internet], 2013, 1, 972028, 2013 [cited 2025 Jan 30]. Available from: https://pubmed.ncbi.nlm.nih.gov/23984424/.
42. Diegelmann, R.F. and Evans, M.C., Wound healing: an overview of acute, fibrotic and delayed healing. *Front. Biosci.* [Internet], 9, 283–9, 2004 [cited 2025 Jan 30]. Available from: https://pubmed.ncbi.nlm.nih.gov/14766366/.
43. Sinno, H. and Prakash, S., Complements and the wound healing cascade: an updated review. *Plast. Surg. Int.* [Internet], 2013, 1–7, 2013 Jul 24 [cited 2025 Jan 30]. Available from: https://pubmed.ncbi.nlm.nih.gov/23984063/.
44. Ayavoo, T., Murugesan, K., Gnanasekaran, A., Roles and mechanisms of stem cell in wound healing. *Stem Cell Investig.* [Internet], 8, 4, 2021 Mar 2 [cited 2025 Jan 30]. Available from: https://pmc.ncbi.nlm.nih.gov/articles/PMC8022285/.
45. Spence, S. and Gallicchio, V.S., Mesenchymal Stem Cell Therapy For Chronic Wound Healing. *J. Stem Cell Res. (Overl Park)*, 3, 1, 1, 2022 Jan 3.
46. Kopan, R. and Ilagan, M.X.G., The canonical Notch signaling pathway: unfolding the activation mechanism. *Cell [Internet]*, 2, 216–33, 2009, cell.com, [cited 2025 Jan 31]. Available from: https://www.cell.com/cell/fulltext/S0092-8674(09)00382-1?cc=y.
47. Otero-Viñas, M. and Falanga, V., Mesenchymal Stem Cells in Chronic Wounds: The Spectrum from Basic to Advanced Therapy. *Adv. Wound Care (New Rochelle) [Internet]*, 5, 4, 149, 2016 Apr 1 [cited 2025 Jan 31]. Available from: https://pmc.ncbi.nlm.nih.gov/articles/PMC4817558/.
48. Cheng, L., Wang, S., Peng, C., Zou, X., Yang, C., Mei, H., *et al.*, Human umbilical cord mesenchymal stem cells for psoriasis: a phase 1/2a, single-arm study. *Signal Transduction Targeted Ther.* [Internet], 7, 1, 1–11, 2022 Aug 5 [cited 2025 Feb 20]. Available from: https://www.nature.com/articles/s41392-022-01059-y.
49. An, L., Kim, D., Donahue, L.R., Mejooli, M.A., Eom, C.Y., Nishimura, N., *et al.*, Sexual dimorphism in melanocyte stem cell behavior reveals combinational therapeutic strategies for cutaneous repigmentation. *Nat. Commun.* [Internet], 15, 1, 1–18, 2024 Jan 27 [cited 2025 Feb 20]. Available from: https://www.nature.com/articles/s41467-024-45034-3.
50. Chen, L., Huang, R., Huang, C., Nong, G., Mo, Y., Ye, L., *et al.*, Cell therapy for scleroderma: progress in mesenchymal stem cells and CAR-T treatment. *Front. Med. (Lausanne)*, 11, 1530887, 2025 Jan 15.
51. Vanikar, A.V., Trivedi, H.L., Patel, R.D., Kanodia, K.V., Modi, P.R., Shah, V.R., Allogenic Hematopoietic Stem Cell Transplantation in Pemphigus Vulgaris: A Single-Center Experience. *Indian J. Dermatol. [Internet]*, 57, 1, 9, 2012 Jan [cited 2025 Mar 4]. Available from: https://pmc.ncbi.nlm.nih.gov/articles/PMC3312672/.
52. Li, A., Guo, F., Pan, Q., Chen, S., Chen, J., Liu, H.F., *et al.*, Mesenchymal Stem Cell Therapy: Hope for Patients With Systemic Lupus Erythematosus. *Front.*

Immunol. [Internet], 12, 728190, 2021 Sep 30 [cited 2025 Feb 20]. Available from: https://pubmed.ncbi.nlm.nih.gov/34659214/.

53. Wang, T., Li, Y., Zhu, Y., Liu, Z., Huang, L., Zhao, H., *et al.*, Anti-aging mechanism of different age donor-matched adipose-derived stem cells. *Stem Cell Res. Ther. [Internet]*, 14, 1, 192, 2023 Dec 1 [cited 2025 Jan 31]. Available from: https://pmc.ncbi.nlm.nih.gov/articles/PMC10394785/.
54. Ding, D.C., Shyu, W.C., Lin, S.Z., Mesenchymal stem cells. *Cell Transplant [Internet]*, 20, 1, 5–14, 2011 [cited 2025 Jan 31]. Available from: https://pubmed.ncbi.nlm.nih.gov/21396235/.
55. El Assaad, N., Chebly, A., Salame, R., Achkar, R., Atme, N.B., Akouch, K., *et al.*, Anti-aging based on stem cell therapy: A scoping review. *World J. Exp. Med. [Internet]*, 14, 3, 97233, 2024 Sep 20 [cited 2025 Jan 31]. Available from: https://pmc.ncbi.nlm.nih.gov/articles/PMC11372738/.
56. Chou, Y., Alfarafisa, N.M., Ikezawa, M., Khairani, A.F., Progress in the Development of Stem Cell-Derived Cell-Free Therapies for Skin Aging. *Clin. Cosmet. Investig. Dermatol. [Internet]*, 16, 3383, 2023 [cited 2025 Jan 31]. Available from: https://pmc.ncbi.nlm.nih.gov/articles/PMC10676866/.
57. Chandrashekar, B., Nandhini, T., Vasanth, V., Sriram, R., Navale, S., Topical minoxidil fortified with finasteride: An account of maintenance of hair density after replacing oral finasteride. *Indian Dermatol. Online J. [Internet]*, 6, 1, 17, 2015 [cited 2025 Jan 31]. Available from: https://pubmed.ncbi.nlm.nih.gov/25657911/.
58. Talebzadeh, A.T. and Talebzadeh, N., Stem Cell Applications in Human Hair Growth: A Literature Review. *Cureus* [Internet], 15, 4, e37439, 2023 Apr 11 [cited 2025 Jan 31]. Available from: https://pmc.ncbi.nlm.nih.gov/articles/PMC10174680/.
59. Ullah, S., Mansoor, S., Ayub, A., Ejaz, M., Zafar, H., Feroz, F., *et al.*, An update on stem cells applications in burn wound healing. *Tissue Cell*, 72, 101527, 2021 Oct 1.
60. Nourian Dehkordi, A., Mirahmadi Babaheydari, F., Chehelgerdi, M., Raeisi Dehkordi, S., Skin tissue engineering: wound healing based on stem-cell-based therapeutic strategies. *Stem Cell Res. Ther.* [Internet], 10, 1, 1–20, 2019 Mar 29 [cited 2025 Jan 31]. Available from: https://pubmed.ncbi.nlm.nih.gov/30922387/.
61. Sharma, R.K. and John, J.R., Role of stem cells in the management of chronic wounds. *Indian J. Plast. Surg.* [Internet], 45, 2, 237–43, 2012 May [cited 2025 Jan 31]. Available from: https://pubmed.ncbi.nlm.nih.gov/23162222/.
62. Liu, J., Ren, J., Su, L., Cheng, S., Zhou, J., Ye, X., *et al.*, Human adipose tissue-derived stem cells inhibit the activity of keloid fibroblasts and fibrosis in a keloid model by paracrine signaling. *Burns* [Internet], 44, 2, 370–85, 2018 Mar 1 [cited 2025 Jan 31]. Available from: https://pubmed.ncbi.nlm.nih.gov/29029852/.

63. Kareem, N.A., Aijaz, A., Jeschke, M.G., Stem Cell Therapy for Burns: Story so Far. *Biologics* [Internet], 15, 379, 2021 [cited 2025 Jan 31]. Available from: https://pmc.ncbi.nlm.nih.gov/articles/PMC8418374/.
64. Miguel, S.P., Moreira, A.F., Correia, I.J., Chitosan based-asymmetric membranes for wound healing: A review. *Int. J. Biol. Macromol.* [Internet], 127, 460–75, 2019 Apr 15 [cited 2025 Jan 31]. Available from: https://pubmed.ncbi.nlm.nih.gov/30660567/.
65. Kathawala, M.H., Ng, W.L., Liu, D., Naing, M.W., Yeong, W.Y., Spiller, K.L., *et al.*, Healing of Chronic Wounds: An Update of Recent Developments and Future Possibilities. *Tissue Eng. Part B Rev.* [Internet], 25, 5, 429–44, 2019 Oct 1 [cited 2025 Jan 31]. Available from: https://pubmed.ncbi.nlm.nih.gov/31068101/.
66. Koch, L., Kuhn, S., Sorg, H., Gruene, M., Schlie, S., Gaebel, R., *et al.*, Laser printing of skin cells and human stem cells. *Tissue Eng. Part C Methods* [Internet], 16, 5, 847–54, 2010 Oct 1 [cited 2025 Jan 31]. Available from: https://pubmed.ncbi.nlm.nih.gov/19883209/.
67. Matai, I., Kaur, G., Seyedsalehi, A., McClinton, A., Laurencin, C.T., Progress in 3D bioprinting technology for tissue/organ regenerative engineering. *Biomaterials* [Internet], 226, 119536, 2020 Jan 1 [cited 2025 Jan 31]. Available from: https://pubmed.ncbi.nlm.nih.gov/31648135/.
68. Fayyazbakhsh, F., Leu MC. A Brief Review on 3D Bioprinted Skin Substitutes. *Procedia Manuf.*, 48, 790–6, 2020 Jan 1.
69. Formigli, L., Paternostro, F., Tani, A., Mirabella, C., Quattrini Li, A., Nosi, D., *et al.*, MSCs seeded on bioengineered scaffolds improve skin wound healing in rats. *Wound Repair Regener.* [Internet], 23, 1, 115–23, 2015 Jan 1 [cited 2025 Mar 4]. Available from: https://pubmed.ncbi.nlm.nih.gov/25571903/.
70. Sivaraj, D., Chen, K., Chattopadhyay, A., Henn, D., Wu, W., Noishiki, C., *et al.*, Hydrogel Scaffolds to Deliver Cell Therapies for Wound Healing. *Front. Bioeng. Biotechnol.* [Internet], 9, 660145, 2021 May 3 [cited 2025 Mar 4]. Available from: https://pmc.ncbi.nlm.nih.gov/articles/PMC8126987/.
71. Zhou, S., Wang, Q., Huang, A., Fan, H., Yan, S., Zhang, Q., Advances in Skin Wound and Scar Repair by Polymer Scaffolds. *Molecules* [Internet], 26, 20, 6110, 2021 Oct 1 [cited 2025 Jan 31]. Available from: https://pmc.ncbi.nlm.nih.gov/articles/PMC8541489/.
72. Sundaramurthi, D., Krishnan, U.M., Sethuraman, S., Electrospun Nanofibers as Scaffolds for Skin Tissue Engineering. *Polym. Rev.* [Internet], 54, 2, 348–76, 2014 Apr 3 [cited 2025 Mar 4]. Available from: https://www.tandfonline.com/doi/abs/10.1080/15583724.2014.881374.
73. Doan, H.E., Inayat, M.S., Gallicchio, V.S., Doan, H.E., Inayat, M.S., Gallicchio, V.S., Stem Cell Use to Treat Dermatological Disorders, in: *Stem Cell Transplantation [Working Title]* [Internet], 2025 Feb 18 [cited 2025 Mar 4]; Available from: https://www.intechopen.com/online-first/1191841.
74. Moioli, E.K., Bolotin, D., Alam, M., Regenerative medicine and stem cells in dermatology. *Dermatol. Surg.* [Internet], 43, 5, 625–34, 2017 [cited 2025

Feb 20]. Available from: https://journals.lww.com/dermatologicsurgery/fulltext/2017/05000/regenerative_medicine_and_stem_cells_in.2.aspx.

75. Mekuye, B. and Abera, B., Nanomaterials: An overview of synthesis, classification, characterization, and applications. *Nano Select* [Internet], 4, 8, 486–501, 2023 Aug 1 [cited 2025 Jan 31]. Available from: https://onlinelibrary.wiley.com/doi/full/10.1002/nano.202300038.
76. Thamarai, P., Karishma, S., Kamalesh, R., Shaji, A., Saravanan, A., Bibi, S., *et al.*, Current advancements in nanotechnology for stem cells. *Int. J. Surg.* [Internet], 110, 12, 7456, 2024 Dec 5 [cited 2025 Jan 31]. Available from: https://pmc.ncbi.nlm.nih.gov/articles/PMC11634102/.
77. Lü, J.M., Wang, X., Marin-Muller, C., Wang, H., Lin, P.H., Yao, Q., *et al.*, Current advances in research and clinical applications of PLGA-based nanotechnology. *Expert Rev. Mol. Diagn.* [Internet], 9, 4, 325, 2009 May [cited 2025 Feb 20]. Available from: https://pmc.ncbi.nlm.nih.gov/articles/PMC2701163/.
78. Raszewska-Famielec, M. and Flieger, J., Nanoparticles for Topical Application in the Treatment of Skin Dysfunctions-An Overview of Dermo-Cosmetic and Dermatological Products. *Int. J. Mol. Sci.* [Internet], 23, 24, 15980, 2022 Dec 1 [cited 2025 Feb 20]. Available from: https://pubmed.ncbi.nlm.nih.gov/36555619/.
79. Niska, K., Zielinska, E., Radomski, M.W., Inkielewicz-Stepniak, I., Metal nanoparticles in dermatology and cosmetology: Interactions with human skin cells. *Chem. Biol. Interact.*, 295, 38–51, 2018 Nov 1.
80. Delouise, L.A., Applications of nanotechnology in dermatology. *J. Invest. Dermatol.* [Internet], 132, 3 Pt 2, 964–75, 2012 [cited 2025 Feb 20]. Available from: https://pubmed.ncbi.nlm.nih.gov/22217738/.
81. Embryonic Stem Cell Research: An Ethical Dilemma [Internet], [cited 2025 Feb 20]. Available from: https://www.eurostemcell.org/embryonic-stem-cell-research-ethical-dilemma.
82. Lo, B. and Parham, L., Ethical Issues in Stem Cell Research. *Endocr. Rev.* [Internet], 30, 3, 204, 2009 May [cited 2025 Feb 20]. Available from: https://pmc.ncbi.nlm.nih.gov/articles/PMC2726839/.
83. Volarevic, V., Markovic, B.S., Gazdic, M., Volarevic, A., Jovicic, N., Arsenijevic, N., *et al.*, Ethical and Safety Issues of Stem Cell-Based Therapy. *Int. J. Med. Sci.* [Internet], 15, 1, 36–45, 2018 Jan 1 [cited 2025 Feb 20]. Available from: https://pubmed.ncbi.nlm.nih.gov/29333086/.
84. Ethical and Social Aspects of Stem Cell Research: Stem Cell Network, NRW [Internet], [cited 2025 Feb 20]. Available from: https://www.stammzellen.nrw.de/en/discover/ethical-and-legal-aspects/ethical-issues.
85. Narsinh, K.H., Plews, J., Wu, J.C., Comparison of Human Induced Pluripotent and Embryonic Stem Cells: Fraternal or Identical Twins? *Mol. Ther.* [Internet], 19, 4, 635, 2011 Apr [cited 2025 Mar 4]. Available from: https://pmc.ncbi.nlm.nih.gov/articles/PMC3070108/.

86. Ben-David, U. and Benvenisty, N., The tumorigenicity of human embryonic and induced pluripotent stem cells. *Nat. Rev. Cancer* [Internet], 11, 4, 268–77, 2011 Apr [cited 2025 Feb 1]. Available from: https://pubmed.ncbi.nlm.nih.gov/21390058/.
87. Ingulli, E., Mechanism of cellular rejection in transplantation. *Pediatr. Nephrol.* [Internet], 25, 1, 61, 2010 Jan [cited 2025 Mar 4]. Available from: https://pmc.ncbi.nlm.nih.gov/articles/PMC2778785/.
88. Meissner, T.B., Schulze, H.S., Dale, S.M., Immune Editing: Overcoming Immune Barriers in Stem Cell Transplantation. *Curr. Stem Cell Rep.* [Internet], 8, 4, 206, 2022 Dec 1 [cited 2025 Mar 4]. Available from: https://pmc.ncbi.nlm.nih.gov/articles/PMC9643905/.
89. Hayat, H., Hayat, H., Francis, B., Gudi, D.M., Bishop, J.O., Wang, P., A Concise Review: The Role of Stem Cells in Cancer Progression and Therapy. *Onco Targets Ther.* [Internet], 14, 2761, 2021 [cited 2025 Feb 1]. Available from: https://pmc.ncbi.nlm.nih.gov/articles/PMC8068480/.
90. Is There a Risk of Immune Rejection From Stem Cell Transplants?, Dana-Farber [Internet], [cited 2025 Feb 20]. Available from: https://blog.dana-farber.org/insight/2023/04/is-there-a-risk-of-immune-rejection-from-stem-cell-transplants/.
91. Short, S., Lewik, G., Issa, F., An Immune Atlas of T Cells in Transplant Rejection: Pathways and Therapeutic Opportunities. *Transplantation* [Internet], 107, 11, 2341–52, 2023 Nov 1 [cited 2025 Mar 4]. Available from: https://journals.lww.com/transplantjournal/fulltext/2023/11000/an_immune_atlas_of_t_cells_in_transplant.13.aspx.
92. Strobl, J., Pandey, R.V., Krausgruber, T., Bayer, N., Kleissl, L., Reininger, B., *et al.*, Long-term skin-resident memory T cells proliferate *in situ* and are involved in human graft-versus-host disease. *Sci. Transl. Med.* [Internet], 12, 570, eabb7028, 2020 Nov 1 [cited 2025 Feb 20]. Available from: https://www.sciencedaily.com/releases/2020/11/201119103039.htm.
93. Nowacki, M., Kloskowski, T., Pietkun, K., Zegarski, M., Pokrywczyńska, M., Habib, S.L., *et al.*, The use of stem cells in aesthetic dermatology and plastic surgery procedures. A compact review of experimental and clinical applications. *Adv. Dermatol. Allergology/Postępy Dermatol. Alergol.* [Internet], 34, 6, 526, 2017 Dec 1 [cited 2025 Feb 20]. Available from: https://pmc.ncbi.nlm.nih.gov/articles/PMC5799755/.
94. Brockmann, I., Ehrenpfordt, J., Sturmheit, T., Brandenburger, M., Kruse, C., Zille, M., *et al.*, Skin-Derived Stem Cells for Wound Treatment Using Cultured Epidermal Autografts: Clinical Applications and Challenges. *Stem Cells Int.* [Internet], 2018, 1, 4623615, 2018 Jan 1 [cited 2025 Feb 20]. Available from: https://onlinelibrary.wiley.com/doi/full/10.1155/2018/4623615.
95. Yael, D., Addressing the High Costs of Stem Cell Therapies. *Stem Cell Res. Regener. Med.* [Internet], 7, 5, 255–6, 2024 Oct 31 [cited 2025 Feb 20]. Available from: https://www.openaccessjournals.com/articles/addressing-the-high-costs-of-stem-cell-therapies-18254.html.

96. Bridging Gaps for Affordable Cell and Gene Therapies: Overcoming Financial and Systematic Obstacles, Inside Precision Medicine [Internet], [cited 2025 Feb 20]. Available from: https://www.insideprecisionmedicine.com/topics/precision-medicine/bridging-gaps-for-affordable-cell-and-gene-therapies-overcoming-financial-and-systematic-obstacles/.
97. Finn, J.D., Smith, A.R., Patel, M.C., Shaw, L., Youniss, M.R., van Heteren, J., *et al.*, A Single Administration of CRISPR/Cas9 Lipid Nanoparticles Achieves Robust and Persistent *In Vivo* Genome Editing. *Cell Rep.*, 22, 9, 2227–35, 2018 Feb.
98. Jinek, M., Chylinski, K., Fonfara, I., Hauer, M., Doudna, J.A., Charpentier E. A Programmable Dual-RNA–Guided DNA Endonuclease in Adaptive Bacterial Immunity. *Science (1979)*, 337, 6096, 816–21, 2012 Aug 17.
99. Zhu, Y., Advances in CRISPR/Cas9. *Biomed Res. Int.*, 2022, 1, 9978571, 2022 Jan 23.
100. Koller, U., Gene therapy advances shine the spotlight on epidermolysis bullosa, bringing hope to patients. *Mol. Ther.*, 31, 7, 1860–1, 2023 Jul.
101. Saad, F.A., Gene Therapy for Skin Aging. *Curr. Gene Ther.*, 25, 1, 2–9, 2025 Feb.
102. Lv, J., Yang, S., Lv, M., Lv, J., Sui, Y., Guo, S., Protective roles of mesenchymal stem cells on skin photoaging: A narrative review. *Tissue Cell*, 76, 101746, 2022 Jun.
103. Chen, K.Y. and Knoepfler, P.S., To Crispr And Beyond: The Evolution Of Genome Editing In Stem Cells. *Regener. Med.*, 11, 8, 801–16, 2016 Dec.
104. Litman, T., Personalized medicine—concepts, technologies, and applications in inflammatory skin diseases. *APMIS*, 127, 5, 386–424, 2019 May 24.
105. Kumar, S., Mahajan, B.B., Kaur, S., Singh, A., Autologous therapies in dermatology. *J. Clin. Aesthet. Dermatol.*, 7, 12, 38–45, 2014 Dec.
106. Tabar, V. and Studer, L., Pluripotent stem cells in regenerative medicine: challenges and recent progress. *Nat. Rev. Genet.*, 15, 2, 82–92, 2014 Feb 17.
107. Köseer, A.S., Di Gaetano, S., Arndt, C., Bachmann, M., Dubrovska, A., Immunotargeting of Cancer Stem Cells. *Cancers (Basel)*, 15, 5, 1608, 2023 Mar 5.
108. Helgadottir, H., Rocha Trocoli Drakensjö, I., Girnita, A., Personalized Medicine in Malignant Melanoma: Towards Patient Tailored Treatment. *Front. Oncol.*, 8, 208, 2018 Jun 12.
109. Rapalli, V.K., Waghule, T., Gorantla, S., Dubey, S.K., Saha, R.N., Singhvi, G., Psoriasis: pathological mechanisms, current pharmacological therapies, and emerging drug delivery systems. *Drug Discov. Today*, 25, 12, 2212–26, 2020 Dec.
110. Kumar, R., Sharma, A., Siddiqui, M.H., Tiwari, R.K., Prediction of Human Intestinal Absorption of Compounds Using Artificial Intelligence Techniques. *Curr. Drug Discov. Technol.*, 14, 4, 244–54, 2017 Oct 27.

111. Kumar, R., Sharma, A., Siddiqui, M.H., Tiwari, R.K., Promises of Machine Learning Approaches in Prediction of Absorption of Compounds. *Mini-Rev. Med. Chem.*, 18, 3, 196–207, 2018 Jan 10.
112. Haenlein, M. and Kaplan, A., A Brief History of Artificial Intelligence: On the Past, Present, and Future of Artificial Intelligence. *Calif. Manage. Rev.*, 61, 4, 5–14, 2019 Aug 17.
113. He, J., Baxter, S.L., Xu, J., Xu, J., Zhou, X., Zhang, K., The practical implementation of artificial intelligence technologies in medicine. *Nat. Med.*, 25, 1, 30–6, 2019 Jan 7.
114. Ara, J. and Khatun, T., A literature review: machine learning-based stem cell investigation. *Ann. Transl. Med.*, 12, 3, 52–52, 2024 Jun.
115. Srinivasan, M., Thangaraj, S.R., Ramasubramanian, K., Thangaraj, P.P., Ramasubramanian, K.V., Exploring the Current Trends of Artificial Intelligence in Stem Cell Therapy: A Systematic Review. *Cureus*, 12, 2021 Dec 1.
116. Wu, H., Yin, H., Chen, H., Sun, M., Liu, X., Yu, Y., *et al.*, A deep learning, image based approach for automated diagnosis for inflammatory skin diseases. *Ann. Transl. Med.*, 8, 9, 581–581, 2020 May.
117. Hussen, B.M., Taheri, M., Yashooa, RK,., Abdullah, G.H., Abdullah, S.R., Kheder, R.K., *et al.*, Revolutionizing medicine: recent developments and future prospects in stem-cell therapy. *Int. J. Surg.*, 110, 12, 8002–24, 2024 Dec.
118. Farini, A., Sitzia, C., Erratico, S., Meregalli, M., Torrente, Y., Clinical Applications of Mesenchymal Stem Cells in Chronic Diseases. *Stem Cells Int.*, 2014, 1–11, 2014.
119. Uppuluri, V.N.V.A., Thukani Sathanantham, S., Bhimavarapu, S.K., Elumalai, L., Polymeric Hydrogel Scaffolds: Skin Tissue Engineering and Regeneration. *Adv. Pharm. Bull.*, 12, 3, 437–48, 2022 May 21.
120. Suamte, L. and Babu, P.J., Electrospun based functional scaffolds for biomedical engineering: A review. *Nano TransMed.*, 100055, 2024 Oct. https://doi.org/10.1016/j.ntm.2024.100055
121. Sethuram, L. and Thomas, J., Therapeutic applications of electrospun nanofibers impregnated with various biological macromolecules for effective wound healing strategy – A review. *Biomed. Pharmacother.*, 157, 113996, 2023 Jan.
122. Olejnik, A., Semba, J.A., Kulpa, A., Dańczak-Pazdrowska, A., Rybka, J.D., Gornowicz-Porowska, J., 3D Bioprinting in Skin Related Research: Recent Achievements and Application Perspectives. *ACS Synth. Biol.*, 11, 1, 26–38, 2022 Jan 21.
123. Li, S., Dan, X., Chen, H., Li, T., Liu, B., Ju, Y., *et al.*, Developing fibrin-based biomaterials/scaffolds in tissue engineering. *Bioact. Mater.*, 40, 597–623, 2024 Oct.
124. Huang, Y.Z., Gou, M., Da, L.C., Zhang, W.Q., Xie, H.Q., Mesenchymal Stem Cells for Chronic Wound Healing: Current Status of Preclinical and Clinical

Studies. *Tissue Eng. Part B Rev.* [Internet], 26, 6, 555–570, 2020 Dec 1 [cited 2025 Feb 20]. Available from: https://pubmed.ncbi.nlm.nih.gov/32242479/.
125. Prodinger, C.M., Reichelt, J., Bauer, J.W., Laimer, M., Current and Future Perspectives of Stem Cell Therapy in Dermatology. *Ann. Dermatol.* [Internet], 29, 6, 667, 2017 Dec 1 [cited 2025 Feb 2]. Available from: https://pmc.ncbi.nlm.nih.gov/articles/PMC5705348/.
126. Brem, H., Adipose-derived Stem Cells to Treat Chronic Wounds, ClinicalTrials.gov. 2023.
127. Coots, K.B., Healing Chronic Venous Stasis Wounds With Autologous Cell Therapy, ClinicalTrials.gov. 2020.
128. Aronowitz, A.J., Adipose Derived Regenerative Cellular Therapy of Chronic Wounds, ClinicalTrials.gov. 2019.
129. Burrows, F. and Armstrong, G.D., Rising Tide - Amniotic Tissue(s) Treatments for Chronic Diabetic Foot Ulcers, ClinicalTrials.gov. 2025.
130. Brenner, M., Stem Cell Transplant to Treat Patients With Systemic Sclerosis, ClinicalTrials.gov. 2020.
131. Georges, G., Allogeneic Hematopoietic Cell Transplantation for Severe Systemic Sclerosis, ClinicalTrials.gov. 2018.
132. Kamen, L.D., Pilot Trial of Mesenchymal Stem Cells for Systemic Lupus Erythematosus, ClinicalTrials.gov. 2019.
133. Schulman, C., Stem Cell Therapy to Improve Burn Wound Healing, ClinicalTrials.gov. 2020.
134. Li, Y., Yan, B., Wang, H., Li, H., Li, Q., Zhao, D., *et al.*, Hair regrowth in alopecia areata patients following Stem Cell Educator therapy. *BMC Med.* [Internet], 13, 1, 1–11, 2015 Dec 12 [cited 2025 Feb 18]. Available from: https://pubmed.ncbi.nlm.nih.gov/25896390/.
135. Verkh, L., Safety Study of Bone Marrow Derived Stem Cells on Patients With Cutaneous Photoaging, ClinicalTrials.gov. 2022.
136. Mitrano, T.I., Grob, M.S., Carrión, F., Nova-Lamperti, E., Luz, P.A., Fierro, F.S., *et al.*, Culture and Characterization of Mesenchymal Stem Cells From Human Gingival Tissue. *J. Periodontol.*, 81, 6, 917–25, 2010 Jun.
137. Wang, S.G., Hsu, N.C., Wang, S.M., Wang, F.N., Successful Treatment of Plaque Psoriasis with Allogeneic Gingival Mesenchymal Stem Cells: A Case Study. *Case Rep. Dermatol. Med.*, 2020, 1–4, 2020 Mar 28.
138. Hong, W.S., Hu, D.N., Qian, G.P., McCormick, S.A., Xu, A.E., Ratio of size of recipient and donor areas in treatment of vitiligo by autologous cultured melanocyte transplantation. *Br. J. Dermatol.*, 3 520–5, 2011 Aug.
139. Seetharaman, R., Mahmood, A., Kshatriya, P., Patel, D., Srivastava, A., Mesenchymal Stem Cell Conditioned Media Ameliorate Psoriasis Vulgaris: A Case Study. *Case Rep. Dermatol. Med.*, 2019, 1–5, 2019 May 2.
140. Mahmood, A., SRKPPDMRSA, Mesenchymal Stem Cell Conditioned Media Induces Hair Regeneration in Alopecia Areata: A Case Study. *Ann. Stem Cell Res. Ther.*, 1, 2021.

141. Kohlman-Trigoboff, D., Lawson, J.H., Murphy, M.P., Stem cell use in a patient with an ischemic foot ulcer: A case study. *J. Vasc. Nurs.*, 24, 2, 56–61, 2006 Jun.
142. Yoon, J., Lee, S., Kim, M.J., Kim, J.H., Brief summary of the regulatory frameworks of regenerative medicine therapies. *Front. Pharmacol.*, 15, 1486812, 2025 Jan 22.
143. Ancans, J., Cell therapy medicinal product regulatory framework in Europe and its application for MSC-based therapy development. *Front. Immunol.* [Internet], 3, Aug, 253, 2012 [cited 2025 Feb 20]. Available from: https://pmc.ncbi.nlm.nih.gov/articles/PMC3418507/.
144. Hayakawa, T., Aoi, T., Umezawa, A., Ozawa, K., Sato, Y., Sawa, Y., *et al.*, A study on ensuring the quality and safety of pharmaceuticals and medical devices derived from the processing of autologous human somatic stem cells. *Regen. Ther.*, 2, 57–69, 2015 Dec 1.
145. Ethical Considerations in Stem Cell Research General remarks.
146. Chakraborty, D. and De, A., Stem-cell therapy in dermatology – Challenges and opportunities. *Indian J. Skin Allergy*, 3, 93, 2024 Aug 5.
147. Aboul-Soud, M.A.M., Alzahrani, A.J., Mahmoud, A., Induced Pluripotent Stem Cells (iPSCs)—Roles in Regenerative Therapies, Disease Modelling and Drug Screening. *Cells*, 10, 9, 2319, 2021 Sep 5.
148. Weng, T., Zhang, W., Xia, Y., Wu, P., Yang, M., Jin, R., *et al.*, 3D bioprinting for skin tissue engineering: Current status and perspectives. *J. Tissue Eng.*, 12, 20417314211028574, 2021 Jan 13.
149. Valenti, M.T., Serena, M., Carbonare, L.D., Zipeto, D., CRISPR/Cas system: An emerging technology in stem cell research. *World J. Stem Cells*, 11, 11, 937–56, 2019 Nov 26.
150. Zhuang, W.Z., Lin, Y.H., Su, L.J., Wu, M.S., Jeng, H.Y., Chang, H.C., *et al.*, Mesenchymal stem/stromal cell-based therapy: mechanism, systemic safety and biodistribution for precision clinical applications. *J. Biomed. Sci.*, 28, 1, 28, 2021 Apr 14.
151. Ortiz, J.M., Rosipal, N.C., Woodworth, G., Morse, E.W., Ann-Yi, S., Barnett, V., *et al.*, Addressing the Unmet Social Needs Of the Aya Stem Cell Transplant Patient. *Biol. Blood Marrow Transplant.*, 15, 2, 148, 2009 Feb.
152. Robinson, T., Bailey, C., Morris, H., Burns, P., Melder, A., Croft, C., *et al.*, Bridging the research–practice gap in healthcare: a rapid review of research translation centres in England and Australia. *Health Res. Policy Syst.*, 18, 1, 117, 2020 Dec 9.
153. Noël, X. and Bechara, A., Bridging the Gap between the Lab and the Clinic: Psychopathology's Grand Challenge. *Front. Psychol.*, 7, 2016 Nov 8.
154. Teachman, B.A., Drabick, D.A.G., Hershenberg, R., Vivian, D., Wolfe, B.E., Goldfried, M.R., Bridging the gap between clinical research and clinical practice: Introduction to the special section. *Psychotherapy*, 49, 2, 97–100, 2012.

10

Clinical Significance and Regulatory Aspects of Stem Cell Therapy

Debasmita Paul, Karan Kr. Das, Abhijit Ghosh, Soukat Chandra and Shaileyee Das*

Bharat Technology, Bahirtafa, Beltala, Uluberia, Howrah, West Bengal, India

Abstract

The potential of stem cell therapy to tackle a wide range of degenerative and chronic diseases is a major step forward in modern medicine. In this chapter, we will take a look at stem cell treatment and its potential to change the face of regenerative medicine. We will see how it might help with autoimmune diseases, heart conditions, neurological issues, and more. In recent years, stem cell therapy—which makes use of human pluripotent stem cells and multipotent mesenchymal stem cells—has become an integral part of regenerative medicine. A delicate balancing act between the expected therapeutic benefits and hazards is required when vital stem cell research is integrated into traditional medicinal uses. While the European Union and United States have passed legislation pertaining to stem cell products, India is still missing key pieces of that puzzle. Global perspectives are shown by examining regulatory frameworks from agencies such as the US Food and Drug Administration, European Medicines Agency, and the Indian Council of Medical Research. Topics covered include the need for thorough clinical trials to verify safety and effectiveness, as well as commercialization and standardization. Despite the revolutionary potential of stem cell treatment, the scientific, regulatory, and ethical communities must work together to bring it to a wider audience.

***Keywords*:** Stem cell therapy, regulatory guidelines, clinical applications, ethical considerations, therapeutic potential

**Corresponding author*: drshaileyeedas@gmail.com

Sumel Ashique, Sanjay Dey, Biplab Debnath, Mohhammad Ramzan and Swarupananda Mukherjee (eds.) Stem Cell Therapy in Dermatological Disorders, (271–300) © 2025 Scrivener Publishing LLC

10.1 Introduction

A number of diseases that were formerly thought to be incurable are now within reach, thanks to stem cell therapy, a revolutionary development in modern medicine. Regenerative medicine has been transformed by stem cells, which can self-renew and differentiate into multiple cell types. These cells hold great promise in treating degenerative illnesses, immune deficiencies, and injuries to important organs. The therapeutic potential of stem cells has been the subject of extensive preclinical and clinical studies, spurred by the current acceleration of stem cell research [1]. Despite these advancements, there are still significant ethical, therapeutic, and regulatory challenges to overcome before stem cell–based medicines may be used safely and effectively. With an eye on the possible impact on healthcare systems throughout the world, this chapter will provide a thorough analysis of the regulatory framework and therapeutic relevance of stem cell treatment. Stem cell therapy makes use of the unique biological properties of stem cells, such as their ability to differentiate into various cell types and promote the regeneration of damaged tissues. Medicinal applications of stem cells are not a new idea. Stem cell treatments have come a long way from the first successful bone marrow transplant in 1957, which occurred in the middle of the 20th century. Interest in induced embryonic stem cells (ESCs), adult stem cells (ASCs) (mesenchymal stem cells [MSCs], etc.), and induced pluripotent stem cells (iPSCs) has grown substantially. The regulatory and therapeutic framework of stem cell therapy is more complex because each category has its own set of advantages and disadvantages [2].

These days, stem cell therapy is widely recognized as a game changer in the fight against serious, long-term illnesses. In neurodegenerative diseases, such as Alzheimer and Parkinson disease, and metabolic and autoimmune disorders, including diabetes and multiple sclerosis, stem cell–based treatments are creating new possibilities for tailored treatment. The capacity to mend injured tissues reinstates normal physiological processes and perhaps inhibits disease development, underscoring the importance of stem cell treatment in medicine.

10.1.1 Clinical Applications and Challenges

Encouraging outcomes in several preclinical and clinical trials have accelerated the shift of stem cell treatment from laboratory investigation to clinical use. In conditions such as rheumatoid arthritis and graft-versus-host disease (GVHD), MSCs have shown promise in regulating immune responses and reducing inflammation. Hematopoietic stem cells (HSCs) have been utilized for an extended period to treat leukemia and various blood diseases, underscoring the clinical efficacy of stem cell therapy.

Despite these advancements, the transition from research to extensive clinical implementation continues to encounter obstacles. Safety, effectiveness, ethical problems, and standardization challenges impede the smooth incorporation of stem cell treatment into conventional medicine. Tumorigenicity, immunological rejection, and the ethical discourse around ESCs are significant issues necessitating stringent regulatory supervision. Moreover, the heterogeneity in procedures, patient reactions, and manufacturing processes necessitates the formulation of explicit clinical translation guidelines [3].

10.1.2 The Role of Regulatory Frameworks

Regulatory agencies play an essential role in ensuring the safety of patients and the effectiveness of stem cell treatment because of the complexities involved. Disparities in healthcare legislation, scientific progress, and ethical perspectives are reflected in the different regulatory measures used by different nations. The Indian Council of Medical Research (ICMR), the European Medicines Agency (EMA), and the Food and Drug Administration (FDA) are among the many agencies that have published guidelines for stem cell therapies to follow.

Intricacies with regulations arise because stem cell therapies are frequently classified as medications, biologics, or advanced therapy medical products (ATMPs). US stem cell products must undergo rigorous evaluations in clinical trials and IND applications since they are classified as "biologics" by the FDA. Similar to how the EMA has tightened regulations on ATMPs, cell-based medications must first pass thorough preclinical and clinical testing before they can be given to patients.

A vital component of regulation is the necessity to address unverified and marketed stem cell treatments. Numerous regions globally have direct-to-consumer stem cell clinics offering unregulated treatments that lack sufficient clinical approval, hence posing considerable hazards to patient safety. This needs robust international cooperation and rigorous policies to address unethical conduct and advocate for scientifically approved therapies.

10.1.3 Objectives and Scope of this Chapter

a) Outline of stem cell therapy principles, emphasizing biological properties and disease treatment mechanisms.

b) Current clinical applications—Investigate stem cell–based treatments for neurodegenerative, metabolic, cardiovascular, and autoimmune disorders.
c) Examine US, European, Indian, and other major nation regulatory systems for stem cell treatment.
d) Handle ethical, commercialization, safety, and efficiency issues in stem cell research and uses.
e) Investigate stem cell therapy, new technologies, and the path to general clinical integration, which will help us to define future directions and translational potential.

This chapter aims to bridge the divide between scientific advancement and legal issues by providing a comprehensive review of the evolving landscape of stem cell treatment. The effective and accountable implementation of stem cell therapies relies on an understanding of their clinical and regulatory characteristics as advancements in regenerative medicine progress [4].

10.2 Definition and Types of Stem Cells

Undifferentiated cells with the potential to develop into several cell types are known as stem cells. They are the building blocks of the body and may mend harm and keep things in balance because of their regenerative and dividing capabilities. These cells are essential for embryonic development, tissue regeneration, and medical procedures. Stem cells' remarkable capacity to develop into a wide variety of cell types has made them a hot topic in the fields of regenerative medicine and disease treatment. Based on their origin and ability to develop into distinct cell types, stem cells can be widely classed. There are mainly two types:

10.2.1 Embryonic Stem Cells

ESCs are produced by the inner cell mass of an embryo in its early stages, known as a blastocyst. As pluripotent entities, they possess the capacity to develop into nearly all cell types. ESCs hold significant potential for regenerative therapy due to their extensive differentiation capabilities. Nonetheless, ethical concerns over the use of ESCs have led to stringent regulations and investigations into alternative stem cell sources [5].

10.2.2 Adult Stem Cells

Brain, skin, bone marrow, and liver are just a few of the many bodily tissues that contain ASCs, which are also called somatic or tissue-specific stem cells. Their ability to differentiate into a limited range of cell types linked to their original tissue is a testament to their multipotency. ASCs are HSCs derived from bone marrow, responsible for the production of blood cells, and MSCs, which differentiate into bone, cartilage, and adipose cells. One therapeutic use of these cells is the treatment of leukemia by bone marrow transplants.

10.2.3 Induced Pluripotent Stem Cells

IPSCs are produced by reprogramming adult somatic cells to become pluripotent. By avoiding the use of ESCs, this innovative technique allows for the production of stem cells tailored to each individual patient. iPSCs have established novel therapeutic avenues for disease modeling, drug screening, and customized therapy.

10.2.4 Perinatal Stem Cells

The stem cells come from many perinatal tissues, including the amniotic fluid, placenta, and umbilical cord. They are multipotent and are thought of as a bridge between ESCs and ASCs. Umbilical cord blood–derived stem cells have been successfully utilized in the treatment of hematological disorders and immunological deficiencies [6].

10.2.5 Cancer Stem Cells

Cancer stem cells (CSCs), a distinct population of stem cells located within tumors, has the ability to self-renew and facilitate tumor progression. They are considered to be factors contributing to cancer recurrence and treatment resistance. Targeting CSCs has become more significant in cancer research to develop more effective therapies. Stem cells provide significant potential for medicinal advancements, ranging from regenerative therapies to insights into disease causes. However, its utilization is fraught with ethical, legal, scientific, and technical challenges. Maximizing their therapeutic potential and addressing ethical difficulties are achievable *via* ongoing study and policy reform.

10.3 Mechanisms of Action of Stem Cell Therapy

Stem cell therapy works through a variety of pathways, including differentiation, paracrine signaling, immunomodulation, and homing. Stem cells develop into specialized cells, which contribute to tissue regeneration [7]. They release bioactive molecules that promote angiogenesis, reduce inflammation, and enhance healing [8]. Stem cells modulate immune responses, reducing autoimmunity and graft rejection [9]. Additionally, they migrate to injury sites through chemotactic signaling [10]. These mechanisms make stem cell therapy a promising approach in regenerative medicine. Table 10.1 shows the list of currently approved stem cell therapies along with their clinical indications and regulatory status across different countries.

10.3.1 Regenerative Potential

Because of its distinct action mechanisms, which involve immunomodulation, paracrine signaling, and differentiation, stem cell therapy has great regeneration potential. Multipotent stem cells (MSCs) can help regenerate damaged tissues by differentiating into many cell types [11]. The regenerative milieu is further enhanced by the bioactive chemicals secreted by stem cells, which promote angiogenesis, reduce apoptosis, and control inflammation [12].

Paracrine signaling is essential, as exosomes and cytokines produced from stem cells promote tissue healing without direct differentiation [13]. The immunomodulatory characteristics of stem cells, including the reduction of proinflammatory cytokines and the enhancement of anti-inflammatory responses, enhance treatment effectiveness in illnesses such as autoimmune disorders and neurodegeneration [14]. Furthermore, iPSCs provide patient-specific restorative treatments, therefore reducing the risks of immunological rejection [15]. Notwithstanding its potential, issues such as tumorigenicity, ethical dilemmas, and delivery mechanisms necessitate more investigation. Progress in gene editing and biomaterial scaffolds seeks to improve stem cell retention and therapeutic efficacy [16]. Stem cell treatment is progressively advancing as a transformative method in regenerative medicine.

10.3.2 Immune Modulation

Stem cell treatment is essential for immune regulation by managing inflammatory responses and facilitating tissue repair. MSCs are extensively researched for their immunomodulatory effects, especially *via* paracrine signaling and cell–cell interactions. MSCs release beneficial chemicals, including transforming growth factor β (TGF-β), interleukin 10 (IL-10),

Clinical significance of stem cell therapy

Table 10.1 Clinical significance of stem cell therapy.

Application area	Specific diseases/ conditions	Description	Clinical trial	Clinical significance	References
Hematopoietic stem cell transplantation (HSCT)	Leukemia, lymphoma, bone marrow failure	**HSCT involves the infusion of hematopoietic stem cells to reestablish normal blood cell production in individuals with hematologic tumors or diseases of the bone marrow.**	**Phase I:** Assesses safety and dose in a limited cohort **Phase II:** Assesses efficacy and supplementary safety in a more extensive cohort **Phase III:** Validates efficacy and monitors adverse effects in a substantial cohort **Phase IV:** Postapproval studies to get further data	HSCT is essential in the treatment of hematological malignancies such as leukemia, lymphoma, and diseases of intrinsic bone marrow failure. It may offer a potential treatment by replenishing healthy stem cells in diseased bone marrow and can serve as a consolidative surgery following chemotherapy, therefore enhancing the likelihood of sustained remission. In conjunction with bone marrow insufficiency.	[27]

(Continued)

Table 10.1 Clinical significance of stem cell therapy. (*Continued*)

Application area	Specific diseases/ conditions	Description	Clinical trial	Clinical significance	References
Bone and cartilage regeneration	Osteoarthritis, bone fractures, cartilage damage	**Stem cells are frequently used in regenerative medicine and tissue engineering to repair damaged tissues, as they may develop into osteoblasts and chondrocytes, contributing to orthopedic therapies.**	**Phase I:** Evaluates the safety and dose of a novel therapy in a limited cohort of healthy volunteers **Phase II:** Assesses treatment efficacy while also examining safety in a larger population **Phase III:** Evaluates the novel treatment against established therapies in a substantial population to validate its efficacy, monitor adverse effects, and gather data to ensure safe application of the medication **Phase IV:** Research to assess the long-term consequences following the approval of the therapy for public use and its application	Bone fractures that heal improperly can result in complications and permanent disability. To improve healing, techniques such as using mesenchymal stem cells (MSCs) to facilitate bone regeneration are essential. Likewise, such damage to cartilage—a wear and tear injury seen frequently in athletes and the elderly—can lead to joint instability and pain. Regenerative therapies seek to restore cartilage and repair it. These have great potentials to alleviate the burden of musculoskeletal disorders and improve patient outcomes.	[28]

(*Continued*)

Table 10.1 Clinical significance of stem cell therapy. (*Continued*)

Application area	Specific diseases/ conditions	Description	Clinical trial	Clinical significance	References
Neurological disorders	Parkinson disease (PD), spinal cord injuries (SCIs), stroke, multiple sclerosis (MS)	**Stem cell therapy has promise for the treatment of neurodegenerative disorders such as PD, Alzheimer disease, and SCIs, owing to its capacity to regenerate damaged neurons and facilitate neuroprotection.**	**Phase I:** Treatment is administered to a limited cohort of healthy volunteers to assess safety and dose **Phase II:** A more extensive cohort of patients is engaged to evaluate the treatment's efficacy and examine its safety **Phase III:** Conducted on bigger populations to validate its efficacy, monitor adverse effects, and compare it with established therapies **Phase IV:** Following US Food and Drug Administration clearance, the treatment's safety is evaluated in the general population	The clinical significance of neurological illnesses, such as SCI, PD, MS, and stroke is profoundly impactful on patients' quality of life. PD is marked by significant motor impairment and cognitive deterioration; SCI leads to profound loss of mobility and sensory capabilities. Stroke causes immediate brain damage and frequently results in enduring disability. Demyelination in MS induces several neurological symptoms and disabilities. These illnesses provide unique obstacles that necessitate innovative therapies, such as stem cell therapy, to improve patients' quality of life.	[29]

(*Continued*)

Table 10.1 Clinical significance of stem cell therapy. (*Continued*)

Application area	Specific diseases/ conditions	Description	Clinical trial	Clinical significance	References
Cardiovascular diseases (CVDs)	Heart failure, myocardial infarction, coronary artery disease (CAD)	**Cardiac progenitor cells and induced pluripotent stem cells are stem cells that regenerate damaged heart tissue after a myocardial infarction, therefore restoring cardiac function.**	**Phase I:** A therapy is evaluated in a limited cohort of healthy volunteers to determine safety and dose **Phase II:** Assembles a more extensive cohort of patients to assess the treatment's efficacy and examine its safety **Phase III:** Research on the treatment's effectiveness, side effects, and comparison to previous conventional therapy constitutes phase III **Phase IV:** Studies to clarify further details about the treatment's risks, advantages, and appropriate community application	CVD is a predominant cause of morbidity and mortality globally. Heart failure results in incapacitating sequelae and recurrent hospitalizations, whereas myocardial infarction causes immediate cardiac injury and lasting sequelae. CAD resulting from atherosclerosis is the predominant cause of human death attributable to diminished blood flow. Modifications in lifestyle, pharmacological interventions, and surgical treatments: a demand for innovative solutions to enhance clinical outcomes and mitigate disease burden enhancing treatment and preventive alternatives provide the opportunity to improve patient outcomes while alleviating a healthcare load.	[30]

(*Continued*)

Table 10.1 Clinical significance of stem cell therapy. (*Continued*)

Application area	Specific diseases/ conditions	Description	Clinical trial	Clinical significance	References
Diabetes (islet cell regeneration)	Type 1 and type 2 diabetes	**Diabetes and insulin therapy for diabetic patients may benefit from such stem cell–derived beta cells, as they can be a renewable source of insulin and may reduce our dependency on insulin therapy.**	**Phase I:** A small group of healthy volunteers or patients is studied to determine the optimal dose and safety of an experimental drug **Phase II:** Evaluation of the treatment's effectiveness and safety in a larger population is the focus of phase II **Phase III:** In the third and final stage, researchers compared the therapy's efficacy and safety to those of standard therapies and a large population **Phase IV:** Evaluates the medication's effectiveness and long-term effects in the broader population through post-marketing surveillance	By restoring the body's ability to produce its own insulin, islet cell transplantation has great therapeutic potential in the treatment of type 1 and type 2 diabetes. If beta cell regeneration in type 1 diabetes is successful, it might help patients overcome the challenges caused by autoimmune destruction, which could lead to a reduction in complications, a possible elimination of insulin need, and an overall improvement in their quality of life. Regenerative approaches seek to improve beta cell activity and insulin secretory capacity; islet cell failure is a component of insulin resistance in type 2 diabetes. Still, new developments in immunomodulation and stem cell therapies provide hope for long-term therapeutic alternatives that can help with glucose management and diabetic complications.	[31]

(*Continued*)

Table 10.1 Clinical significance of stem cell therapy. (*Continued*)

Application area	Specific diseases/ conditions	Description	Clinical trial	Clinical significance	References
Skin and wound healing	Chronic wounds, burns, ulcers	**Stem cells help with wound healing by generating new skin cells, speeding up tissue repair in burns, ulcers and other skin injuries.**	**Phase I:** evaluates a novel medicine or therapy in a limited cohort of healthy volunteers (20–100) to assess its safety, establish a safe dose range, and identify adverse effects **Phase II:** In this phase, the medication is given to a larger cohort of persons (100–300) afflicted by the targeted ailment to ascertain the drug's efficacy and to examine potential adverse effects more comprehensively **Phase III:** This step is conducted on a wider scale (1000–3000 participants) to validate the treatment's efficacy, observe adverse effects, and compare it with standard therapies **Phase IV:** These studies are performed subsequent to the approval and commercialization of a medication. Phase IV trials enable researchers to gather further data about the long-term hazards and benefits of the therapy within a general population	Repairing damaged skin and facilitating wound healing are of the utmost importance when dealing with burns, ulcers, and chronic wounds. Infections, amputations, and other complications from chronic wounds such as diabetic and venous ulcers cause substantial morbidity, which in turn increases healthcare costs and reduces quality of life. Severe burns may cause permanent scarring, so it is important to take the right steps to treat them. Regenerating damaged tissues and speeding up the healing process after an injury could be possible with stem cell therapy	[32]

(*Continued*)

Table 10.1 Clinical significance of stem cell therapy. (*Continued*)

Application area	Specific diseases/ conditions	Description	Clinical trial	Clinical significance	References
Other diseases	Autoimmune diseases, muscular dystrophy, liver cirrhosis, blindness	**Stem cell research explores to liver disease, autoimmune disorders, and genetic disorders, providing hope for regenerative therapies in various medical fields.**	**Phase I:** Assesses safety and dose in a limited cohort (20–100) of healthy volunteers or patients **Phase II:** Evaluates effectiveness and ongoing safety in a limited cohort (100–300) of patients **Phase III:** Validates efficacy; monitors adverse effects in a broader demographic (1000–3000) **Phase IV:** Evaluates long-term effects and safety following the approval of therapy for public use	Stem cell therapy has great promise as a treatment for autoimmune diseases like MS since it can reduce inflammation and improve neuroprotection by administering MSCs. Although not stated directly, stem cells have the potential to aid in muscle regeneration in muscular dystrophy. Although there is some emerging research on stem cells and their potential use in liver cirrhosis, there is currently little evidence to support this claim. Treatment of visual impairment using stem cells: although it is still in its early stages, stem cell therapy is being considered as a potential future solution for retinal diseases that might restore eyesight. To confirm its efficacy and safety across different contexts, more studies are needed.	[33]

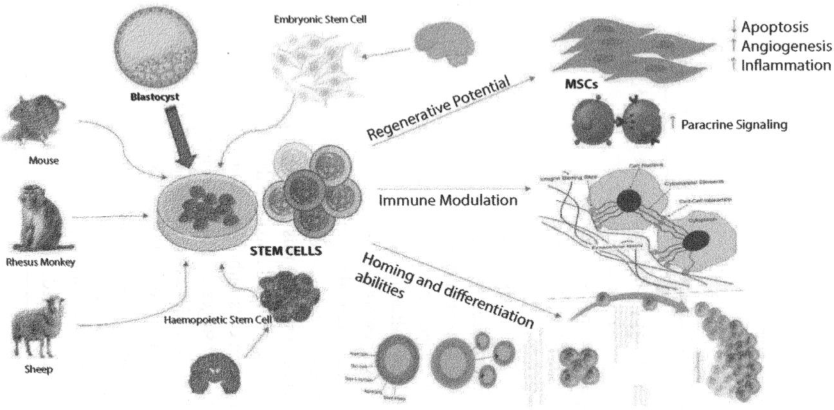

Figure 10.1 Clinical significance, regulatory aspects of stem cell therapy.

and prostaglandin E_2, which inhibit proinflammatory cytokines such as interferon γ and tumor necrosis factor α (TNF-α) [17]. Dendritic cells, T cells, B cells, and macrophages are all immune cells whose activity stem cells control [18]. MSCs stimulate the generation of regulatory T cells while suppressing T helper 17 cells, thus facilitating immunological tolerance [19]. Furthermore, MSCs induce macrophage polarization toward an anti-inflammatory M2 phenotype, hence mitigating tissue damage [20].

Exosomes originating from stem cells facilitate immune regulation by transporting microRNAs and proteins that govern immune responses [21]. The immunomodulatory characteristics of stem cell therapy provide a prospective treatment for autoimmune illnesses, GVHD, and inflammatory disorders. Figure 10.1 shows the evidence of the therapeutic potential of stem cell therapy and Clinical significance, regulatory aspects of stem cell therapy.

10.3.3 Homing and Differentiation Abilities

Stem cell treatment depends on the capacity of stem cells to migrate to damage locations and specialize into certain cell types, facilitating tissue repair and regeneration. Homing is facilitated by chemotactic signals, such as stromal-derived factor 1 and its receptor CXCR4, which direct stem cells to injured tissues [22]. Moreover, inflammatory cytokines, including TNF-α and IL-6, promote stem cell migration [23].

Stem cells respond to cues from their surroundings when they are localized, causing them to differentiate. Osteoblasts, chondrocytes, and adipocytes are differentiated from MSCs by distinct signaling pathways such Wnt, Notch, and TGF-β [24]. HSCs replenish blood lineages *via* niche-dependent signals [25].

In addition to direct differentiation, paracrine actions are crucial in tissue healing through the secretion of growth hormones, including vascular endothelial growth factor and insulin-like growth factor 1 [26]. Improving homing and differentiation efficacy is a primary objective in refining stem cell–based regenerative treatments.

10.3.4 Emerging Therapies and Innovations

a) **Organoid development**
   Organoid technology providing novel ways to addressing disease, therapeutic strategies, the development of human organs, etc., through three-dimensional (3D) cell culture systems are developing from PSCs or ASCs [34]. When it comes to complicated disorders such as glioblastoma (GBM), which require more advanced treatment procedures, organoid creation will pave the way for stem cell treatments to be used in translational medicine. Using human pluripotent stem cells (hPSCs) derived from a reprogrammed fibroblast, allowing for exploration of GBM initiation and progression in artificial brain organoids that recapitulate the tumor microenvironment [35].
b) **Gene-editing applications with stem cells**
   Stem cell therapy and gene editing technologies, notably CRISPR-Cas9. One example of a genetic illness that has shown promise in the treatment of Wolfram syndrome using CRISPR-Cas9 technology is hereditary disorders such as Down syndrome, which are caused by mutations in specific areas of the genome that may be edited in stem cells. One benefit of iPSCs is that they reduce the likelihood of immunological rejection because they are made from the patient's own cells. Regarding individualized medication, this is very helpful. In order to precisely fix genetic errors, the homology-directed repair technique makes use of a homologous DNA template [36].
c) **Personalized medicine and disease modeling**
   Stem cell therapy and novel treatments facilitate personalized medicine and disease modeling. Utilizing MSCs and hPSCs may result in tailored therapies that consider genetic

profiles and illness causes, hence enhancing the efficacy of therapeutic protocols. Recent advancements in regenerative medicine demonstrate the potential of using stem cells for diseases previously lacking viable treatment choices, hence revolutionizing medical procedures and improving healing capabilities [37].

10.3.5 Challenges in Stem Cell Therapy

Problems with stem cell therapy include improved methods for cell production and delivery [38], the possibility of immune rejection of implanted cells [39], poor cell survival after transplantation [40], inconsistent product quality [41], the necessity of proving treatment safety and effectiveness [42], and proof of immune rejection or adverse events. Moreover, the complexity of these neurological disorders makes it difficult to develop effective therapeutic strategies [38]. Clinical outcomes can be further modulated by factors such as cryopreservation effects and the need for optimized application methods, creating a need for better engineering strategies for more efficacious models in a device [39].

10.3.6 Regulatory Aspects of Stem Cell Therapy

a) **Global regulatory frameworks**
Stem cell treatment has shown to be a hopeful method for regenerative medicine and illustrates personalized medicine, disease modeling, and drug development in the United States, Europe, and Asia. Nevertheless, there are many legal restrictions imposed by national borders that influence the advancement and use of these methods [40].

b) **United States (FDA regulations)**
   - **i.** Such regulations important for human cells, tissues, and products, "Minimal Manipulation and Homologous Use: Guidance for Industry and Food and Drug Administration Staff" (2020) [40].
   - **ii.** Draft Guidance: "Assuring Potency for Cellular and Gene Therapy Products (2023)" [40].
   - **iii.** "Guidance for the Use of Human and Animal-Derived Material Intended for Manufacturing Cellular and Gene Therapy and Tissue Engineered Medical Products" (2024) [40].

c) **Europe (EMA guidelines)**
   i. "Reflection Paper on Stem Cell–Based Medicinal Products" (2009) [40, 41].
   ii. "Potency and Quality of Non-clinical Data for Certification of Advanced Therapy Medicinal Products" (2010) [40, 41].
   iii. "Quality, Non-clinical and Clinical Requirements for Investigational Advanced Therapy Medicinal Products in Clinical Trials" (2024) [40, 41].

d) **Asia**
   As in other regions, there is a diversity of countries and regulatory agencies in Asia.

a. **Japan (Ministry of Health, Labour, and Welfare Guidelines)**
   i. "Guidelines for Clinical Research Involving Human Stem Cells." MHLW Notification No. 425 (2006) [40, 43].
   ii. The technical recommendations about the quality, non-clinical, and clinical research of regenerative medicinal goods (human cell–processed products). PSEHB/MDED Administrative Notice No. 0614043 (2016) [40, 43].
   iii. Factors that the certified special committee should contemplate in the evaluation of tumorigenicity assessment within the regenerative medicine provision plans utilizing hPSCs are as follows: 2021 [40].

b. **India (National Stem Cell Guidelines for Research)**
   i. In 2007, the ICMR and the Department of Biotechnology collaborated in preparation [44].
   ii. The paper was originally titled "Guidelines for Stem Cell Research and Therapy," but was subsequently revised in 2013/2017 to [National] Guidelines for stem cell Research (2007) [44].
   iii. There were no sanctioned clinical situations for which stem cells, including primary cells and minimally modified stem cells such as bone marrow or peripheral blood stem cells, have been utilized in homologous applications of peripheral blood stem cells (2017) [44].

c. **Other regions**
   In other regions such as Africa, stem cell therapy is in use—but its popularity generally suffers because there are not any concrete firm legal guidelines in place [45].

10.3.7 Quality Control and Safety Requirements

a) **GMP (Good Manufacturing Practices) standards**
MSCs used in treatments need to be made following GMP—that is, the process is tightly controlled and can be audited, and generally speaking, it helps lower the risk of transferring pathogens. Storage and distribution sometimes get less attention, even though they really matter in how the whole manufacturing process comes together. In most cases, tests for sterility, potency, purity, and identity happen earlier on and are then reviewed by quality assurance before the cells are cleared for use. What this setup does is keeping the procedure in check and minimizes the chance of viral contamination, especially in the culture medium. Basically, when you stick with a GMP method, MSCs expand easily and tend to keep both their usual appearance and genetic makeup intact [46, 47].

b) **Preclinical and clinical trial phases**
Initial testing is necessary prior to doing a thorough assessment of safety and effectiveness in preclinical studies. For these preclinical studies, *in vitro* and *in vivo* models could be used. Investigations on the safety of stem cell therapy should cover a wide range of topics, including cellular characterization, tumorigenicity testing, biodistribution studies, adjunct treatments, long-term safety assessments, and possible toxicological uses of stem cells. Before beginning research and animal trials, we will conduct efficacy studies that include evidence of effectiveness. Future reviews, approvals, and continuing monitoring of all research, including clinical trials using stem cell–based therapies, must be conducted by an independent committee on human subjects. Everything from risk assessments to testing goals to subject selection to capacity evaluations to reports on event discussions to long-term observations falls under this category [48].

c) **Risk assessment and pharmacovigilance**
Several factors affect the potential risks of stem cell–based medicinal products. These include the stem cells' type, their differentiation status, their proliferation capacity, the method of administration, the area to be treated, any manipulation processes (*in vitro* culture, etc.), the treatment's irreversibility, the need or possibility of concurrent tissue regeneration

in cases of irreversible tissue loss, and the viability of the engrafted cells in the long run. The identified risks or potential hazards include tumor growth, adverse immunological responses, and the transfer of unintended agents. At now, there appears to be a lack of clinical experience using pluripotent stem cells. Treatments utilizing MSCs in regenerative therapy programs have so far shown no severe adverse effects in small-scale clinical studies, which raises the possibility that MSCs are generally safe. On the other hand, major side effects have been recorded in certain clinical studies; thus, there has to be further research, especially into the molecular processes and the long-term safety [49].

Legal and policy challenges

Stem cell therapy holds significant promise in regenerative medicine, but its advancement is intertwined with complex regulatory challenges, particularly concerning patent and intellectual property (IP) issues, the proliferation of unregulated "stem cell tourism," and the pressing need for international harmonization of standards.

a) **Patent and IP issues**
   In order to safeguard inventions and guarantee ethical considerations, strong IP frameworks are required for the commercialization of stem cell treatments [50]. When it comes to human ESCs in particular, there are ethical questions that arise when stem cell–based innovations are considered for patenting. It is still a difficult balancing act to ensure the accessibility of therapies while also protecting private innovations [51]. To understand the complexities of patent eligibility and scope in this dynamic industry, clear standards are required.
b) **Combatting unregulated "stem cell tourism"**
   Patients are increasingly turning to nations with less stringent rules for stem cell treatments, a practice known as medical tourism [52]. Exposure to therapies that have not been well tested and may even be harmful is one of the major dangers posed by these phenomena. Clinics in India, for example, have provided stem cell therapies without strong preclinical data, which points to a lack of regulation or unclear jurisdiction [53]. Both patient safety and genuine scientific advancement are undermined by such tactics.

c) **Need for international harmonization**
The inconsistencies in regulatory systems among nations enable clinics providing unverified cures to leverage legal ambiguities and differing enforcement practices [54]. The International Society for Stem Cell Research (ISSCR) has denounced these methods and supports the establishment of uniform criteria to ensure stringent control and openness [55]. Attaining global harmonization is essential to avert unethical activities and guarantee that stem cell therapies are effective, safe, and accessible.

10.3.8 Role of Regulatory Bodies in Innovation

a) **Supporting research and development**
When it comes to stem cell therapy, ISSCR has been a huge help. Important safety and effectiveness results from well-planned, correctly executed, and strictly controlled clinical trials provide credence to this. By working together, the ISSCR has improved public understanding of stem cell therapy and launched programs to ensure patients' safety; this has benefited research subjects, patients, their families, lawmakers, and other political groups, and it has helped distinguish between problematic and practical research. A number of ISSCR-related initiatives, such as gastric research and the creation of global standards for the clinical translation of stem cell treatment, have come to fruition. Scientists from many countries are working on developing international guidelines for stem cell research. Healthcare providers, government officials, bioethicists, social scientists, and academics from a wide range of scientific fields all had a hand in developing this strategy [56].

b) **Fast-tracking promising therapies**
Novel approaches to treating diseases and conditions that were previously thought to have no hope of recovery are becoming available because of advancements in stem cell research. The use of stem cells, and more specifically hPSCs and MSCs, is gaining prominence in the field of regenerative medicine. The unique ability of hPSCs to self-renew and differentiate into all human cell types is what makes them so special. The International Society for Cell and Gene Therapy defines MSCs as multipotent progenitor cells that

can differentiate into mesenchymal lineages and self-renew (although to a limited extent *in vitro*). hPSCs and MSCs derived from fat, bone marrow, or umbilical cord have recently found use in clinical trials to treat a wide range of human diseases and conditions [57].

10.4 Future Perspectives

New therapies and potential remedies are emerging as a result of advancements in stem cell research. As research into stem cell treatment continues to advance, there is great promise in combining it with emerging technologies. The potential of CRISPR-Cas9 gene editing, synthetic biology, 3D bioprinting, regulatory hurdles, and stem cell implications for healthcare systems and aging populations are all areas that stem cell research is looking toward in the future [58].

10.4.1 CRISPR-Cas9 and Genetic Engineering

Applying CRISPR-Cas9 to precisely modify genes is a huge step forward in stem cell research. Thanks to CRISPR-Cas9, genome editing has been a game changer, allowing for the targeted correction of harmful mutations in stem cells. This finding has major ramifications for how genetic diseases including muscular dystrophy, cystic fibrosis, and sickle cell anemia are treated. Gene editing in iPSCs has decreased the risk of immune rejection and opened up new possibilities for tailored treatments. While CRISPR-Cas9 gene editing shows promise, it also raises concerns about safety and ethics. Unwanted consequences, such as cancer, might arise from off-target effects, which are unintentional changes to DNA [59]. Additional research is necessary to improve CRISPR methods by reducing risks, increasing therapeutic efficacy, and improving specificity using base and prime editing technologies.

10.4.2 Synthetic Biology and 3D Bioprinting

Synthetic biology and 3D bioprinting are revolutionizing stem cell–based regenerative medicine. Synthetic biology enables the creation of personalized biological systems, such as engineered stem cells with improved regenerative abilities. These bioengineered cells can be programmed to respond to specific biological signals, which increase their therapeutic efficacy [60]. In contrast, 3D bioprinting makes it easier to create complex tissue structures from stem cells and biomaterials. Bioprinting advancements have helped the fabrication of functional tissues such as skin, cartilage, and vascularized organoids [61]. Future research requires scaling up bioprinting technology for whole organ development, which can address the world's organ donor shortage. There are still challenges in ensuring the vascularization and functional integration of printed tissues in the body.

10.4.3 Balancing Innovation with Safety

As stem cell technologies advance, balancing innovation and safety remains a challenge. Although novel therapies have great potential, it is important to maintain their long-term safety and efficacy. Clinical trials are assessing the risks and benefits of stem cell–based treatments, but the complexity of these therapies results in longer regulatory approval timelines [62]. Unregulated stem cell clinics that offer unproven treatments pose another challenge. These clinics frequently operate outside of regulatory oversight, promoting therapies that lack scientific validation. To protect patients from potential harm while also encouraging responsible innovation in the field, regulations must be strengthened and ethical standards enforced.

10.4.4 Bridging Gaps in Regulatory Frameworks

Stem cell treatments face a complex and regionally diverse regulatory environment. Many differences persist despite the fact that several countries' procedures for authorizing stem cell–based treatments exist, including the United States, European Union, and Japan [5]. A unified global regulatory framework could improve approval processes and encourage international collaboration. Another new approach is to use adaptive regulatory pathways, which allow for faster approval of promising therapies while ensuring postmarket surveillance. Using risk-based regulatory models, in which therapies with higher risks face more stringent evaluations, can help balance innovation and patient safety [63].

10.4.5 Role in Aging Populations

The potential to cure age-related degenerative diseases has been made possible by stem cell therapy. The potential of stem cells to promote healing in conditions such as osteoarthritis, Alzheimer disease, and Parkinson disease has prompted research into this field [64]. MSCs and iPSCs have promise for reversing cellular senescence and facilitating tissue repair. A crucial domain for future study is the advancement of rejuvenation medicines aimed at addressing the fundamental causes of aging. Senolytic agents, when used with stem cell treatment, have the capacity to enhance lifespan and quality of life. Moreover, the utilization of autologous stem cells, sourced from the patient's own body, may mitigate immunogenic issues while enhancing therapeutic results.

10.4.6 Integration into Healthcare Systems

For extensive clinical use, stem cell treatments must be included into healthcare systems. The exorbitant expenses and intricate manufacturing procedures presently restrict access to stem cell–based therapies. Enhancing production capacity while maintaining cost efficiency presents a significant challenge for future research [65]. Standardized protocols for transportation, stem cell banking, and administration must be established to enable clinical application. Collaboration among doctors, biotechnologists, and politicians is essential to bridge the divide between research and clinical practice. Public knowledge and education of the advantages and constraints of stem cell therapy enable patients and healthcare providers to make more educated decisions.

The future of stem cell research holds tremendous potential, encompassing CRISPR-Cas9 genetic modifications, bioprinted organs, and rejuvenation therapy for the elderly. Nonetheless, it is essential to balance these advancements with stringent safety testing and regulatory supervision for their effective integration into clinical practice. To realize stem cell therapy's full potential, we must resolve issues with regulatory frameworks, ethical concerns, and healthcare integration. The next era of regenerative medicine will be shaped by the continuous research and cooperation of scientists, legislators, and healthcare practitioners [66].

10.5 Conclusions and Future Directions

The promise of stem cell therapy to cure a wide range of inherited, immunological, and degenerative diseases has prompted many to hail it as a game-changing approach to regenerative medicine. Stem cell research has come a long way in the last several decades, with promising outcomes in both the laboratory and the clinic. The therapeutic importance of stem cell treatment is evidenced by its utilization across several medical fields, such as neurology, cardiology, orthopedics, and cancer. Notwithstanding its promise, the discipline encounters several hurdles, encompassing ethical dilemmas, safety problems, and regulatory intricacies [67].

Tumorigenicity, immune rejection, and uncontrolled differentiation are the key problems with stem cell therapy. To ensure that stem cell–based treatments are safe and successful, extensive preclinical research and clinical trials are required. Some countries have very strict regulations regarding stem cell therapy, whereas others have far more relaxed policies. Guidelines for the development and delivery of stem cell–based medicines have been established to ensure patient safety and therapeutic efficacy by regulatory bodies such as the FDA and the EMA [68]. However, stem cell product commercialization and clinical practice uniformity are impeded by differences in worldwide regulatory frameworks.

Ethical issues surrounding ESCs have generated increased interest in iPSCs and ASCs, which provide alternatives devoid of ethical dilemmas. iPSCs have potential in personalized medicine as they may be sourced from a patient's own cells, hence reducing the dangers of immunological rejection. Issues such as genomic instability and insufficient reprogramming necessitate more examination [69, 70].

Future research should focus on enhancing stem cell transport techniques, refining differentiation processes, and establishing uniform manufacturing standards to guarantee repeatability and safety. The application of artificial intelligence and machine learning in stem cell research can enhance data analysis. Advancements in biomaterials and tissue engineering provide prospects to enhance the results of stem cell transplants. Seventy regulatory harmonization among nations would facilitate international cooperation and expedite the licensing process for stem cell–based medicines. Policymakers want to establish a worldwide framework that guarantees safety and accessibility while promoting innovation in the sector.

In conclusion, although stem cell treatment possesses transformational potential in medicine, it is essential to tackle existing problems *via* comprehensive research, ethical considerations, and regulatory progress. Future initiatives should focus on enhancing treatment methodologies, augmenting patient safety, and establishing a regulatory framework that reconciles innovation with safety. Ongoing collaboration among academics, physicians, politicians, and regulatory bodies will be essential to fully actualize the potential of stem cell treatment in clinical applications.

AI disclosure statement: The author acknowledges the use of AI-based tools for minor language editing or grammatical correction. However, all intellectual contributions, analyses, and conclusions presented in this chapter are entirely the author's own.

References

1. Le Blanc, K. and Ringdén, O., Immunomodulation by mesenchymal stem cells and clinical experience. *J. Intern. Med.*, 262, 5, 509–25, 2007.
2. Marks, P.W., Witten, C.M., Califf, R.M., Clarifying Stem-Cell Therapy's Benefits and Risks. *N. Engl. J. Med.*, 376, 11, 1007–9, 2017.
3. Marote, A., Teixeira, F.G., Mendes-Pinheiro, B., Salgado, A.J., MSCs-Derived Exosomes: Cell-Secreted Nanovesicles with Regenerative Potential. *Front. Pharmacol.*, 7, 231, 2016.
4. Thomas, E.D., Lochte, H.L., Jr, Lu, W.C., Ferrebee, J.W., Intravenous infusion of bone marrow in patients receiving radiation and chemotherapy. *N. Engl. J. Med.*, 257, 11, 491–6, 1957.
5. Trounson, A. and McDonald, C., Stem Cell Therapies in Clinical Trials: Progress and Challenges. *Cell Stem Cell*, 17, 1, 11–22, 2015.
6. Zhao, Y., Glesne, D., Huberman, E., A human peripheral blood monocyte-derived subset acts as pluripotent stem cells. *Proc. Natl. Acad. Sci. U.S.A.*, 100, 5, 2426–31, 2003.
7. Zakrzewski, W., Dobrzyński, M., Szymonowicz, M., Rybak, Z., Stem cells: past, present, and future. *Stem Cell Res. Ther.*, 10, 1–22, 2019 Dec.
8. Caplan, A.I. and Dennis, J.E., Mesenchymal stem cells as trophic mediators. *J. Cell. Biochem.*, 98, 5, 1076–84, 2006 Aug 1.
9. Le Blanc, K. and Mougiakakos, D., Multipotent mesenchymal stromal cells and the innate immune system. *Nat. Rev. Immunol.*, 12, 5, 383–96, 2012 May.
10. Karp, J.M. and Teo, G.S., Mesenchymal stem cell homing: the devil is in the details. *Cell Stem Cell*, 4, 3, 206–16, 2009 Mar 6.
11. Uccelli, A., Moretta, L., Pistoia, V., Mesenchymal stem cells in health and disease. *Nat. Rev. Immunol.*, 8, 9, 726–36, 2008 Sep.
12. Caplan, A.I. and Correa, D., The MSC: an injury drugstore. *Cell Stem Cell*, 9, 1, 11–5, 2011 Jul 8.
13. Phinney, D.G. and Pittenger, M.F., Concise review: MSC-derived exosomes for cell-free therapy. *Stem Cells*, 35, 4, 851–8, 2017 Apr 1.
14. Wang, J., Neutrophils in tissue injury and repair. *Cell Tissue Res.*, 371, 531–9, 2018 Mar.
15. Takahashi, K. and Yamanaka, S., Induction of pluripotent stem cells from mouse embryonic and adult fibroblast cultures by defined factors. *Cell*, 126, 4, 663–76, 2006 Aug 25.
16. González, S., Chen, L., Deng, S. X., Comparative Study of Xenobiotic-Free Media for the Cultivation of Human Limbal Epithelial Stem/Progenitor Cells. *Tissue Engineering Part C: Methods*, 23, 4, 219-227, 2017.
17. Shi, Y., Wang, Y., Li, Q., Liu, K., Hou, J., Shao, C., Wang, Y., Immunoregulatory mechanisms of mesenchymal stem and stromal cells in inflammatory diseases. *Nat. Rev. Nephrol.*, 14, 8, 493–507, 2018 Aug.

18. Vandermeulen, M., Erpicum, P., Weekers, L., Briquet, A., Lechanteur, C., Detry, O., Beguin, Y., Jouret, F., Mesenchymal stromal cells in solid organ transplantation. *Transplantation*, 104, 5, 923–36, 2020 May 1.
19. Galipeau, J. and Sensébé, L., Mesenchymal stromal cells: clinical challenges and therapeutic opportunities. *Cell Stem Cell*, 22, 6, 824–33, 2018 Jun 1.
20. Wu, X., Jiang, J., Gu, Z., Zhang, J., Chen, Y., Liu, X., Mesenchymal stromal cell therapies: immunomodulatory properties and clinical progress. *Stem Cell Res. Ther.*, 11, 1, 345, 2020 Aug 8.
21. Kordelas, L., Rebmann, V., Ludwig, A.K., Radtke, S., Ruesing, J., Doeppner, T.R., Epple, M., Horn, P.A., Beelen, D.W., Giebel, B., MSC-derived exosomes: a novel tool to treat therapy-refractory graft-versus-host disease. *Leukemia*, 28, 4, 970–3, 2014 Apr.
22. Jiqian, Z., Ming, L., Tongxin, W., Wende, T., Jianqing, J., Hao, X., Association between visceral adiposity index and all-cause and cardiovascular mortality in the non-elderly adults, *Front. Endocrinol.*, 16, 2025.
23. Ayala-Cuellar, A.P., Kang, J.-H., Jeung, E.-B., Choi, K.-C., Roles of mesenchymal stem cells in tissue regeneration and immunomodulation. *Biomol. Ther.*, 27, 1, 25–33, 2018.
24. Pittenger, M.F., Mackay, A.M., Beck, S.C., Jaiswal, R.K., Douglas, R., Mosca, J.D., Moorman, M.A., Simonetti, D.W., Craig, S., Marshak, D.R., Multilineage potential of adult human mesenchymal stem cells. *Science*, 284, 5411, 143–147, 1999.
25. Orkin, S.H. and Zon, L.I., Hematopoiesis: An Evolving paradigm for stem cell biology. *Cell*, 132, 4, 631–644, 2008.
26. Wu, L., Mesenchymal stem cells as trophic mediators in cartilage regeneration, 2012.
27. Siegal, D., Keller, A., Xu, W., Bhuta, S., Kim, D.H., Kuruvilla, J., Lipton, J.H., Messner, H., Gupta, V., Central nervous system complications after allogeneic hematopoietic stem cell transplantation: incidence, manifestations, and clinical significance. *Biol. Blood Marrow Transplant.*, 13, 11, 1369–79, 2007 Nov 1.
28. Filardo, G., Perdisa, F., Roffi, A., Marcacci, M., Kon, E., Stem cells in articular cartilage regeneration. *J. Orthop. Surg. Res.*, 11, 1–5, 2016 Dec.
29. Rahman, M.M., Islam, M.R., Islam, M.T., Harun-Or-Rashid, M., Islam, M., Abdullah, S., Uddin, M.B., Das, S., Rahaman, M.S., Ahmed, M., Alhumaydhi, F.A., Stem cell transplantation therapy and neurological disorders: current status and future perspectives. *Biology*, 11, 1, 147, 2022 Jan 7.
30. Bakinowska, E., Kiełbowski, K., Boboryko, D., Bratborska, A.W., Olejnik-Wojciechowska, J., Rusiński, M., Pawlik, A., The role of stem cells in the treatment of cardiovascular diseases. *Int. J. Mol. Sci.*, 25, 7, 3901, 2024 Mar 31.
31. Rech Tondin, A. and Lanzoni, G., Islet Cell Replacement and Regeneration for Type 1 Diabetes: Current Developments and Future Prospects. *BioDrugs*, 39, 1–20, 2025 Feb 7.

32. Farabi, B., Roster, K., Hirani, R., Tepper, K., Atak, M.F., Safai, B., The efficacy of stem cells in wound healing: a systematic review. *Int. J. Mol. Sci.*, 25, 5, 3006, 2024 Mar 5.
33. Karussis, D., Petrou, P., Kassis, I., Clinical experience with stem cells and other cell therapies in neurological diseases. *J. Neurol. Sci.*, 324, 1–2, 1–9, 2013 Jan 15.
34. Chang, L., Li, L., Han, Y., Cheng, H., Yang, L., Organoids in Haematologic Research: Advances and Future Directions. *Cell Prolif.*, 26, 6, 2573, 2025.
35. Correia, C.D., Calado, S.M., Matos, A., Esteves, F., De Sousa-Coelho, A.L., Campinho, M.A., Fernandes, M.T., Advancing Glioblastoma Research with Innovative Brain Organoid-Based Models. *Cells*, 14, 4, 292, 2025 Feb 16.
36. Abusalah, M.A., Abd Rahman, E.N., Choudhary, O.P., Evolving trends in stem cell therapy: an emerging and promising approach against various diseases. *Int. J. Surg.*, 110, 11, 6862–8, 2024 Nov 1.
37. Hoang, D.M., Pham, P.T., Bach, T.Q., Ngo, A.T., Nguyen, Q.T., Phan, T.T., Nguyen, G.H., Le, P.T., Hoang, V.T., Forsyth, N.R., Heke, M., Stem cell-based therapy for human diseases. *Signal Transduction Targeted Ther.*, 7, 1, 272, 2022 Aug 6.
38. Rahimi Darehbagh, R., Seyedoshohadaei, S.A., Ramezani, R., Rezaei, N., Stem cell therapies for neurological disorders: current progress, challenges, and future perspectives. *Eur. J. Med. Res.*, 29, 1, 386, 2024 Jul 25.
39. Mei, R., Wan, Z., Yang, C., Shen, X., Wang, R., Zhang, H., Yang, R., Li, J., Song, Y., Su, H., Advances and clinical challenges of mesenchymal stem cell therapy. *Front. Immunol.*, 15, 1421854, 2024 Jul 19.
40. Song, S.J., Nam, Y., Rim, Y.A., Ju, J.H., Sohn, Y., Comparative analysis of regulations and studies on stem cell therapies: focusing on induced pluripotent stem cell (iPSC)based treatments. *Stem Cell Res. Ther.*, 15, 447, 1-24, 2024.
41. Committee for Advanced Therapies (CAT), Reflection paper on stem cell-based medicinal products, European Medicines Agency, pp. 1–14, 2011.
42. Ministry of Health, Labour and Welfare, Guidelines on clinical research using human stem cells, pp. 1–41, 2010.
43. Tiwari, S.S. and Desai, P.N., Unproven Stem Cell Therapies in India: Regulatory Challenges and Proposed Paths Forward. *Cell Stem Cell*, 23, 649–652, November 1, 2018.
44. Gaobotse, G., Stem Cell Research in Africa: Legislation and Challenges. *J. Regener. Med.*, 7, 1, 1–3, 2018.
45. Hanley, P.J., Therapeutic Mesenchymal Stromal Cells: Where We Are Headed. *Methods Mol. Biol.*, 1283, 1–11, 2015.
46. Pham, P.V. and Phan, N.K., Production of Good Manufacturing Practice-Grade Human Umbilical Cord Blood-Derived Mesenchymal Stem Cells for Therapeutic Use. *Methods Mol. Biol.*, 1283, 73–85, 2015.
47. International Society for Stem Cell Research (ISSCR), ISSCR Guidelines *for Stem Cell Research and Clinical Translation*, 1-73, May, 2021.

48. Herberts, C.A., Kwa, M.S.G., Hermsen, H.P.H., Risk factors in the development of stem cell Therapy. *J. Transl. Med.*, 9, 29, 1–14, 2011.
49. Papageorgiadis, N. and Sharma, A., Intellectual property rights and innovation: A panel analysis. *Econ. Lett.*, 141, 70–2, Apr 1, 2016.
50. Patent Politics: Life Forms, Markets and the Public Interest in the United States and Europe (review). *Am. J. Sociol.*, 123, 6, 1854–56, 2018.
51. Gharaibeh, B., Anderson, J.E., Deasy, B.M., Combating the threat of stem cell tourism through patient education and government regulation. *Innovation Entrepreneurship Health*, 2016:3, 15–24, Jan 27, 2016.
52. Tiwari, S.S. and Raman, S., Governing stem cell therapy in India: regulatory vacuum or jurisdictional ambiguity. *New Genet. Soc.*, 33, 4, 413–433, Oct 2, 2014, doi: 10.1080/14636778.2014.970269. Epub 2014 Oct 29. PMID: 25431534; PMCID: PMC4226321.
53. Petricciani, J., Hayakawa, T., Stacey, G., Trouvin, J.H., Knezevic, I., Scientific considerations for the regulatory evaluation of cell therapy products. *Biologicals*, 50, 20–6, Nov 1, 2017.
54. Stem-cell society condemns medical tourism. *Nature*, 453, 969, 2008, https://doi.org/10.1038/453969e.
55. Turner, L., ISSCR's Guidelines for Stem Cell Research and Clinical Translation: Supporting development of safe and efficacious stem cell-based interventions. *Stem Cell Rep.*, 16, 1394–1397, June 8, 2021.
56. Bolideei, M., Barzigar, R., Gahrouei, R.B. et al. Applications of Gene Editing and Nanotechnology in Stem Cell-Based Therapies for Human Diseases. *Stem Cell Rev. and Rep.*, 21, 905–934, 2025.
57. Altyar, A.E., Sayed, A.E., Abdeen, A., Piscopo, M., Mousa, S.A., Najda, A., Abdel-Daim, M.M., Future regenerative medicine developments and their therapeutic applications. *Biomed. Pharmacother.*, 158, 114131, 1–18, 2023.
58. Bianco, P. and Robey, P.G., Stem cells in tissue engineering. *Nature*, 414, 6859, 118–21, 2001.
59. Kosicki, M., Tomberg, K., Bradley, A., Repair of double-strand breaks induced by CRISPR-Cas9 leads to large deletions and complex rearrangements. *Nat. Biotechnol.*, 36, 8, 765–71, 2018.
60. Nielsen, A.A., Der, B.S., Shin, J., Vaidyanathan, P., Paralanov, V., Strychalski, E.A., *et al.*, Genetic circuit design automation. *Science*, 352, 6281, aac7341, 2016.
61. Ozbolat, I.T. and Hospodiuk, M., Current advances and future perspectives in extrusion-based bioprinting. *Biomaterials*, 76, 321–43, 2016.
62. Gluckman, E., Broxmeyer, H.A., Auerbach, A.D., Friedman, H.S., Douglas, G.W., Devergie, A., *et al.*, Hematopoietic reconstitution in a patient with Fanconi's anemia by means of umbilical-cord blood from an HLA-identical sibling. *N. Engl. J. Med.*, 321, 17, 1174–8, 1989.
63. Marks, P. and Gottlieb, S., Balancing Safety and Innovation for Cell-Based Regenerative Medicine. *N. Engl. J. Med.*, 378, 10, 954–9, 2018.

64. Kirkland, J.L. and Tchkonia, T., Senolytic drugs: from discovery to translation. *J. Intern. Med.*, 288, 5, 518–36, 2020.
65. Mason, C. and Dunnill, P., A brief definition of regenerative medicine. *Regen. Med.*, 3, 1, 1–5, 2008.
66. Park, J.R., Kim, E., Yang, J., Lee, H., Hong, S.H., Woo, H.M., *et al.*, Isolation of human dermis derived mesenchymal stem cells using explants culture method: expansion and phenotypical characterization. *Cell Tissue Bank.*, 16, 2, 209–18, 2015.
67. Reya, T., Morrison, S.J., Clarke, M.F., Weissman, I.L., Stem cells, cancer, and cancer stem cells. *Nature*, 414, 6859, 105–11, 2001.
68. Mendicino, M., Bailey, A.M., Wonnacott, K., Puri, R.K., Bauer, S.R., MSC-based product characterization for clinical trials: an FDA perspective. *Cell Stem Cell*, 14, 2, 141–5, 2014.
69. Song, B., Niclis, J. C., Alikhan, M. A., Sakkal, S., Sylvain, A., Kerr, P. G., Laslett, A. L., Bernard, C. A., Ricardo, S. D., Generation of Induced Pluripotent Stem Cells from Human Kidney Mesangial Cells. *J. Am. Soc. Nephrol.* 22, 7, 1213–20, July 2011.
70. Marei, M.K. and El Backly, R.M., Dental Mesenchymal Stem Cell-Based Translational Regenerative Dentistry: From Artificial to Biological Replacement. *Front. Bioeng. Biotechnol.*, 6, 49, 2018

11

Translational Research for Dermatological Stem Cell Therapy

Mishra Bharti[1], Khan Anam[1,2], Gothalwal Ragini[2], Jaiswal Prashansa[3] and Kotnis Ashwin[1*]

[1]*Department of Biochemistry, All India Institute of Medical Sciences, Bhopal, India*
[2]*Department of Biotechnology, Barkatullah University, Bhopal, India*
[3]*Department of Dermatology, All India Institute of Medical Sciences, Bhopal, India*

Abstract
Stem cell therapy offers significant potential in dermatology for skin regeneration, wound healing, and treating various skin conditions. This review highlights the types of stem cells used in dermatology, such as epidermal and mesenchymal stem cells, and explores their mechanisms in promoting healing and maintaining skin homeostasis. It examines current applications in treating conditions such as vitiligo, psoriasis, and cosmetic dermatology. Challenges in translating stem cell research to clinical practice, including safety concerns, ethical issues, and regulatory requirements, are discussed. The review also covers patient acceptance, emerging technologies such as three-dimensional bioprinting, and future prospects for stem cell therapies in dermatology.

Keywords: Stem cell, dermatology, skin homeostasis, wound healing, epidermal, mesenchymal, tumorigenicity, skin regeneration

11.1 Introduction

The field of stem cell dermatology speaks to an energetic and quickly advancing zone researchers inquire about, fueled by the potential to revolutionize the treatment of various skin conditions [1]. At the center of this field lies the exceptional capacity of stem cells to distinguish into

**Corresponding author*: Ashwin.biochemistry@aiimsbhopal.edu.in
***H-index: 16**

Sumel Ashique, Sanjay Dey, Biplab Debnath, Mohhammad Ramzan and Swarupananda Mukherjee (eds.) Stem Cell Therapy in Dermatological Disorders, (301–320) © 2025 Scrivener Publishing LLC

specialized skin cell sorts, empowering the recovery of harmed or infected tissue [2]. This regenerative potential is especially critical in dermatology, where the skin's steady presentation to natural components and natural maturing forms frequently leads to an assortment of weakening conditions [2, 3]. Embryonic stem cells, while advertising pluripotency, raise moral concerns and are less often as possible utilized in clinical settings. Grown-up stem cells, determined from sources such as bone marrow, fat tissue, and hair follicles, offer a more promptly available and morally less disputable alternative [4, 5].

11.1.1 Types of Stem Cells Used in Dermatology

Stem cell treatment could be a quickly advancing field in dermatology, advertising promising arrangements for different skin clutters, wound recuperating, and hair recovery. Diverse sorts of stem cells are being examined for their one-of-a-kind regenerative capabilities, immunomodulatory properties, and potential applications in skin repair and restoration [6].

11.1.1.1 Epidermal Stem Cells (ESCs)

Epidermal stem cells (ESCs) are fundamentally found within the basal layer of the epidermis and inside the hair follicle bulge, where they keep up skin keenness by creating keratinocytes for nonstop recharging and repair. These cells are basic for supplanting the external layers of skin that are routinely misplaced or harmed due to natural variables such as UV introduction, mechanical contact, or damage. Their quick turnover and versatile reaction to cellular push guarantee that the skin remains strong and utilitarian. Clinically, ESCs are tackled in skin joining strategies to treat extreme burns, incessant ulcers, and other nonhealing wounds, where their capacity to advance re-epithelialization essentially quickens the mending prepare. Additionally, their regenerative capacity is crucial in reestablishing the skin's obstruction work, in this manner, decreasing contamination dangers and improving recuperation. These highlights make ESCs a foundation in skin repair methodologies, and their clinical effect is really significant [7].

11.1.1.2 Mesenchymal Stem Cells (MSCs)

MSCs are flexible multipotent cells gotten from different tissues such as bone marrow, fat tissue, umbilical line blood, and indeed the dermis. These cells can separate into numerous cell sorts, including fibroblasts, which are fundamental for collagen generation and extracellular lattice arrangement. MSCs, moreover, emit a wide range of development components and cytokines that tweak the immune reaction and encourage tissue repair, subsequently diminishing irritation and advancing mending. In dermatological applications, MSCs are utilized to downplay scar arrangement and upgrade skin flexibility, while contributing to hair follicle incitement in cases of diminishing hair. Their combined capacity to both supplant harmed cells and organize regenerative forms through paracrine signaling makes them an important resource in skin treatment. This double usefulness has driven to their expanding utility in negligibly intrusive strategies pointed at reviving and repairing skin tissue. Their strong regenerative potential makes them irreplaceable in clinical hone in fact [8].

11.1.1.3 Hair Follicle Stem Cells (HFSCs)

HFSCs dwell within the bulge locale of hair follicles, a specialized specialty fundamentally to both the hair development cycle and the repair of the overlying epidermis. By moving to harm locales, HFSCs separate into keratinocytes and other cell sorts that reestablish the skin's auxiliary judgment. Their energetic part in hair follicle recovery makes them especially noteworthy in tending to conditions such as alopecia areata and androgenetic alopecia, where hair misfortune may be an essential concern. In dermatological medicines, HFSCs are progressively recognized for their capacity to progress hair thickness and back wound mending, subsequently advertising double benefits in both corrective and helpful applications. Their exact localization and characteristic regenerative capacity guarantee incite enactment in reaction to harm, which is basic for successful skin recuperation. This special usefulness underscores the guarantee of HFSCs focused on skin and hair rebuilding techniques, without a doubt [9, 10].

11.1.2 Methods of Stem Cell Isolation and Cultivation

Stem cell separation and development are basic forms in regenerative medication and investigation, including differing techniques custom fitted to the particular stem cell sort and inquiries about targets. Segregation starts with sourcing stem cells from different tissues, including bone marrow, fat tissue, umbilical rope blood, or initiated pluripotent stem cells (iPSCs) determined from separated cells [11]. Strategies such as thickness slope centrifugation, fluorescence-activated cell sorting, or magnetic-activated cell sorting are utilized to isolated stem cells from heterogeneous cell populaces based on their one-of-a-kind surface markers or physical properties. This development includes carefully controlled situations, including temperature, pH, and stickiness, at the side specialized culture media containing development components and cytokines that advance stem cell proliferation and self-renewal [12, 13]. For instance, mesenchymal stem cells can be cultured in media supplemented with fetal bovine serum, whereas hematopoietic stem cells may require a cocktail of growth factors like stem cell factor and thrombopoietin. Three-dimensional culture frameworks, such as hydrogels or platforms, are progressively utilized to imitate the common tissue microenvironment and upgrade stem cell practicality and usefulness. Besides, progressions in bioreactor innovation allow for large-scale stem cell generation, significant for clinical applications. Normal observation of stem cell characteristics, such as reasonability, multiplication rate, and separation potential, is fundamental to guarantee the quality and consistency of the developed cells [14, 15].

11.2 Mechanism of Stem Cell Action in Skin Regeneration and Healing

The component of stem cell activity in skin recovery and mending is characterized by a multifaceted approach including cellular separation, paracrine signaling, immunomodulation, and extracellular framework (extracellular matrix [ECM]) balance. Stem cells, strikingly MSCs, display pluripotency, empowering separation into key cutaneous cell ancestries, including keratinocytes, fibroblasts, and endothelial cells, subsequently reconstituting harmed tissue engineering. Concurrently, these cells lock in paracrine signaling, discharging a complex milieu of development components and cytokines that invigorate angiogenesis, fibroblast expansion, and collagen blend, basic for wound closure and tissue remodeling. Besides, stem cells apply immunomodulatory impacts, moderating incendiary reactions and avoiding over-the-top scar arrangement, vital in unremitting wound administration. They too impact ECM flow, controlling its union and remodeling to reestablish tissue keenness. Endogenous ESCs, dwelling inside the epidermis and hair follicles, contribute to re-epithelialization amid wound repair, guaranteeing persistent epidermal recharging. Collectively, these forms emphasize the helpful potential of stem cells in differing dermatological pathologies (Figure 11.1) [16, 17].

How stem cell promote wound healing: Stem cells play an essential part in organizing exceedingly facilitated regenerative cascade basic for successful wound recuperating [18]. Their one-of-a-kind capacities of self-renewal,

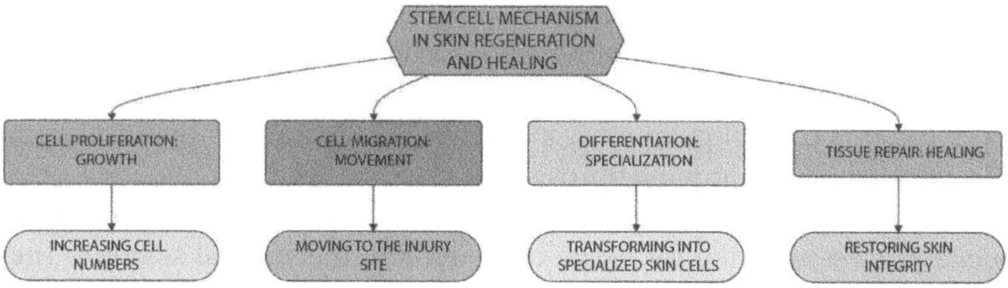

Figure 11.1 Schematic representation of stem cell mechanism in skin regeneration and healing.

multipotency, and paracrine signaling empower them to direct different stages of repair preparation [19].

At the introductory stage of wound mending, stem cells start a vigorous paracrine reaction by emitting a differing collection of development components, cytokines, and chemokines. These bioactive atoms, such as vascular endothelial growth factor (VEGF), transforming growth factor β (TGF-β), epidermal growth factor, and platelet-derived growth factor, fortify the movement, multiplication, and enactment of inhabitant skin cells, including fibroblasts and keratinocytes [20]. This interaction advances the arrangement of granulation tissue, quickens re-epithelialization, and upgrades ECM testimony basic steps in wound closure [20, 21].

Past paracrine signaling, stem cells also show exceptional versatility, separating into specialized cell sorts that are basic for basic and utilitarian tissue rebuilding. Fibroblastic separation bolsters collagen testimony and ECM rebuilding, fortifying the ductile quality of the recently shaping tissue. Endothelial separation contributes to angiogenesis, a significant handle that guarantees satisfactory supplement and oxygen conveyance to the mending tissue. Moreover, epithelial separation helps in reestablishing the epidermal barrier, protecting the wound from infections and environmental insults [22, 23]. Stem cells contribute to ECM remodeling by synthesizing and discharging basic components, giving back basic and encouraging tissue cohesion. This multifaceted approach, including paracrine signaling, cellular separation, angiogenesis, immunomodulation, and ECM remodeling, underscores the basic part of stem cells in advancing productive and viable wound recuperating, eventually driving to utilitarian tissue rebuilding [24]. A basic angle of stem cell–mediated wound mending is their capacity to balance the incendiary microenvironment, avoiding inveterate aggravation that might disable tissue repair. Stem cells accomplish this through immunomodulatory impacts, including the direction of macrophage polarization, concealment of over-the-top proinflammatory cytokine discharge (e.g., tumor necrosis factor α [TNF-α], interleukin 6 [IL-6]), and improvement of anti-inflammatory cytokines (e.g., IL-10, TGF-β). This control guarantees a controlled safe reaction that underpins recovery, while minimizing fibrosis and scarring [25, 26].

Besides, stem cells actively contribute to extracellular matrix remodeling, a process essential for tissue cohesion and mechanical stability. They also help regulate the inflammatory microenvironment, preventing excessive inflammation that could impair healing. They accomplish this by directing resistant cell movement and cytokine discharge, making a favorable environment for tissue recovery.

Role of stem cells in skin homeostasis: The stem cells serve as principal controllers of skin homeostasis, guaranteeing nonstop recovery of epidermal and dermal structures to preserve tissue astuteness, versatility, and boundary work. Their energetic exchange between self-renewal and lineage-specific differentiation enables the skin to resist aging and recover from minor wounds [27].

Within the epidermal compartment, basal layer ESCs (EpSCs) recharge keratinocytes, which shape the defensive peripheral skin boundary. These stem cells are directed by signaling pathways such as Wnt score and BMP, which arrange their multiplication and separation to support epidermal turnover. ESCs also react to neighboring tissue harm by rapidly mobilizing to locales of harm, where they encourage re-epithelialization [28].

Within the dermal compartment, MSCs contribute to fibroblast population mindful for ECM generation, collagen union, and dermal auxiliary bolster. These MSCs keep up the mechanical quality of the skin by balancing collagen homeostasis and reacting to biochemical prompts from the microenvironment, such as fibroblast growth factor and TGF-β [29].

Past basic support, stem cells also contribute to vascular arrangement solidness and resistant observation inside the skin. By emitting angiogenic variables such as VEGF, they regain endothelial cell survival and microvascular remodeling, guaranteeing nonstop oxygen and supplement supply. Also, they connect with resistant cells to control aggravation and improve antimicrobial defense components, providing protection against contaminations and natural aggressors [18, 30]. The profoundly coordinated movement of these particular stem cell populations is indispensable for keeping up skin homeostasis all through an individual's lifespan, allowing for adjustment to physiological changes and outside challenges while avoiding untimely tissue degeneration [31, 32].

11.3 Current Applications of Stem Cell Therapy in Dermatology

Stem cell treatment may be a quick advancing field in dermatology, leveraging the regenerative potential of stem cells to advance tissue repair, tweak safe reactions, and upgrade stylish results [19]. Current clinical and exploratory applications basically center on wound mending, immune system and fiery skin disarrangements, and corrective dermatology. MSCs and adipose-derived stem cells (ADSCs) have garnered significant consideration due to their capacity to distinguish into different cell sorts and discharge bioactive variables that balance cellular capacities [33].

11.3.1 Role of Stem Cells in Wound Healing

Wound healing and scar reduction: Stem cell–based treatments are being broadly investigated for their part in upgrading wound recuperating, especially in persistent nonhealing wounds related to conditions such as diabetes and venous lacking. MSCs and ADSCs advance re-epithelialization, upgrade angiogenesis, and tweak the fiery reaction, subsequently quickening wound closure and diminishing fibrosis. Their application amplifies scar modification, where their regenerative and immunomodulatory properties progress both the stylish and utilitarian results of cutaneous scarring [34].

Treatment of dermatological conditions: Stem cell treatment is being assessed for its viability in treating different skin infections, including vitiligo, psoriasis, pemphigus, systemic lupus erythematosus (SLE), scleroderma, epidermolysis bullosa, and alopecia. Each of these conditions presents special pathophysiological challenges, requiring focus on stem cell mediations as shown in Table 11.1.

It is significant that treatment choices are profoundly individualized, figuring in illness seriousness, persistent comorbidities, and potential treatment dangers and benefits. The administration of complex dermatological disarranges such as vitiligo, psoriasis, SLE, etc., requires a multifaceted and patient-centric approach, as these illnesses have assorted pathophysiological instruments requiring custom-fitted helpful methodologies. Determination of treatment modalities depends on illness seriousness; patient-specific contemplations such as age, hereditary inclination, safety status, and comorbid conditions; and an exhaustive risk–benefit examination to optimize viability while minimizing antagonistic impacts [44]. Restorative mediations incorporate topical medications such as corticosteroids, calcineurin inhibitors, and vitamin D analogs; phototherapy using

Table 11.1 Current therapies and potential stem cell-based applications for various conditions and diseases.

Condition/ disease	Current therapies	Stem cell applications	References
Wound healing	Dressings, debridement, growth factors	Mesenchymal stem cells (MSCs)/ adipose-derived stem cells (ADSCs) for enhanced healing, reduced fibrosis	[35]
Vitiligo	Corticosteroids, phototherapy, melanocyte transplantation	MSC-based immuno-modulation and melanocyte regeneration	[36]
Psoriasis	Biologic agents, phototherapy	MSCs to reduce inflammation and keratinocyte proliferation	[37]
Pemphigus	Immuno-suppressants, corticosteroids	Hematopoietic stem cell transplantation for immune reset and disease remission	[38]
Systemic lupus erythematosus	Antimalarials, immuno-suppressants	MSC therapy for immune modulation	[39]
Scleroderma	Vasodilators, antifibrotics	Autologous stem cell transplantation for severe cases	[40]
Epidermolysis bullosa	Wound care, gene therapy	Stem cell transplantation for tissue regeneration	[41]
Alopecia	Minoxidil, corticosteroids	ADSCs for hair follicle regeneration	[42]
Cosmetic dermatology	Fillers, lasers, chemical peels	Stem cells for skin rejuvenation and scar repair	[43]

bright (UV) light; systemic immunomodulation with corticosteroids; biologics (TNF inhibitors, IL-17/IL-23 inhibitors); small atom inhibitors (JAK inhibitors); and imaginative treatments such as stem cell transplantation and gene-based intercessions [45]. Despite noteworthy progresses, current treatments fundamentally provide symptomatic help instead of tending to fundamental atomic drivers of illness, requiring proceeded inquiry about novel immunomodulatory specialists, quality treatments utilizing CRISPR-based approaches, stem cell and regenerative medication, and exceedingly targeted small molecules to refine exactness treatment. The longer term of dermatological therapeutics lies in personalized medication, where genomic and proteomic profiling will direct individualized treatment procedures, progressing adequacy while lessening unfavorable impacts. Integration of accuracy medication with artificial intelligence-driven diagnostics and rising restorative developments are set to revolutionize dermatology, advertising long-term malady control and potential healing arrangements for complex skin clutters [46].

11.3.2 Use of Stem Cells in Cosmetic Dermatology

The burgeoning field of stem cell applications in corrective dermatology speaks to a critical flight from conventional stylish mediations, grasping regenerative standards to address a range of cutaneous concerns. Central to this approach is the abuse of MSCs and adipose-derived stem cells (ADSCs), famous for their multipotent separation capacity and powerful paracrine signaling. These cells, when presented into target tissues, organize a complex cascade of natural occasions, fundamentally through the emission of development variables, cytokines, and extracellular vesicles. This expansion of the dermal framework deciphers to unmistakable stylish benefits, including upgraded skin versatility and lessened rhytides, and made strides in, by and large, skin surface and tone. Moreover, these cells advance angiogenesis and the arrangement of modern blood vessels, which serves to upgrade tissue perfusion and supplement conveyance, cultivating a revitalized cutaneous environment [47].

The clinical applications of stem cell treatments in restorative dermatology are differing and advancing. Skin revival conventions point to check the inborn and outward maturing forms by recharging exhausted cellular components and reestablishing dermal framework keenness, coming about in a more young and dynamic appearance. Scar amendment strategies use the immunomodulatory and regenerative properties of stem cells to balance fibroblast action, advancing organized collagen testimony and minimizing the arrangement of hypertrophic or keloidal scars. Hair recovery techniques investigate the potential of stem cells to fortify torpid

HFSCs, advertising a promising road for tending to conditions such as androgenetic alopecia. Volumetric reclamation strategies utilize ADSCs in fat transfer techniques to restore volume loss in areas such as the face, providing a natural and long-lasting enhancement [48, 49].

In any case, the interpretation of stem cell treatments from seat to bedside requires fastidious logical thoroughness. Long-term security and adequacy information, inferred from vigorous clinical trials, are fundamental to set up the helpful potential of these mediations and to moderate potential dangers [50]. Besides, moral contemplations encompassing stem cell control, including the potential for tumorigenesis and off-target impacts, must be completely tended to. The field of stem cell–based restorative dermatology is balanced to revolutionize stylish medication by tackling the body's intrinsic regenerative capacity, advertising a naturally grounded approach to accomplishing characteristic and persevering stylish upgrades [49].

The application of stem cells inside corrective dermatology speaks to a wilderness of regenerative pharmaceutical, advertising a flight from conventional, symptom-focused medicines. By leveraging the inalienable regenerative capabilities of cells such as MSCs and ADSCs, clinicians point to address the root causes of stylish maturing and tissue corruption. The paracrine signaling components, advancing collagen and elastin union, nearby angiogenesis, give an experimentally grounded premise for watched enhancements in skin quality, scar diminishment, and hair recovery. Future advance pivots on thorough logical request, including standardized conventions, long-term security evaluations, and strong clinical trials [51]. Besides, constant consideration to moral contemplations is vital to guarantee dependable and secure usage. Eventually, the integration of stem cell advances into restorative dermatology holds the guarantee of conveying characteristic, persevering stylish improvements through the body's regenerative potential, stamping a noteworthy headway within the interest of cutaneous revival [49, 51].

11.4 Safety and Challenges in Translating Stem Cell Research to Clinical Practice

The method of turning stem cell disclosures into dependable clinical treatments is complex. This move is not simply approximately accomplishing a logical point of reference; it requires assembly and strict security benchmarks and administrative prerequisites that adjust development with understanding assurance [52]. Preeminent among security concerns is the characteristic chance of uncontrolled cellular multiplication, which can lead to the arrangement of tumors [53]. Besides, safe reactions, especially in cases where giver cells are utilized, require fastidious administration to anticipate dismissal. Interpretation into clinical hone also faces critical obstacles, counting the requirement for standardized cell generation and thorough quality control to guarantee reliable helpful results [54–57].

Ethical considerations and public awareness: Clinical interpretation of stem cell investigation faces moral and open mindfulness obstacles. Embryonic stem cell utilization evokes ethical talk about respect to developing life status. Potential innovation abuse, like cloning, requires strict moral rules. Educated assent in trials, particularly for defenseless bunches, is critical. Equitable getting to exorbitant treatments could be a concern. Open mindfulness is ruined by deception, sensationalized media, and a need of understanding obstructing educated choices. Recognizing dubious treatments is imperative. Exploring moral complexities and cultivating educated talk are fundamental for capable headway [58, 59].

Regulatory requirements for stem cell–based therapies: Administrative prerequisites for stem cell treatments prioritize understanding security and efficacy through thorough controls. Rigid item characterization, including character, immaculateness, and power, is obligatory, nearby adherence to GMP measures for reliable generation and careful testing for contaminants. Comprehensive preclinical and staged clinical trials, conducted beneath GCP rules, are basic for assessing security and adequacy some time recently and amid human utilize [60]. Administrative organizations manage improvement, tending to cell sourcing, handling, and organization, with particular rules for diverse stem cell sorts. Postmarket reconnaissance guarantees continuous checking of long-term impacts, with vigorous detailing frameworks for antagonistic occasions, emphasizing the adjustment between advancement and persistent assurance [55–60].

Immune rejection and use of autologous versus allogeneic stem cells: Resistant dismissal poses a major impediment in stem cell treatments, directing benefactor cell determination. Autologous cells dodge dismissal but may be restricted or unacceptable. Allogeneic cells, although more readily available, carry the risk of graft-versus-host disease and immune responses, often requiring immunosuppression. Mitigation strategies envelop HLA coordinating, T-cell consumption, and hereditary adjustment to reduce immunogenicity. Developing procedures, such as actuated resistance and immunomodulatory cell utilization, point to diminished dependence on immunosuppressants and upgrade lifespan [17].

Tumorigenicity and other safety concerns: Clinical interpretation of stem cell treatments is restricted by tumorigenesis hazard, emerging from left-over undifferentiated cells possibly driving to neoplasia. Key security concerns incorporate undifferentiated cell determination postdifferentiation, genomic precariousness from delayed culture, vector-mediated insertional mutagenesis, safe dismissal, and ectopic tissue arrangement. Relief requires thorough cell line characterization, comprehensive tumorigenicity testing, and posttransplantation observation [56].

Emerging technologies in stem cell research for dermatology: Dermatological stem cell investigation utilizes advancing procedures to reinforce security and adequacy. Refinements in iPSC reconstruction minimize genomic insecurity and dispense with vector dangers, checking tumorigenesis. Coordinated reconstruction bypasses pluripotency, diminishing this chance. Biomaterials and three-dimensional (3D) bioprinting empowers exact spatial control of separation and tissue arrangement, constraining ectopic tissue. CRISPR-Cas9 encourages focus on quality redress and expression control, improving cell work and lessening off-target impacts. Single-cell RNA sequencing and progressed imaging give point-by-point cell characterization, empowering exacting quality control and posttransplantation checking, tending to both tumorigenicity and broader security concerns [46, 61].

3D bioprinting and organoids for skin regeneration: 3D bioprinting and organoids progress skin recovery, showing particular security contemplations. Bioprinting permits exact layering of cells, biomaterials, and components, reproducing skin structure and restricting ectopic tissue. Be that as it may, vascularization and innervation of constructs pose challenges to graft survival. Organoids, self-organizing tissue societies, help skin improvement and illness, but standardization and reproducibility are crucial. The potential for unintended cell sorts inside organoids requires thorough characterization to relieve dangers such as uncontrolled development and sometime recently clinical application [48].

11.5 Conclusion and Prospects for the Future

The progression of dermatological stem cell treatments requires a strong translational inquiry about worldview. To realize the clinical potential, a center on robotic explanation in preclinical models is vital, taken by thorough clinical approval. Standardized strategies for cell obtainment, separation, and organization are pivotal for guaranteeing information reproducibility and comparative examination [62]. The integration of advanced imaging modalities and biomarker profiling will encourage exact appraisal of helpful reactions and longitudinal results. Collaborative, multidisciplinary endeavors, including fundamental science, clinical hone, and administrative oversight, are irreplaceable for exploring the inborn complexities and quickening the secure and compelling clinical interpretation of these restorative intercessions [63, 64].

Future role of stem cells in dermatology: The dermatological therapeutics is progressively interwoven with the translational potential of stem cell science, promising to revolutionize clinical hone by bridging the hole between crucial inquiry and persistent care. Current examinations are centered on overcoming characteristic confinements in cell-based treatments, including the optimization of stem cell sourcing, extension, and focus on conveyance to cutaneous tissues. A basic area of improvement includes the application of gene-editing technologies, such as CRISPR-Cas9, to engineer stem cells with enhanced therapeutic properties, including increased growth factor secretion, improved homing capabilities, and reduced immunogenicity, thereby addressing complex pathologies like chronic wounds and genetic skin disorders [60–62]. The interest of personalized medication requires the refinement of conventions for utilizing patient-derived iPSCs or MSCs, empowering customized medicines that maximize viability and minimize unfavorable resistant responses. Concurrently, the integration of stem cells with biocompatible biomaterials, such as hydrogels and nanofibers, is being investigated to form 3D frameworks that mirror the local extracellular lattice, advancing cell survival, separation, and tissue recovery. Moreover, the advancement of advanced imaging modalities, including multiphoton microscopy and optical coherence tomography, encourages noninvasive checking of stem cell behavior and tissue remodeling *in vivo*, giving important experiences on treatment viability. The foundation of standardized clinical trial conventions and thorough result measures is basic to improve the security and viability of these rising treatments, clearing the way for their far-reaching appropriation in clinical dermatology and changing the administration of obstinate skin conditions [57, 63].

AI disclosure statement: The author confirms that no artificial intelligence (AI)-generated content or AI-assisted writing tools were used in the preparation of this chapter (including summaries and the abstract). All text, analysis, and conclusions are the result of the author's original work.

References

1. Park, G., Kwon, H.H., Seok, J., Yang, S.H., Lee, J., Park, B.C., *et al.*, Efficacy of combined treatment with human adipose tissue stem cell-derived exosome-containing solution and microneedling for facial skin aging: A 12-week prospective, randomized, split-face study. *J. Cosmet. Dermatol.*, 22, 12, 3418–26, 2023 Dec.
2. Trubiani, O., Marconi, G.D., Pierdomenico, S.D., Piattelli, A., Diomede, F., Pizzicannella, J., Human Oral Stem Cells, Biomaterials and Extracellular Vesicles: A Promising Tool in Bone Tissue Repair. *Int. J. Mol. Sci.*, 20, 20, 4987, 2019 Oct 9.
3. Zhang, B., Yan, W., Zhu, Y., Yang, W., Le, W., Chen, B., *et al.*, Nanomaterials in Neural-Stem-Cell-Mediated Regenerative Medicine: Imaging and Treatment of Neurological Diseases. *Adv. Mater.*, 30, 17, 1705694, 2018 Apr.
4. Ghasemi, N., Azizi, H., Skutella, T., Unraveling the Significance of Nanog in the Generation of Embryonic Stem-like Cells from Spermatogonia Stem Cells: A Combined *In Silico* Analysis and *In Vitro* Experimental Approach. *Int. J. Mol. Sci.*, 25, 9, 4833, 2024 Apr 29.
5. Meng, S., Whitt, A.G., Eaton, J.W., Yaddanapudi, K., Li, C., The efficacy of an embryonic stem cell-based vaccine for lung cancer prevention depends on the undifferentiated state of the stem cells. *Sci. Rep.*, 14, 1, 32127, 2024 Dec 30.
6. Zare, S., Kurd, S., Rostamzadeh, A., Nilforoushzadeh, M.A., Types of Stem Cells in Regenerative Medicine: A Review. *J. Skin Stem Cell [Internet]*, 1, 3, 2014 Dec 20 [cited 2025 Feb 20]. Available from: https://brieflands.com/articles/jssc-14658.html.
7. [PDF] Epidermal Stem Cells in Wound Healing and Regeneration, Semantic Scholar, [cited 2025 Feb 21]; Available from: https://www.semanticscholar.org/paper/Epidermal-Stem-Cells-in-Wound-Healing-and-Yang-Wang/e330b500e11e573d25ad25a92e39477a95c0484f.
8. Wu, S., Sun, S., Fu, W., Yang, Z., Yao, H., Zhang, Z., The Role and Prospects of Mesenchymal Stem Cells in Skin Repair and Regeneration. *Biomedicines*, 12, 4, 743, 2024 Mar 27.
9. [PDF] Short-Term Clinical Assessment of Treating Female Androgenetic Alopecia with Autologous Stem Cells Derived from Human Hair Follicles, Semantic Scholar, [cited 2025 Feb 21]; Available from: https://www.semanticscholar.org/paper/Short-Term-Clinical-Assessment-of-Treating-Female-Krefft-Trzciniecka-Pi%C4%99towska/a38fe89b7ae8537aa6188649bbf411fe04e71c68.
10. Fatehi, A., Sadat, M., Fayyad, M., Tang, J., Han, D., Rogers, I.M., *et al.*, Efficient Generation of Pancreatic Progenitor Cells from Induced Pluripotent Stem Cells Derived from a Non-Invasive and Accessible Tissue Source—The Plucked Hair Follicle. *Cells*, 13, 12, 1010, 2024 Jun 10.

11. Wang, X., Li, F., Wu, S., Xing, W., Fu, J., Wang, R., *et al.*, Research progress on optimization of *in vitro* isolation, cultivation and preservation methods of dental pulp stem cells for clinical application. *Front. Bioeng. Biotechnol.*, 12, 1305614, 2024 Apr 3.
12. Janssens, S., Schotsaert, M., Manganaro, L., Dejosez, M., Simon, V., García-Sastre, A., *et al.*, FACS-Mediated Isolation of Neuronal Cell Populations From Virus-Infected Human Embryonic Stem Cell–Derived Cerebral Organoid Cultures. *Curr. Protoc. Stem Cell Biol.*, 48, 1, e65, 2019 Feb.
13. [PDF] Purification of Pig Muscle Stem Cells Using Magnetic-Activated Cell Sorting (MACS) Based on the Expression of Cluster of Differentiation 29 (CD29), Semantic Scholar [Internet], [cited 2025 Feb 21]. Available from: https://www.semanticscholar.org/paper/Purification-of-Pig-Muscle-Stem-Cells-Using-Cell-on-Choi-Kim/072d1fa64cc2616ec542188de9d399034f248675.
14. Chaicharoenaudomrung, N., Kunhorm, P., Noisa, P., Three-dimensional cell culture systems as an *in vitro* platform for cancer and stem cell modeling. *World J. Stem Cells*, 11, 12, 1065–83, 2019 Dec 26.
15. [PDF] Recent Advances in Three-Dimensional Stem Cell Culture Systems and Applications, Semantic Scholar, [cited 2025 Feb 21]; Available from: https://www.semanticscholar.org/paper/Recent-Advances-in-Three-Dimensional-Stem-Cell-and-Wu-Su/fa157c8168d2568fc81d043c0800aca14966c9da.
16. Ohnsorg, M.L., Hushka, E.A., Anseth, K.S., Photoresponsive Chemistries for User-Directed Hydrogel Network Modulation to Investigate Cell–Matrix Interactions. *Acc. Chem. Res.*, 58, 1, 47–60, 2025 Jan 7.
17. Aghlara-Fotovat, S., Nash, A., Kim, B., Krencik, R., Veiseh, O., Targeting the extracellular matrix for immunomodulation: applications in drug delivery and cell therapies. *Drug Deliv. Transl. Res.*, 11, 6, 2394–413, 2021 Dec.
18. Zou, M., Liu, S., Sun, Z., Wu, J., Yuan, Z., Teng, Y., *et al.*, Insights into the role of adipose-derived stem cells: Wound healing and clinical regenerative potential. *J. Cell. Physiol.*, 236, 4, 2290–7, 2021 Apr.
19. Arwert, E.N., Hoste, E., Watt, F.M., Epithelial stem cells, wound healing and cancer. *Nat. Rev. Cancer*, 12, 3, 170–80, 2012 Mar.
20. Kamentseva, R.S., Kharchenko, M.V., Gabdrahmanova, G.V., Kotov, M.A., Kosheverova, V.V., Kornilova, ES. EGF, TGF-α and Amphiregulin Differently Regulate Endometrium-Derived Mesenchymal Stromal/Stem Cells. *Int. J. Mol. Sci.*, 24, 17, 13408, 2023 Aug 29.
21. Zakaria, S., Elshazly, A.M., Alaa, R., Elsebaey, S., Dantrolene and coenzyme Q10 as a suggested combination therapy to statin-induced myopathy in high fat diet rats: A possible interference with ROS/TGF-β/Smad4 signaling pathway. *Toxicol. Appl. Pharmacol.*, 485, 116900, 2024 Apr.
22. Guo, Z., Su, W., Zhou, R., Zhang, G., Yang, S., Wu, X., *et al.*, Exosomal MATN3 of Urine-Derived Stem Cells Ameliorates Intervertebral Disc Degeneration by Antisenescence Effects and Promotes NPC Proliferation

and ECM Synthesis by Activating TGF-β. *Oxid. Med. Cell Longev.*, 2021, 1, 5542241, 2021 Jan. Yang S, editor.

23. Mao, X., Zhang, X., Zheng, X., Chen, Y., Xuan, Z., Huang, P., Curcumin suppresses LGR5(+) colorectal cancer stem cells by inducing autophagy and *via* repressing TFAP2A-mediated ECM pathway. *J. Nat. Med.*, 75, 3, 590–601, 2021 Jun.
24. [PDF] Therapy Resistance, Cancer Stem Cells and ECM in Cancer: The Matrix Reloaded, Semantic Scholar, [cited 2025 Feb 28]; Available from: https://www.semanticscholar.org/paper/Therapy-Resistance%2C-Cancer-Stem-Cells-and-ECM-in-Kesh-Gupta/7fa724011a5e0c4647773bc384335643742ee649.
25. Amiri, S., The effect of parvovirus B19 on expression levels of IL6, IL-1β, TNF-α and NF-κB genes in human bone marrow mesenchymal stem cells (hBM-MSCs) *in vitro*, 2019 [cited 2025 Feb 28]. Available from: https://www.semanticscholar.org/paper/The-effect-of-parvovirus-B19-on-expression-levels-Amiri/47387324879f909857ec196e043a84590686f4e6.
26. Zhao, C., Pu, Y., Zhang, H., Hu, X., Zhang, R., He, S., *et al.*, IL10-modified Human Mesenchymal Stem Cells inhibit Pancreatic Cancer growth through Angiogenesis Inhibition. *J. Cancer*, 11, 18, 5345–52, 2020.
27. Benitah, S.A., Epidermal stem cells in skin homeostasis and cutaneous carcinomas. *Clin. Transl. Oncol.*, 9, 12, 760–6, 2007 Dec.
28. Zhu, K., Liu, Y., Fan, C., Zhang, M., Cao, H., He, X., *et al.*, Etv5 safeguards trophoblast stem cells differentiation from mouse EPSCs by regulating fibroblast growth factor receptor 2. *Mol. Biol. Rep.*, 47, 12, 9259–69, 2020 Dec.
29. [PDF] Derivation of Cell-Engineered Nanovesicles from Human Induced Pluripotent Stem Cells and Their Protective Effect on the Senescence of Dermal Fibroblasts, Semantic Scholar, [cited 2025 Feb 28]; Available from: https://www.semanticscholar.org/paper/Derivation-of-Cell-Engineered-Nanovesicles-from-and-Lee-Cha/bb68a4fd1eb9409de5bdb3e4ef023f76559536c2.
30. Kumar, S., Mahajan, S., Kale, D., Chourasia, N., Khan, A., Asati, D., *et al.*, Insights into the gut microbiome of vitiligo patients from India. *BMC Microbiol.*, 24, 1, 440, 2024 Oct 28.
31. Li, X., Wu, A., Han, C., Chen, C., Zhou, T., Zhang, K., *et al.*, Bone marrow-derived mesenchymal stem cells in three-dimensional co-culture attenuate degeneration of nucleus pulposus cells. *Aging*, 11, 20, 9167–87, 2019 Oct 30.
32. Ding, J., Zhang, R., Li, H., Ji, Q., Cheng, X., Thorne, R.F., *et al.*, ASIC1 and ASIC3 mediate cellular senescence of human nucleus pulposus mesenchymal stem cells during intervertebral disc degeneration. *Aging*, 13, 7, 10703–23, 2021 Apr 15.
33. Van Pham, P., Adipose stem cells in the clinic. *Biomed. Res. Ther.*, 1, 2, 11, 2014 May.

34. Park, S.Y. and Yi, K.H., Exosome-mediated Advancements in Plastic Surgery: Navigating Therapeutic Potential in Skin Rejuvenation and Wound Healing. *Plast. Reconstr. Surg. Glob. Open*, 12, 8, e6021, 2024 Aug.
35. Guo, E., Wang, L., Wu, J., Chen, Q., Exosomes from MicroRNA-125b-Modified Adipose-Derived Stem CellsPromote Wound Healing of Diabetic Foot Ulcers. *Curr. Stem Cell Res. Ther.*, 20, 4, 409–20, 2025 May.
36. Huang, L., Zuo, Y., Li, S., Li, C., Melanocyte stem cells in the skin: Origin, biological characteristics, homeostatic maintenance and therapeutic potential. *Clin. Transl. Med.*, 14, 5, e1720, 2024 May.
37. Chang, W., Liang, N., Cao, Y., Xing, J., Li, J., Li, J., *et al.*, The effects of human dermal–derived mesenchymal stem cells on the keratinocyte proliferation and apoptosis in psoriasis. *Exp. Dermatol.*, 30, 7, 943–50, 2021 Jul.
38. Gillard, G.O., Proctor, J.L., Brooks, M.L., Lamothe, T.L., Hyzy, S.L., Mikse, O., *et al.*, A Novel Targeted Approach to Achieve Immune System Reset: CD45-Targeted Antibody Drug Conjugates Enable Autologous HSCT, Ameliorate Disease in Autoimmune Models, Potently Kill Human Immune Cells from Normal Donors and MS Patients, and Achieve Immune Depletion in Non-Human Primates (NHP). *Blood*, 134, Supplement_1, 3208–3208, 2019 Nov 13.
39. Ginting, A.R., Munir, D., Amin, M.M., Darlan, D.M., Putra, A., Rusda, M., *et al.*, Mesenchymal stem cells for immune modulation in systemic lupus erythematosus: From bench research to clinical applications. *Narra J.*, 4, 3, e994, 2024 Oct 25.
40. Sullivan, K.M., Goldmuntz, E.A., Keyes-Elstein, L., McSweeney, P.A., Pinckney, A., Welch, B., *et al.*, Myeloablative Autologous Stem-Cell Transplantation for Severe Scleroderma. *N. Engl. J. Med.*, 378, 1, 35–47, 2018 Jan 4.
41. Raoufinia, R., Rahimi, H.R., Keyhanvar, N., Moghbeli, M., Abdyazdani, N., Rostami, M., *et al.*, Advances in Treatments for Epidermolysis Bullosa (EB): Emphasis on Stem Cell-Based Therapy. *Stem Cell Rev. Rep.*, 20, 5, 1200–12, 2024 Jul.
42. Tang, H., Wang, F., Yang, R., Zhao, Z., Zhang, Y., Yang, L., *et al.*, Baricitinib-loaded EVs promote alopecia areata mouse hair regrowth by reducing JAK-STAT-mediated inflammation and promoting hair follicle regeneration. *Drug Discov. Ther.*, 18, 6, 368–74, 2024 Dec 31.
43. The effect of three-dimensional cultured adipose tissue-derived mesenchymal stem cell–conditioned medium and the antiaging effect of cosmetic products containing the medium, Semantic Scholar, [cited 2025 Feb 28]; Available from: https://www.semanticscholar.org/paper/The-effect-of-three-dimensional-cultured-adipose-of-Kim-Kim/ec32cf53776c02b44f56284b082dd340c42f35c4.
44. Reis, C., Gospodarev, V., Reis, H., Wilkinson, M., Gaio, J., Araujo, C., *et al.*, Traumatic Brain Injury and Stem Cell: Pathophysiology and Update on Recent Treatment Modalities. *Stem Cells Int.*, 2017, 1–13, 2017.

45. [PDF] Therapeutic Applications of Mesenchymal Stem Cell Loaded with Gold Nanoparticles for Regenerative Medicine, Semantic Scholar, [cited 2025 Feb 28]; Available from: https://www.semanticscholar.org/paper/Therapeutic-Applications-of-Mesenchymal-Stem-Cell-Cheng-Yang/55428ea281a1a5cdaad72560f905c42582807384.
46. Chakraborty, D. and De, A., Stem-cell therapy in dermatology – Challenges and opportunities. *Indian J. Skin Allergy*, 3, 93–105, 2024 Aug 5.
47. Shen, X., Song, S., Chen, N., Liao, J., Zeng, L., Stem cell-derived exosomes: A supernova in cosmetic dermatology. *J. Cosmet. Dermatol.*, 20, 12, 3812–7, 2021 Dec.
48. Nistor, G., Poole, A., Draelos, Z., Lupo, M., Tzikas, T., Liu, J.H., *et al.*, Human Stem Cell-Derived Skin Progenitors Produce Alpha 2-HS Glycoprotein (Fetuin): A Revolutionary Cosmetic Ingredient. *J. Drugs Dermatol. JDD [Internet]*, 2016 May 1 [cited 2025 Feb 28]. Available from: https://www.semanticscholar.org/paper/Human-Stem-Cell-Derived-Skin-Progenitors-Produce-A-Nistor-Poole/8d238e69b3f029d36f19086949c30cc5cbd9aa73.
49. Boddu, S., Hashim, P., Nia, J., Horowitz, R., Farberg, A.S., Goldenberg, G., Regenerative medicine in cosmetic dermatology. *Cutis [Internet]*, 2018 [cited 2025 Feb 28]. Available from: https://www.semanticscholar.org/paper/Regenerative-medicine-in-cosmetic-dermatology.-Boddu-Hashim/9d1177ebd26a64a23603bd5f8313ce0c4a910508.
50. Eshraghi, A.A., Jung, H.D., Mittal, R., Recent Advancements in Gene and Stem Cell-Based Treatment Modalities: Potential Implications in Noise-Induced Hearing Loss. *Anat. Rec.*, 303, 3, 516–26, 2020 Mar.
51. Kim, H.J., Jung, M.S., Hur, Y.K., Jung, A.H., A study on clinical effectiveness of cosmetics containing human stem cell conditioned media. *Biomed. Dermatol.*, 4, 1, 9, 2020 Dec.
52. Stem cell-based combinatorial therapies for spinal cord injury: A narrative review of current research and future directions, PubMed [Internet], [cited 2025 Mar 20]. Available from: https://pubmed.ncbi.nlm.nih.gov/37554849/.
53. [PDF] Using Defined Neural Cell Populations as a Possible Solution for Challenges in Neural Stem Cell Therapy, Semantic Scholar, [cited 2025 Mar 20]; Available from: https://www.semanticscholar.org/paper/Using-Defined-Neural-Cell-Populations-as-a-Possible-Azari/5ac9a888af6227c4e0eea08c956cbcd4f4623642.
54. Exploring the Pivotal Association of AI in Cancer Stem Cells Detection and Treatment, Semantic Scholar, [cited 2025 Mar 20]; Available from: https://www.semanticscholar.org/paper/Exploring-the-Pivotal-Association-of-AI-in-Cancer-Abubakar/df2ff244d51427e52a891dab35bf4d0a33adde72.
55. [PDF] miR-150 inhibits proliferation and tumorigenicity *via* retarding G1/S phase transition in nasopharyngeal carcinoma, Semantic Scholar, [cited 2025 Mar 20]; Available from: https://www.semanticscholar.org/paper/

miR-150-inhibits-proliferation-and-tumorigenicity-S-Li-Liu/d957972873c3d32d2d54e09bca1494e0a7648b8a.
56. Investigation of immune responses contributing to the pathogenesis of load-induced heart failure and the rejection of stem cell grafts, Semantic Scholar [Internet], [cited 2025 Mar 20]. Available from: https://www.semanticscholar.org/paper/Investigation-of-immune-responses-contributing-to-Hamann/3bbb9a186d2794005f270209ec2e6c7b7cf68c0e.
57. Fundamental Principles of Stem Cell Banking. Semantic Scholar, [cited 2025 Mar 20]; Available from: https://www.semanticscholar.org/paper/Fundamental-Principles-of-Stem-Cell-Banking.-Sun-Yue/6a92987fe736658addb0d15b435d02d250343627.
58. Stem Cell Research: Ethical Considerations, Semantic Scholar [Internet], [cited 2025 Mar 20]. Available from: https://www.semanticscholar.org/paper/Stem-Cell-Research%3A-Ethical-Considerations-Arda-G%C3%BCvercin/f4c48dd1b745cbbb4178bf896d66819f5ed5fc90.
59. [PDF] An overview on ethical considerations in stem cell research in Iran and ethical recommendations: A review, Semantic Scholar, [cited 2025 Mar 20]; Available from: https://www.semanticscholar.org/paper/An-overview-on-ethical-considerations-in-stem-cell-Farajkhoda/41d3bf2855e11c8b3471e8ad55680d5ca6137b18.
60. Comparative Regulatory Requirements for the Stem Cells and Stem Cells Based Products in India, United States and European Union, Semantic Scholar, [cited 2025 Mar 20]; Available from: https://www.semanticscholar.org/paper/Comparative-Regulatory-Requirements-for-the-Stem-in-Ruhela-Bhatia/4896085aecfda88c03beb8283af776d6000d01e4.
61. Anam, K., Ananyan, S., Rishabh, M., Dinesh, A., Ashwin, K., Zinc, copper, and selenium levels in vitiligo: A systematic review and meta-analysis. *Sci. Rep.*, 14, 1, 23700, 2024 Oct 10.
62. [PDF] Current Status and Future Prospects of Stem Cell Therapy for Infertile Patients with Premature Ovarian Insufficiency, Semantic Scholar, [cited 2025 Mar 20]; Available from: https://www.semanticscholar.org/paper/Current-Status-and-Future-Prospects-of-Stem-Cell-Kim-Kim/f2712631c1a721e0424f719fc3e294ace96fe67a.
63. Revolutionizing medicine: recent developments and future prospects in stem-cell therapy, Semantic Scholar, [cited 2025 Mar 20]; Available from: https://www.semanticscholar.org/paper/Revolutionizing-medicine%3A-recent-developments-and-Hussen-Taheri/ba3c7b91d187aba1b07d0390867935ed5c059b63.
64. Wong, J.K.U., Mehta, A., Vũ, T.T., Yeo, G.C., Cellular modifications and biomaterial design to improve mesenchymal stem cell transplantation. *Biomater. Sci.*, 11, 14, 4752–73, 2023 Jul 12.

12

In Vitro and *In Vivo* Model of Stem Cell Therapy in Dermatological Diseases

Sneha Das[1†], Pradip Jana[2†], Mohini Mondal[2†], Mainak Chatterjee[2†] and Moitreyee Chattopadhyay[3*]

[1]*Department of Pharmacology, BCDA College of Pharmacy & Technology, Campus 2, Kolkata, West Bengal, India*
[2]*Department of Pharmaceutical Technology, Bharat Technology, Uluberia, Howrah, West Bengal, India*
[3]*Department of Pharmaceutical Technology, Maulana Abul Kalam Azad University Technology, Haringhata, Nadia, West Bengal, India*

Abstract
Skin is the complex and a vast organ system of the body. When study of the diseases comes into play, we need to consider a complex structure involved in the skin model. So, while establishing the skin models in screening of various drugs, the necessity arises on the layer of the skin involved. The mechanistic pathway for the development of disease is given importance. In this chapter, we have attempted to give a clear idea about the various models both *in vitro* and *in vivo* that are used. Moreover, the future of this models and the role of stem cells have been discussed thoroughly.

***Keywords*:** Dermatological disorders, stem cells, conventional models, *in vitro* models, *in vivo* models

Highlights: Conventional approaches, current and future perspectives of stem cell therapy in dermatology, traditional skin models, mesenchymal stem cells (MSCs) models, induced pluripotent stem cell (iPSC) models, *in vivo* models—mice, rat, and pigs

**Corresponding author*: moit179@rediffmail.com
†Equal contribution

Sumel Ashique, Sanjay Dey, Biplab Debnath, Mohhammad Ramzan and Swarupananda Mukherjee (eds.) Stem Cell Therapy in Dermatological Disorders, (321–340) © 2025 Scrivener Publishing LLC

12.1 Introduction

Dermatological disease: Dermatological conditions are widespread and increasingly frequent as the population ages. Skin serves as a steady but versatile defensive obstruction against hurtful natural components. It embraces three fundamental layers: the hypodermis (the foremost diminished layer), the dermis (the center layer), and the epidermis (the outside layer) [1].

Life-threatening skin infections and conditions

Skin cancer arises from a rapid and unusual increase in cells within the epidermis. This condition develops due to unhealed damage to the DNA, leading to mutations and excessive proliferation of skin cells, eventually creating malignant tumors. The primary factor contributing to skin cancer is prolonged exposure to harmful ultraviolet (UV) rays. Although skin cancer is the most widespread of significant skin conditions, several rare but equally severe conditions can lead to life-altering consequences.

12.1.1 Dermatology

Dermatology is the medical field focused on the skin. Dermatologists are skin surgeons and physicians who concentrate on assessing, diagnosing, and treating issues related to the skin, hair, and nails. Specialists in dermatology encompass the following:

- **Dermatopathologist**—This expert combines dermatology with pathology, diagnosing skin diseases by analyzing biopsy samples from the skin.
- **Mohs surgeon**—Mohs surgery is a method used for treating skin cancer. This technique involves removing and examining thin layers of skin until no cancer cells remain detectable.
- **Pediatric dermatologist**—This specialist focuses on the treatment of infants and children suffering from severe or uncommon skin disorders [2].

Digital dermatology

Like many other medical disciplines, dermatology is continually advancing, largely due to technological advancements. Mobile health applications play a crucial role in digital dermatology. Notably, there are now 526 dermatology-related mobile apps, reflecting an increase of 80.8% since

2014. Additionally, the proportion of teledermatology in the market rose from 11.0% in 2014 to 20.1% in 2017 and 24% in 2024. These applications are used globally.

12.1.2 Conventional Approaches

Diagnosis of skin and transmitted diseases today involves combination of laboratory techniques, visual technologies, and modern solutions.

a. **Laboratory techniques** encompass serological tests and polymerase chain reaction (PCR). Serological tests detect antibodies against infectious agents, making them particularly important for diagnosing syphilis and HIV. Due to its high sensitivity and specificity, especially in cases with mild or no symptoms, PCR is often the preferred method for identifying pathogens like *Chlamydia trachomatis* and *Neisseria gonorrhoeae.*
b. **Recent advances**: Natural substances have been utilized for skin wound treatment for numerous years because of their healing properties, which encompass inflammatory, antimicrobial, and cell-boosting effects. In addition to significant advancements in new extraction techniques, purification methods, quality control measures, and treatment guidelines, more investigation is required to understand the precise mechanisms of action, potential side effects, and safety profiles of these compounds.
c. **Critical issues**: The healing of skin wounds is among the most intricate biological processes in humans, unfolding through a well-coordinated series of overlapping biochemical and cellular activities, so both traditional treatments and natural remedies have shown encouraging results. Although these options tend to be more affordable than contemporary therapies, they can be influenced by factors such as geographic location and seasonal changes. Additionally, variations from batch to batch can lead to unforeseen allergic reactions, side effects, and inconsistent clinical outcomes.
d. **Future directions**: The scientific research supporting the use of traditional therapies for wound healing shows positive effects on various types of injuries. Nonetheless, there are still unresolved challenges. To enhance the effectiveness and application of natural substances in wound management,

collaborative efforts are essential to validate the safety of these products, explore their side effects, and establish standardized controlled trials. The establishment of good manufacturing practices and regulatory frameworks plays a crucial role in facilitating the use of traditional therapies by healthcare providers. Present trends focus on creating innovative wound care solutions that merge traditional healing methods with modern products and techniques, such as nanofibers infused with silver nanoparticles, *Aloe vera* incorporated into alginate hydrogels, propolis used in dressing films, hydrogel sheets, etc. [3, 4].

12.1.3 Stem Cell Biology and Technology

Stem cells exist in nearly all body tissues and play a crucial role in tissue maintenance and healing after injuries. There are numerous sorts of stem cells (SCs). The fetus SC is considered predominant adaptability.

Embryonic SCs: These are made from embryos for almost 3 to 5 days. These SCs are flexible, which deliver numerous SCs or separated cells in any sort of cell.
Adult SCs: These SCs exist in compelled whole internal parts mostly grown-up tissues, tallying bone marrow and fat. Embryonic stem cells are pluripotent cells derived from the inner cell mass of a blastocyst, an early-stage embryo [5, 6].
Altered adult SCs: Exploration has encouraged altering over standard grown-up cells into SCs through intrinsic retrying. By adjusting the qualities of these cells, analysts can begin them to operate as embryonic SCs that are hinted to as induced pluripotent SCs (iPSCs). This imaginative approach might empower the utility of replicated cells as an elective to embryonic SCs, conceivably lessening the chance of secure removal.
Perinatal SCs: Analysts have originated SCs in both amniotic fluid and umbilical rope blood. These SCs have the capacity to create specialized cell sorts. Amniotic fluid is a clear, slightly yellowish liquid that envelops the fetus during pregnancy. Contained within the amniotic sac, it plays a vital role in fetal development and protection.

SC innovation is an advancing zone that brings together the work of cell analysts, geneticists, and healthcare masters, advancing guarantee for viable arrangements for unmistakable pernicious and nonmalignant conditions.

- SCs are trailblazer cells with the capacity to both copy themselves and make into diverse sorts of cells.
- These cells are committed to treating a wide range of cancer-like and non–cancer-like diseases.
- Peripheral blood SCs are commonly utilized in both autologous and allogeneic bone marrow transplants.
- Introducing genes into hematopoietic SCs could enable the treatment of both genetic and acquired disorders.
- In the future, embryonic SCs may be cultivated in the laboratory to create intricate organs.
- Neuronal SCs are right now being connected for neuron substitution in neurodegenerative maladies such as Parkinson and Huntington infections [7–10].

12.2 Current and Future Perspectives of SC Therapy in Dermatology by *In Vitro* and *In Vivo* Models

SCs having the huge regenerative property are getting major attention in the field of dermatology as it opens a new arena of treatment for skin conditions such as burns, chronic wounds, scars, autoimmune diseases, etc. [11–13]. This treatment can both replace and repair damaged skin tissues, aiding in the treatment of destructive skin conditions and making it antiaging as well [14]. In the following, both the current scenario and future perspectives of SC therapy have been discussed with reference to the *in vitro* and *in vivo* models.

12.2.1 Current Use of SCs in Dermatology

12.2.1.1 Skin Regeneration and Wound Healing

Severe burns, accidental skin injuries, and chronic wounds due to diabetic ulcers are major concerns of dermatology where skin regeneration can be an effective remedy. SCs are being used in the enormous research being done on skin regeneration globally. Among various SCs, mesenchymal stem cells (MSCs), iPSCs, and epidermal SCs are of utmost importance due to their skin regenerating ability [15]. Increased collagen production, increased angiogenesis, suppressed inflammation, and enhanced skin proliferation and differentiation have been observed in the SC-mediated wound healing process [16]. The creation of SC-based *in vitro/in vivo* models with an emphasis on the production of keratinocytes (KCs), fibroblasts (FBs), and melanocytes, all of which are critical for the function and appearance of skin. Skin organoids and three-dimensional (3D) skin equivalents are the *in vitro* models that are popularly used [17]. These models help to simulate the properties and functions of native skin and test the SC-mediated growth of new skin tissues. Mouse models with fully grown epidermal wounds are often used as an *in vivo* model to test the efficiency of SCs in wound healing and skin tissue regeneration [18].

12.2.1.2 Treatment of Skin Cancers

SCs have shown promising activity in melanoma and nonmelanoma skin cancer treatment. Genetically modified SCs are often used as carriers to target the cancerous cells that promote the tumor growth and metastasis. They are also used as mediators to deliver chemotherapeutic agents to the target sites [19]. These treatment strategies are usually tested using *in vitro* (e.g., 3D tumor spheroids) and *in vivo* models (e.g., mouse xenografts) [20, 21]. These models are used to evaluate the interaction between tumor cells and SCs, assess toxicity of the said treatment, and monitor the tumor regression.

12.2.1.3 Disease Modeling

Diseases such as epidermolysis bullosa and ichthyosis are due to genetic disorder, and in these, skin is severely damaged. Blisters appear on the skin, mouth, and stomach. SC therapies are being tested to manage these conditions for correcting genetic defects. Gene-editing techniques such as CRISPR-cas9 are being used to correct the mutation in SCs and regenerate the damaged skin [22]. Skin cell culture is one of the *in vitro* models that are crucial for assessing gene-editing techniques and determining how well they produce healthy skin tissue. Genetically engineered mice and patient-derived xenografts are examples of *in vivo* models that aid in evaluating the possible hazards and long-term effects of such treatments. *In vitro* models also play a crucial role in understanding other dermatological diseases such as eczema, vitiligo, and psoriasis. Besides these, personalized disease models can be created using patient-derived xenografts to investigate medication responses, disease causes, and possible therapies [23].

12.2.2 Future Perspectives of SC Therapy in Dermatology

12.2.2.1 Personalized Medicine

The future of any particular disease condition lies in the personalized medicine, and scientists are trying to develop the same for genetic skin conditions and other skin-related disorders. iPSCs are being advanced for their ability to derive patient-specific SCs and for developing customized skin care regimens [15]. The use of the patient's own SCs, i.e., autologous SC therapy, reduces the possibility of immunological rejection and enhances therapeutic outcomes. The personalized medicine also permits targeted therapies by identifying the specific genetic and environmental factors that influence the development of skin diseases. Patient-derived skin organoids are generally used as an *in vitro* model to test the personalized medicine approach using SCs [24].

12.2.2.2 Antiaging and Hair Restoration

SCs are increasingly being investigated as possible remedies for skin aging and hair restoration. The aging of skin is generally due to decreased regenerative capacity, thinning of the epidermis, and loss of elasticity. SC therapy can rejuvenate the aging skin by enhancing the SC population and by promoting the synthesis of collagen and other fibers of extracellular matrix (ECM) [25]. In hair restoration, SC injections into the bald scalp are used in clinical trials, and hair follicle organoids are used as *in vivo* models for SC therapy.

12.2.2.3 3D Skin Models and Bioprinting

In order to improve disease modeling and medication testing, future developments will concentrate on enhancing 3D skin models to more closely resemble the architecture of actual skin. More intricate elements, such as the dermis, epidermis, and the skin's supporting structures including sweat glands and hair follicles, may be included in these models. 3D bioprinting can be an effective technology for creating complex skin grafts from SCs. These skin grafts can be applied for the treatment of tertiary-burn patient and severe wound therapy. 3D skin grafts can also be used as *in vitro* models to test new therapies for different dermatological conditions before *in vivo* and clinical trials. SC-based *in vitro* models of the skin have already been created and used [26], but an *in vivo* model needs to be developed for advanced trials.

12.2.3 Challenges and Considerations

Safety concerns: The potential of immunogenicity, unchecked cell development, and tumor formation are among the main obstacles to SC therapy for dermatology. Before SC-derived therapies may be extensively used in clinical settings, it is imperative that their safety be guaranteed.
Regulatory approval: SC-based treatments must go through a strict and thorough regulatory approval process.
Ethical issues: In some places, the use of embryonic SCs has sparked ethical questions. However, because iPSCs may be produced without the moral dilemmas surrounding embryos, advancements in iPSC technology have alleviated some of these worries.

12.3 *In Vitro* Models for SC Therapy in Dermatological Diseases

The strategies for developing the screening models include the involvement of *in vitro* and *in vivo* studies. Skin is a complex organ, so the framework should be analyzed critically before implementation, and the steps to be followed are given in Figure 12.1.

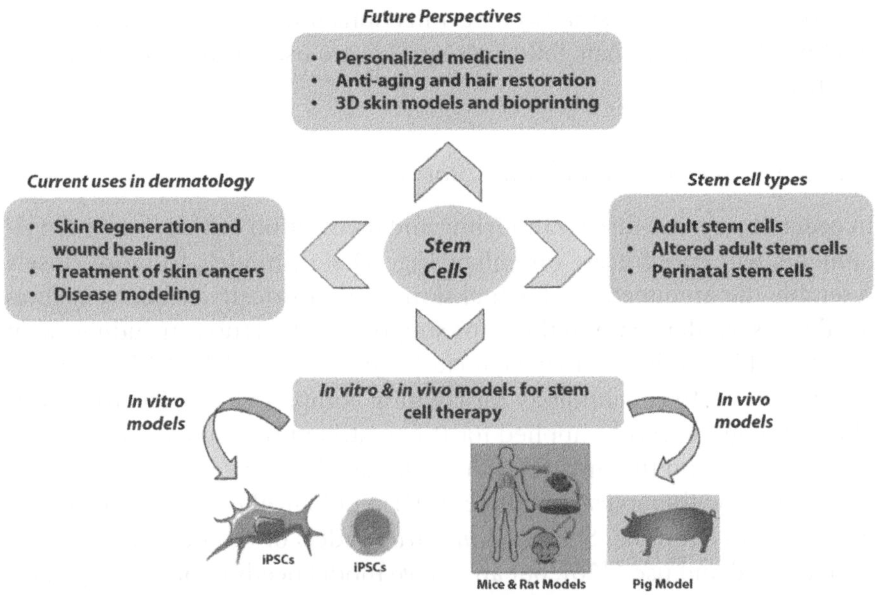

Figure 12.1 Framework for developing dermatological disease model.

In vitro models are widely explored for finding newer therapeutic approaches for dermatological diseases as they mimic different pathologic features of epidermis. These models help to analyze different skin features and related gene expressions and observation of changes in epidermal barrier related to skin disorders and allow understanding the pharmacological and toxicological aspects.

For dermatological diseases, two-dimensional (2D) *in vitro* models are well established, which generally includes FB feeder cells and primary human KCs monolayer culture [27]. Skin disorders such as atopic dermatitis (AD), which involves immune dysfunction for the disease progression, have been modeled by incubating KCs with cytokines [28]. Further, immune cells such as lymphocytes, mast cells, and eosinophils have also been used with KCs in 2D models for their incorporation in inflammatory skin conditions such as psoriasis and AD [29, 30]. These 2D models for skin diseases are simple and easy to produce, but these models lack considerations of the intricacy and connections with human skin and are unable to develop efficient skin barrier. Thus, these 2D models offer a limited view of skin disorders.

To overcome limitations of 2D skin disease models, 3D *in vitro* models are introduced that mimic human skin features including histology, differentiation, and barrier function more efficiently similar to *in vivo* [31, 32]. 3D *in vitro* models generally comprise FBs supported on a collagen matrix mimicking the dermis layer, covered by epidermal KCs, which is facing the air–liquid interface. These 3D cell cultures have been modified time to time using several human epidermal equivalents and human skin equivalents for different purposes, including study mechanism of skin disorders and drug development and testing. Different 3D skin models have also been widely developed by integrating macrophages, dermal FBs, Langerhans cells, endothelial cells, melanocytes, etc., to evaluate drug permeation and toxicity studies [33–37].

12.3.1 Traditional Skin Models

Traditionally, skin models are made of KCs and FBs on a dermal component, which are used for investigation of proliferation, differentiation, dermal remodeling, and epithelial attachments [38]. *In vitro* skin models made up of KCs in the presence of a dermal equivalent along with or without FBs facing the air–liquid interface lead to a multilayered differentiated epidermis, which mimics the native skin. When FB is present, it produces a more precise construction of epidermal layer with increased level of ECM proteins mimicking *in vivo* skin. The common dermal equivalents used while constructing skin models include collagen lattices contracted using dermal FBs and de-epidermalized dermis (DED). DED represents a natural dermis containing basement membrane and ECM component. The DED is advantageous as it offers reduced antigenicity and good durability [39].

12.3.2 MSC Models

MSCs are collected from different tissues such as umbilical cord blood, skin, adipose tissue, placenta, amniotic fluid, etc. MSCs from diverse sources represent similar types of features, but these are not identical in nature. Irrespective of their source, MSCs showcase a wide range of differentiating capability and plasticity and generate cells of mesodermal (e.g., osteocytes, adipocytes, myoblasts, etc.), ectodermal (melanocytes, neurocytes), and endodermal (hepatocytes, thyroid cells) origin [40].

MSCs developed from bone marrow produce different mesenchymal-derived tissues/cells both *in vitro* and *in vivo* settings [41]. SC-based therapies are one of the recent advanced avenues for wound healing, which includes skin equivalents by bioengineering, negative-pressure wound therapy, and growth factors to regain the tissue integrity and functions of damaged tissues in wound [42, 43]. Multicomponent MSCs can produce many tissue lineages, which is involved in wound healing and thus evolved as a promising candidate for wound healing as cell-based bioengineered product. Ojesh and Navsaria hypothesized the use of MSCs replacing FBs for wound healing. They used a DED skin model to compare mesenchymal interaction of cultured MSCs and cultured FBs with respect to epidermal proliferation, differentiation, and ECM protein expression. They also explored the possibility of MSCs to transdifferentiate into KCs using this model. Different groups including unseeded DED, DED with MSCs, and DED with FBs at an air–liquid interface were used to culture KCs to bring epidermal differentiation. KC–MSC skin model and KC–FB skin model both exhibited epidermal phenotype similar to *in vivo* epidermis. Both these models showed promising expression of proliferation, differentiation, and ECM proteins, and thus, they establish the possibility of MSC use for the treatment of wounds [44].

Further, different studies also explored human adipose-derived MSCs (AD-MSCs) for the wound healing ability. AD-MSCs have the advantages including easy isolation in high quantity and good recovery potency [45].

12.3.3 iPSC Models

Conversion of somatic cells into iPSCs enables the facility to produce a broad spectrum of cellular phenotypes useful for different therapeutic interventions [46]. Skin cells including KCs, FB, melanocytes, and adult adipose SCs are the sources for this method [47]. Further, the differentiation of iPSCs into KCs, melanocytes, and FB has already been reported, and thus, this iPSC technology has great potential in the area of dermatological diseases [48–50]. SC therapies are well-known in genetic diseases, even in rare diseases such as in recessive dystrophic epidermolysis bullosa, where genes for synthesis of collagen VII are absent in patients. For *in vitro* models, iPSCs are effectively developed from healthy (allogeneic) individuals or gene-corrected (autologous) RDRB FBs. This can further differentiate into MSCs and hematopoietic SCs and stay in the mucocutaneous blistering areas, which subsequently differentiate into FBs and KCs [51]. Besides, iPSCs can lead to autologous grafting of 3D skin equivalents to develop stratified epidermis *in vitro* as well as *in vivo* [52].

12.4 Various Animal Model for *In Vivo* in Dermatological Diseases

12.4.1 Mice Models

Okamura *et al.* have conducted a study on wild-type C57BL/6, B10. D2 (H-2d) and BALB/c (H-2d) mice were used to explore the potential of using adipose-derived stromal/SCs (ASCs) as a treatment for systemic sclerosis that causes fibrosis (scarring) of the skin and lungs by inducing two models (bleomycin-induced scleroderma and sclerodermatous chronic graft-versus-host disease) by administering ASCs intravenously, and it was found out that ASCs significantly reduced both skin and lungs fibrosis, suppressed immune cell infiltration ($CD4^+$ T cells, $CD8^+$ T cells, and macrophages), and also reduced the expression of fibrogenic factors (IL-6 and IL-13). Although results were promising, further study is needed [53].

Dalirfardouei *et al.* isolated and characterized exosomes from menstrual blood MSCs and incorporated on male C57BL/6 mice by creating diabetic full-thickness wound to assess wound healing and found out that exosomes reduce inflammation by converting proinflammatory M1 macrophages to anti-inflammatory M2 macrophages. They also increased the expression of vascular endothelial growth factor A (VEGF-A) and activation of NF-κB p65, which helps in cell regrowth, reduces scar formation by decreasing the ratio of type 1 collagen to type 3 collagen (Col1:Col3), and can be used as a treatment for difficult-to-treat scars in the future [54].

Psoriasis is a chronic skin inflammatory disease, and to treat that with a newer approach, Zhang *et al.* used fluorescent mesenchymal stem/stromal cell (MSC) exosomes topically on male Balb/c mice induced by imiquimod. Although the results showed limited exosome penetration (most of the exosomes remain on stratum corneum), they still significantly reduced inflammatory factors IL-17, C5b-9, maybe through complement activation, and C5b-9 formation in the stratum corneum [55].

12.4.2 Rat Models

Mahmood *et al.* examine the possibilities of highly regenerative neonatal SCs from the placenta, specifically amniotic epithelial cells for KCs and MSCs for FBs by inducing severe burn injuries on Sprague–Dawley rats. After incorporating bioengineered skin substitute on wounds and evaluating wound closure, they observed better skin structure, faster re-epithelialization, and reduced apoptosis, and most importantly, placenta SCs are excellent candidates for developing "off-the-shelf" artificial skin substitutes for future clinical use [56].

Zafari *et al.* investigated the effectiveness of using rat CD93 hematopoietic stem cells (HSCs) and a polycaprolactone (PCL)–gelatin scaffold, alone or in combination, to improve diabetic wound healing in Wistar rats. They isolated CD93 HSCs from rat bone marrow using fluorescent-activated cell sorting and assessed wound healing at days 7, 14, and 28 posttreatment through histological analysis, reverse transcriptase–PCR, and Western blot. The most significant wound healing, evidenced by the thickest epidermal layer and highest growth rate of skin appendages, was observed in the group treated with a combination of CD93 HSCs and a PCL–gelatin scaffold. This enhanced healing was accompanied by a surge in VEGF expression, indicative of robust early angiogenesis, observed on day 7 in both the HSCs alone and HSCs + scaffold groups, mirroring levels seen in the sham group. Furthermore, the combination therapy, along with HSCs alone and the scaffold alone, effectively minimized DAPK-1 gene expression, suggesting a significant reduction in cell death compared to the control group, which displayed the highest DAPK-1 levels. These findings collectively demonstrate that the synergistic effect of CD93 HSCs and the

PCL-gelatin scaffold promotes superior diabetic wound healing through enhanced vascularization and reduced cellular apoptosis [57].

Autografts (skin taken from the patient's own body) are the gold standard for severe burns, but there is often not enough healthy skin available, which calls for alternative treatments, and to target that issue, Nazempour *et al.* explored cell transplantation and tissue engineering. They combined acellular dermal matrix (ADM), a processed skin tissue, with human Wharton jelly stem cells (hWJSCs), SCs derived from Wharton jelly, the connective tissue of the umbilical cord on third-degree burns induced on 40 male Sprague–Dawley rats. Their findings showed that hWJSCs exhibited typical SC characteristics and demonstrated robust growth in culture. Although ADM alone resulted in mild healing, hWJSCs seeded onto ADM significantly enhanced wound repair, indicating a synergistic effect and demonstrating moderate healing. These findings suggest that allogeneic hWJSCs on ADM improve burn wound healing, likely through therapeutic and anti-inflammatory mechanisms, highlighting the potential for this approach as a valuable clinical treatment for burn injuries [58].

12.4.3 Other (Pig) Models

Irons *et al.* also investigated a 28-day study to understand the potential of using ASCs and their related products to improve wound healing in diabetic patients but used Yorkshire pig model by creating a full-thickness wound on their back and saw that ASCs, endothelial differentiated ASCs, and conditioned media from ASCs or endothelial cells significantly accelerate wound healing by increasing angiogenesis and immunomodulation [59].

Murphy *et al.* explored the potential of using amnion-derived products, specifically a hydrogel and a powdered form, as effective and cost-efficient wound treatments on Yorkshire pigs. Both amnion hydrogel and amnion powder treatments effectively promoted significant and rapid wound healing, primarily through new epithelialization, resulting in superior skin regeneration compared to closure by contraction. Histological analysis confirmed the formation of a mature epidermis and dermis, closely resembling healthy skin, with the less processed amnion powder demonstrating comparable efficacy, highlighting its potential for cost-effective manufacturing and clinical application [60].

12.5 Analytical and Numerical Solutions to the Various Models

12.5.1 Parameter Estimation and Interpretability

The validated mathematical model was designed to predict the percutaneous diffusion of a nerve agent to predict the conditions—occlusion and dilution using guinea pig and pig skin. The skin hydration and permeability and other conditions usually vary with different skin types. The mathematical models proved advantageous without invasive methods for estimating the chemical concentration of drugs within skin layers in the presence of factors such as blood circulation and sweating [61].

12.5.2 Stability of Various Models

In vivo animal models, particularly mice, have been instrumental in advancing our understanding of skin biology and disease pathogenesis. While *in vitro* cell culture, utilizing patient-derived skin cells, offers a complementary approach, standard 2D cultures lack the complex microenvironment of native skin, potentially leading to altered cell behavior and requiring validation. To bridge this gap, 3D organotypic and organ culture techniques provide a more realistic, tissue-level environment [62].

12.5.3 Application of Various Models for the Treatment of Dermatological Diseases

Future *in vitro* skin substitutes should incorporate skin appendages such as sweat glands and hair follicles for better *in vivo* realism. Developing vascularized models remains challenging due to difficulties in integrating vascular-like structures with external systems that maintain physiological shear, hindering maturation [63].

Machine learning (ML) holds significant potential in dermatology for diagnosis and treatment prediction, requiring dermatologists to understand its function and clinical application. Physician interpretation remains crucial, despite ML's power, and awareness of potential biases within algorithms is essential, particularly regarding skin of color. Transparent research, including public access to algorithms and datasets, and rigorous clinical trials are necessary. Increased dermatologist involvement in ML development and testing is vital for clinically relevant technology [64].

12.6 Conclusions and Future Directions

The chapter highlights the significant advancements and potential of various screening models that are in practice for searching the treatment of the dermatological diseases. Moreover, the influence of the SCs in research has been highlighted. The futuristic approaches that may widen the research in dermatological diseases have been taken up in this chapter to make the views clear about the disease models.

AI disclosure statement: The author confirms that no artificial intelligence (AI)-generated content or AI-assisted writing tools were used in the preparation of this chapter (including summaries and the abstract). All text, analysis, and conclusions are the result of the author's original work.

References

1. Al Abbadi, N.K., SaadiDahir, N., AL-Dhalimi, M.A., Restom, H., Psoriasis Detection Using Skin Color and Texture Features. *J. Comput. Sci.*, 6, 6, 648–652, 2010.
2. Sreekala, K., Rajkumar, N., Sugumar, R., Daya Sagar, K.V., Shobarani, R., Parthiban Krishnamoorthy, K., Saini, A.K., Palivela, H., Yeshitla, A., Skin Diseases Classification Using Hybrid AI Based Localization Approach. *Comput. Intell. Neurosci.*, 3, 7–10, 2022.
3. Thapar, P., Rakhra, M., Cazzato, G., Hossain, M.S., A novel hybrid deep learning approach for skin lesion segmentation and classification. *J. Healthcare Eng.*, 21, 2022.
4. Lim, H.W., *et al.*, The burden of skin disease in the United States. *J. Am. Acad. Dermatol.*, 76, 5, 958–972, 2017.
5. Guy, G.P., Jr, *et al.*, Prevalence and costs of skin cancer treatment in the U.S. *Am. J. Preventative Med.*, 48, 2, 183–187, 2015.
6. Kessinger, H.H., Armitage, J.O., Smither, D.M., Landmark, J.D., Bierman, P.J., Weisenburger, D.D., High dose chemotherapy and autologous peripheral blood stem cell transplantation for patients with lymphoma. *Blood*, 74, 1260–1265, 1989.
7. Morrison, S.J., Hemmati, H.D., Wandycz, A.M., Weissman, I.L., The purification and characterization of fetal liver hematopoietic stem cells. *Proc. Natl. Acad. Sci. U.S.A.*, 92, 10302–10306, 1995.
8. Rafii, S., Shapiro, F., Rimarachin, J., Nachan, R.L., Ferris, B., Weksler, B., *et al.*, Isolation and characterization of human bone marrow microvascular endothelial cells: hematopoietic progenitor cell adhesion. *Blood*, 84, 10–19, 1994.

9. Sutherland, H.J., Lansdorp, P.M., Henkelman, D.H., Eaves, A.C., Eaves, C.J., Functional characterization of individual human hematopoietic stem cells cultured at limiting dilution on supportive marrow stromal layers. *Proc. Natl. Acad. Sci. U.S.A.*, 87, 3584–3588, 2021.
10. Thomson, A.W., Lu, L., Murase, N., Demetris, A.J., Rao, A.S., Starzl, T.E., Microchimerism, dendritic cell progenitors and transplantation tolerance. *Stem Cells*, 13, 622–639, 2021.
11. Klar, A.S., Zimoch, J., Biedermann, T., Skin tissue engineering: application of adipose-derived stem cells. *Biomed Res. Int.*, 2017, 9747010, 2017.
12. Sierra-Sánchez, A., Montero-Vilchez, T., Quiñones-Vico, M.I., *et al.*, Current advanced therapies based on human mesenchymal stem cells for skin diseases. *Front. Cell Dev. Biol.*, 9, 643125, 2021.
13. Montero-Vilchez, T., Sierra-Sánchez, A., Sanchez-Diaz, M., *et al.*, Mesenchymal stromal cell- conditioned medium for skin diseases: a systematic review. *Front. Cell Dev. Biol.*, 9, 654210, 2021.
14. Bellei, B., Migliano, E., Picardo, M., Therapeutic potential of adipose tissue-derivatives in modern dermatology. *Exp. Dermatol.*, 31, 12, 1837–52, 2022 Dec.
15. Hu, C. and Li, L., Preconditioning influences mesenchymal stem cell properties *in vitro* and *in vivo*. *J. Cell. Mol. Med.*, 22, 3, 1428–42, 2018 Mar.
16. Damayanti, R.H., Rusdiana, T., Wathoni, N., Mesenchymal stem cell secretome for dermatology application: a review. *Clin. Cosmet. Invest. Dermatol.*, 5, 1401–12, 2021 Oct.
17. Sun, H., Zhang, Y.X., Li, Y.M., Generation of skin organoids: Potential opportunities and challenges. *Front. Cell Dev. Biol.*, 9, 709824, 2021 Nov 4.
18. Borena, B.M., Martens, A., Broeckx, S.Y., Meyer, E., Chiers, K., Duchateau, L., Spaas, J.H., Regenerative skin wound healing in mammals: state-of-the-art on growth factor and stem cell based treatments. *Cell. Physiol. Biochem.*, 36, 1, 1–23, 2015 Apr 27.
19. Hoang, D.M., Pham, P.T., Bach, T.Q., Ngo, A.T., Nguyen, Q.T., Phan, T.T., Nguyen, G.H., Le, P.T., Hoang, V.T., Forsyth, N.R., Heke, M., Stem cell-based therapy for human diseases. *Signal Transduct. Target. Ther.*, 7, 1, 272, 2022 Aug 6.
20. Ohguro, H., Watanabe, M., Sato, T., Hikage, F., Furuhashi, M., Okura, M., Hida, T., Uhara, H., 3D spheroid configurations are possible indictors for evaluating the pathophysiology of melanoma cell lines. *Cells*, 12, 5, 759, 2023 Feb 27.
21. Müller, I. and Kulms, D., A 3D organotypic melanoma spheroid skin model. *J. Visualized Exp. JoVE*, 135, 57500, 2018 May 18.
22. Li, T., Yang, Y., Qi, H., Cui, W., Zhang, L., Fu, X., He, X., Liu, M., Li, P.F., Yu, T., CRISPR/Cas9 therapeutics: progress and prospects. *Signal Transduct. Target. Ther.*, 8, 1, 36, 2023 Jan 16.
23. Aparicio, S., Hidalgo, M., Kung, A.L., Examining the utility of patient-derived xenograft mouse models. *Nat. Rev. Cancer*, 15, 5, 311–6, 2015 May.

24. Berardesca, E. and Cameli, N., Perspectives of Regenerative Medicine in Dermatology and Cosmetology. *Cosmetics*, 11, 6, 188, 2024 Oct 30.
25. Taub, A.F. and Pham, K., Stem cells in dermatology and anti-aging care of the skin. *Facial Plast. Surg. Clin.*, 26, 4, 425–37, 2018 Nov 1.
26. Vijayavenkataraman, S., Lu, W.F., Fuh, J.Y., 3D bioprinting of skin: a state-of-the-art review on modelling, materials, and processes. *Biofabrication*, 8, 3, 032001, 2016 Sep 7.
27. Khurana, P., Kolundzic, N., Flohr, C., Ilic, D., Human pluripotent stem cells: An alternative for 3D *in vitro* modelling of skin disease. *Exp. Dermatol.*, 30, 11, 1572–87, 2021.
28. Omori-Miyake, M., Yamashita, M., Tsunemi, Y., Kawashima, M., Yagi, J., *In vitro* assessment of IL-4-or IL-13-mediated changes in the structural components of keratinocytes in mice and humans. *J. Invest. Dermatol.*, 134, 5, 1342–50, 2014.
29. Jiao, D., Wong, C.-K., Qiu, H.-N., Dong, J., Cai, Z., Chu, M., *et al.*, NOD2 and TLR2 ligands trigger the activation of basophils and eosinophils by interacting with dermal fibroblasts in atopic dermatitis-like skin inflammation. *Cell. Mol. Immunol.*, 13, 4, 535–50, 2016.
30. Martin, G., Guérard, S., Fortin, M.-M.R., Rusu, D., Soucy, J., Poubelle, P.E., *et al.*, Pathological crosstalk *in vitro* between T lymphocytes and lesional keratinocytes in psoriasis: necessity of direct cell-to-cell contact. *Lab. Invest.*, 92, 7, 1058–70, 2012.
31. Carlson, M.W., Alt-Holland, A., Egles, C., Garlick, J.A., Three-dimensional tissue models of normal and diseased skin. *Curr. Protoc. Cell Biol.*, 41, 1, 19–9, 2008.
32. Stark, H.-J., Szabowski, A., Fusenig, N.E., Maas-Szabowski, N., Organotypic cocultures as skin equivalents: a complex and sophisticated *in vitro* system. *Biol. Proced. Online*, 6, 55–60, 2004.
33. Linde, N., Gutschalk, C.M., Hoffmann, C., Yilmaz, D., Mueller, M.M., Integrating macrophages into organotypic co-cultures: a 3D *in vitro* model to study tumor-associated macrophages. *PLoS One*, 7, 7, e40058, 2012.
34. Facy, V., Flouret, V., Régnier, M., Schmidt, R., Reactivity of Langerhans cells in human reconstructed epidermis to known allergens and UV radiation. *Toxicol. In Vitro*, 19, 6, 787–95, 2005.
35. Li, L., Fukunaga-Kalabis, M., Herlyn, M., The three-dimensional human skin reconstruct model: a tool to study normal skin and melanoma progression. *J. Visualized Exp. JoVE*, 54, 2937, 2011.
36. Ponec, M., Ghalbzouri, A.E., Dijkman, R., Kempenaar, J., Gvd, P., Koolwijk, P., Endothelial network formed with human dermal microvascular endothelial cells in autologous multicellular skin substitutes. *Angiogenesis*, 7, 295–305, 2004.
37. Itoh, M., Umegaki-Arao, N., Guo, Z., Liu, L., Higgins, C.A., Christiano, A.M., Generation of 3D skin equivalents fully reconstituted from human induced pluripotent stem cells (iPSCs). *PLoS One*, 8, 10, e77673, 2013.

38. Krejci, N.C., Cuono, C.B., Langdon, R.C., McGuire, J., *In vitro* reconstitution of skin: fibroblasts facilitate keratinocyte growth and differentiation on acellular reticular dermis. *J. Invest. Dermatol.*, 97, 5, 843–8, 1991.
39. Ghosh, M.M., Boyce, S., Layton, C., Freedlander, E., Mac Neil, S., A comparison of methodologies for the preparation of human epidermal-dermal composites. *Ann. Plast. Surg.*, 39, 4, 390–404, 1997.
40. Marfia, G., Navone, S.E., Di Vito, C., Ughi, N., Tabano, S., Miozzo, M., *et al.*, Mesenchymal stem cells: potential for therapy and treatment of chronic non-healing skin wounds. *Organogenesis*, 11, 4, 183–206, 2015.
41. Pittenger, M.F., Mackay, A.M., Beck, S.C., Jaiswal, R.K., Douglas, R., Mosca, J.D., *et al.*, Multilineage potential of adult human mesenchymal stem cells. *Science*, 284, 5411, 143–7, 1999.
42. Griffiths, M., Ojeh, N., Livingstone, R., Price, R., Navsaria, H., Survival of Apligraf in acute human wounds. *Tissue Eng.*, 10, 7–8, 1180–95, 2004.
43. Khanolkar, M.P., Bain, S.C., Stephens, J.W., The diabetic foot. *QJM: Int. J. Med.*, 101, 9, 685–95, 2008.
44. Ojeh, N.O. and Navsaria, H.A., An *in vitro* skin model to study the effect of mesenchymal stem cells in wound healing and epidermal regeneration. *J. Biomed. Mater. Res. Part A*, 102, 8, 2785–92, 2014.
45. Hassan, W.U., Greiser, U., Wang, W., Role of adipose-derived stem cells in wound healing. *Wound Repair Regener.*, 22, 3, 313–25, 2014.
46. Takahashi, K. and Yamanaka, S., Induction of pluripotent stem cells from mouse embryonic and adult fibroblast cultures by defined factors. *Cell*, 126, 4, 663–76, 2006.
47. Vangipuram, M., Ting, D., Kim, S., Diaz, R., Schüle, B., Skin punch biopsy explant culture for derivation of primary human fibroblasts. *J. Visualized Exp. JoVE*, 77, 3779, 2013.
48. Bilousova, G., Chen, J., Roop, D.R., Differentiation of mouse induced pluripotent stem cells into a multipotent keratinocyte lineage. *J. Invest. Dermatol.*, 131, 4, 857–64, 2011.
49. Ohta, S., Imaizumi, Y., Okada, Y., Akamatsu, W., Kuwahara, R., Ohyama, M., *et al.*, Generation of human melanocytes from induced pluripotent stem cells. *PLoS One*, 6, 1, e16182, 2011.
50. Hewitt, K.J., Shamis, Y., Hayman, R.B., Margvelashvili, M., Dong, S., Carlson, M.W., *et al.*, Epigenetic and phenotypic profile of fibroblasts derived from induced pluripotent stem cells. *PLoS One*, 6, 2, e17128, 2011.
51. Itoh, M., Kiuru, M., Cairo, M.S., Christiano, A.M., Generation of keratinocytes from normal and recessive dystrophic epidermolysis bullosa-induced pluripotent stem cells. *Proc. Natl. Acad. Sci.*, 108, 21, 8797–802, 2011.
52. Sebastiano, V., Zhen, H.H., Haddad, B., Bashkirova, E., Melo, S.P., Wang, P., *et al.*, Human COL7A1-corrected induced pluripotent stem cells for the treatment of recessive dystrophic epidermolysis bullosa. *Sci. Transl. Med.*, 6, 264, 264ra163–264ra163, 2014.

53. Okamura, A., Matsushita, T., Komuro, A., Kobayashi, T., Maeda, S., Hamaguchi, Y., Takehara, K., Adipose-derived stromal/stem cells successfully attenuate the fibrosis of scleroderma mouse models. *Int. J. Rheum. Dis.*, 23, 2, 216–25, 2020 Feb.
54. Dalirfardouei, R., Jamialahmadi, K., Jafarian, A.H., Mahdipour, E., Promising effects of exosomes isolated from menstrual blood-derived mesenchymal stem cell on wound-healing process in diabetic mouse model. *J. Tissue Eng. Regener. Med.*, 13, 4, 555–68, 2019 Apr.
55. Zhang, B., Lai, R.C., Sim, W.K., Choo, A.B., Lane, E.B., Lim, S.K., Topical application of mesenchymal stem cell exosomes alleviates the imiquimod induced psoriasis-like inflammation. *Int. J. Mol. Sci.*, 22, 2, 720, 2021 Jan 13.
56. Mahmood, R., Mehmood, A., Choudhery, M.S., Awan, S.J., Khan, S.N., Riazuddin, S., Human neonatal stem cell-derived skin substitute improves healing of severe burn wounds in a rat model. *Cell Biol. Int.*, 43, 2, 147–57, 2019 Feb.
57. Zafari, F., Shirian, S., Sadeghi, M., Teimourian, S., Bakhtiyari, M., CD93 hematopoietic stem cells improve diabetic wound healing by VEGF activation and downregulation of DAPK-1. *J. Cell. Physiol.*, 235, 3, 2366–76, 2020 Mar.
58. Nazempour, M., Mehrabani, D., Mehdinavaz-Aghdam, R., Hashemi, S.S., Derakhshanfar, A., Zare, S., Zardosht, M., Moayedi, J., Vahedi, M., The effect of allogenic human Wharton's jelly stem cells seeded onto acellular dermal matrix in healing of rat burn wounds. *J. Cosmet. Dermatol.*, 19, 4, 995–1001, 2020 Apr.
59. Irons, R.F., Cahill, K.W., Rattigan, D.A., Marcotte, J.H., Fromer, M.W., Chang, S., Zhang, P., Behling, E.M., Behling, K.C., Caputo, F.J., Acceleration of diabetic wound healing with adipose-derived stem cells, endothelial-differentiated stem cells, and topical conditioned medium therapy in a swine model. *J. Vasc. Surg.*, *68*, 6, 115S–125S, 2018.
60. Murphy, S.V., Skardal, A., Nelson Jr., R.A., Sunnon, K., Reid, T., Clouse, C., Kock, N.D., Jackson, J., Soker, S., Atala, A., Amnion membrane hydrogel and amnion membrane powder accelerate wound healing in a full thickness porcine skin wound model. *Stem Cells Transl. Med.*, 9, 1, 80–92, 2020 Jan 1.
61. Jenner, J., Dalton, C.H., Graham, S.J., A simple mathematical model of skin diffusion of nerve agent VX *in vitro*. *Comput. Toxicol.*, 17, 100137, 2021 Feb 1.
62. Atwood, S.X. and Plikus, M.V., Fostering a healthy culture: Biological relevance of *in vitro* and *ex vivo* skin models. *Exp. Dermatol.*, 30, 3, 298, 2021 Feb 10.
63. Groeber, F., Holeiter, M., Hampel, M., Hinderer, S., Schenke-Layland, K., Skin tissue engineering—*in vivo* and *in vitro* applications. *Adv. Drug Deliv. Rev.*, 63, 4–5, 352–66, 2011 Apr 30.
64. Chan, S., Reddy, V., Myers, B., Thibodeaux, Q., Brownstone, N., Liao, W., Machine learning in dermatology: current applications, opportunities, and limitations. *Dermatol. Ther.*, 10, 365–86, 2020 Jun.

13

Preclinical and Clinical Trials and Case Studies Involved in Stem Cell Therapy in Dermatological Diseases

Rony Mandal*, Somali Gorai and Sajal Kumar Jha

Faculty of Pharmacology, Bengal College of Pharmaceutical Technology, Dubrajpur, West Bengal, India

Abstract

Stem cell treatment evolved as a revolutionary methodology in dermatology, offering potential solutions for chronic inflammatory skin diseases and reconstructive procedures. Immunomodulatory and regenerative properties of mesenchymal stem cells (MSCs) have shown favorable results in preclinical and clinical studies for conditions such as atopic dermatitis and other inflammatory dermatoses. Advanced techniques, including cell priming and gene modification, are enhancing MSC therapeutic efficacy, paving the way for next-generation regenerative treatments. In addition to inflammatory skin conditions, MSC-based interventions have demonstrated significant potential in reconstructive surgery, particularly in improving skin flap survival. Distal necrosis remains a major challenge in flap procedures, necessitating novel treatment strategies. Studies indicate that stem cell therapy enhances angiogenesis, upregulates vascular endothelial growth factor, and upsurges blood vessel density, contributing to improved flap viability and reduced necrosis. The transition of stem cell therapies from experimental research to clinical application requires adherence to stringent regulatory guidelines set by organizations such as the Food and Drug Administration and European Medicines Agency. This review offers a complete overview of the preclinical and clinical trials, as well as case studies, which support the inclusion of stem cell therapy in dermatology. By exploring latest innovations and regulatory considerations, this study highlights the potential of regenerative medicine in transforming treatment paradigms for complex dermatological diseases.

**Corresponding author*: mandalrony93@gmail.com

Sumel Ashique, Sanjay Dey, Biplab Debnath, Mohhammad Ramzan and Swarupananda Mukherjee (eds.) Stem Cell Therapy in Dermatological Disorders, (341–370) © 2025 Scrivener Publishing LLC

Keywords: Mesenchymal stem cells, atopic dermatitis, distal necrosis, vascular endothelial growth factor, regenerative medicine

13.1 Introduction

13.1.1 Skin Flap Necrosis

Skin necrosis occurs when skin cells die due to insufficient blood supply, often resulting from trauma, infection, prolonged pressure, or medical conditions such as diabetes and vascular disease. The affected area may become dark, firm, and ulcerated, potentially progressing to open wounds. If untreated, necrosis can cause serious impediments such as infections and tissue destruction. Treatment typically involves removing dead tissue through debridement, enhancing circulation, and addressing the underlying cause [1]. In some cases, skin grafting or restoration procedure may be prerequisite to restore both function and appearance. Preventive strategies include promoting good circulation, proper wound management, and controlling chronic health conditions. Skin flap necrosis is a complication that occurs when a surgically created skin flap fails in supplying blood, leading to tissue damage. Factors, such as excessive tension, inadequate vascularization, or conditions such as diabetes and smoking, impair circulation and can contribute to its development [2]. Symptoms often include discoloration, slow healing, and tissue breakdown, which may increase the risk of infection if left untreated. Management includes wound care, debridement of necrotic tissue, and, in severe cases, surgical intervention to reconstruct viable tissue. Preventive measures such as meticulous surgical planning and ensuring sufficient blood flow are essential in minimizing the risk of flap necrosis.

13.1.2 Atopic Dermatitis

Atopic dermatitis (AD) is a long-term skin condition that causes inflammation, dryness, and intense itching. It is influenced by genetic factors, an overactive immune system, and environmental triggers such as allergens, harsh soaps, stress, and weather changes. The condition is common in children but can persist into adulthood, often appearing in individuals with several allergic disorders such as asthma or hay fever. Indications includes red, scaly patches, skin thickening, and, in severe cases, cracked or oozing skin [2]. Managing AD involves keeping the skin moisturized, avoiding known irritants, and using treatments such as topical corticosteroids or antihistamines to reduce inflammation and discomfort. For severe cases, advanced therapies such as immune suppressants or biologic medications may be necessary. Preventive care, including gentle skincare routines and a healthy lifestyle, can help minimize flare-ups and maintain skin health [3]. The factors that trigger the cause of AD are allergens, irritants, environmental factors, stress, and skin infections.

13.2 Forms of Stem Cell Therapy

13.2.1 Allogeneic Stem Cells

Allogeneic stem cells are the ones that are sourced from a donor and transplanted into a recipient. These cells are derived from medulla ossium, peripheral blood cells, or blood from funiculus umbilicalis. Since they come from a genetically different individual, compatibility is determined through human leukocyte antigen to reduce the likelihood of immune rejection. Allogeneic stem cell transplantation is usually used in treating conditions such as leukemia, lymphoma, and certain genetic disorders [4]. They help replace damaged or diseased cells with healthy ones, supporting the regeneration of blood and immune cells. However, these transplants carry risks, comprising graft-versus-host disease (GVHD), attacking the recipient's tissues with donor's immune cells. To minimize complexity, patients often receive immunosuppressive therapy. Despite the risks, allogeneic stem cell therapy remains a vital treatment option for various life-threatening diseases.

The process of allogeneic stem cell transplantation involves several key steps to ensure the safe transfer of healthy stem cells to recipients from donors. The following is how it is carried out:

***Donor selection and matching*:** A compatible donor is identified, usually a sibling or an unrelated individual with a closely matched HLA profile [4]. Matching of HLA is crucial to lessen the risk of immune rejection and GVHD.

***Stem cell collection*:** Donor's stem cells are collected from one of three sources: bone marrow stem cells are drawn out from the pelvic bone under anesthesia.

***Peripheral blood*:** The donor is given medication (granulocyte colony-stimulating factor) to increase stem cells in the bloodstream, which are then collected through apheresis.

Umbilical cord blood: After childbirth, this is collected from the placenta and umbilical cord, stored in cord blood banks for future use.
Conditioning therapy: Before transplantation, the recipient undergoes chemotherapy, radiation, or immunosuppressive treatment to destroy diseased cells, suppressing the immune system, and prepares the body to accept the donor stem cells.
Stem cell infusion: The collected stem cells are delivered into the recipient's bloodstream through an intravenous (IV) infusion, similar to a blood transfusion [5, 9]. These cells travel to the bone marrow and begin producing healthy blood cells. In the following weeks, the donor stem cells integrate into the recipient's bone marrow and start generating new blood cells. Patients are carefully observed for complications such as infections, GVHD, or transplant rejection.
Posttransplant care: The recipient may require medications, including immune suppressants, to prevent GVHD and support immune function. Frequent follow-ups are necessary to track recovery and detect any complications early. This process is essential in treating blood-related cancers, immune disorders, and genetic diseases, offering patients a chance for long-term remission and improved health [5].

13.2.2 Mesenchymal Stem Cells

Mesenchymal stem cells (MSCs) are adaptable, multipotent cells capable of developing into different cell types, including bone, cartilage, fat, and muscle. Although they are mainly found in bone marrow, they are also present in adipose tissue, the umbilical cord, placenta, and dental pulp [5]. These cells have attracted considerable interest in restorative medicine due to their capacity to support tissue repair, regulate immune responses, and alleviate inflammation [5, 38]. Their unique properties make them valuable for treating various medical conditions, including degenerative diseases, autoimmune disorders, and tissue injuries. The key characteristics are as follows:

Multipotency: Can differentiate into osteocytes (bone cells), chondrocytes (cartilage cells), and adipocytes.
Immunomodulatory properties: Help regulate immune responses, making them useful for treating autoimmune diseases [6, 38].
Paracrine effects: Release bioactive molecules that aid in tissue repair and reduce inflammation.

Dysfunction of MSCs has been linked in the development of dermatitis and other skin-related disorders. MSCs play a pivotal role in skin homeostasis, wound healing, and immune regulation, and any impairment in their function can contribute to chronic inflammation, impaired tissue repair, and skin barrier dysfunction. The contributing factors are as follows:

Impaired skin regeneration: MSCs support skin repair by differentiating into fibroblasts and other skin cells [9, 14]. Dysfunctional MSCs reduce collagen production and slow down wound healing, leading to persistent skin damage.
Weakened immunomodulation: MSCs regulate immune responses by suppressing excessive inflammation. Their dysfunction can cause hyperactive immune reactions, increasing susceptibility to AD, psoriasis, and contact dermatitis.
Disrupted skin barrier function: MSCs influence keratinocyte function, which maintains the skin barrier. Dysfunctional MSCs fail to support proper epidermal renewal, leading to dryness, irritation, and susceptibility to allergens [7, 11].
Inflammatory cytokine imbalance: Healthy MSCs release anti-inflammatory cytokines to balance immune activity. When MSCs malfunction, cytokine proinflammatory mediators such as tumor necrosis factor α (TNF-α) and interleukin 6 (IL-6) increase, contributing to chronic skin inflammation seen in dermatitis.
Oxidative stress and aging: MSC dysfunction can lead to oxidative stress, accelerating skin aging and making it more prone to eczema and inflammatory skin diseases.

13.2.3 Embryonic Stem Cells

Embryonic stem cells (ESCs) are pluripotent cells obtained from the inner cell mass of a blastocyst, an early-stage embryo that develops around 5 to 7 days after fertilization. These cells have the remarkable ability to transform into any cell type in the body, making them crucial for studies in regenerative medicine, disease modeling, and drug discovery. Characteristics of ECS are as follows:

Pluripotency: ESCs can develop into all three germ layers—ectoderm (skin and nervous system), mesoderm (muscles, bones, and blood), and endoderm (internal organs such as the liver and pancreas).
Self-renewal: They can divide indefinitely under controlled laboratory conditions, maintaining their undifferentiated state.

***High plasticity*:** ESCs can be directed to differentiate into specialized cells, making them a promising tool for regenerative therapies [5]. ESCs have shown promise in treating dermatitis due to their ability to regenerate damaged skin, regulate immune responses, and restore skin barrier function.

Since dermatitis is often associated with chronic inflammation, skin barrier defects, and immune system dysfunction, ESC-based therapies offer potential solutions for repairing damaged tissues and modulating immune activity. Embryonic stem cells perform the job role to cure dermatitis in the following manner:

***Regeneration of skin cells*:** ESCs are pluripotent and can be distinguished into keratinocytes (skin cells), fibroblasts, and other essential skin components. This can help repair damaged or inflamed skin, restoring normal structure and function in AD patients, contact dermatitis, and psoriasis [5, 11].
***Restoration of the skin barrier*:** ESC-derived keratinocytes improve the skin's ability to retain moisture and protect against allergens, irritants, and microbial infections. This is particularly beneficial in eczema, where the skin barrier is compromised.
***Anti-inflammatory and immunomodulatory Effects*:** Cells derived from ESCs help reduce inflammation by releasing anti-inflammatory cytokines such as IL-10 and transforming growth factor β (TGF-β), which regulate and control excessive immune responses. They can also regulate T-cell activity, preventing immune system overactivation that contributes to chronic skin inflammation [8].
***Wound healing and tissue repair*:** ESCs promote angiogenesis (development of new blood vessels), accelerating wound healing in patients with severe dermatitis or skin ulcers. They enhance the construction of collagen and elastin, improving skin texture and reducing scarring.

13.2.4 Induced Pluripotent Stem Cells

Induced pluripotent stem cells (iPSCs) are generated by reprogramming adult somatic cells, such as skin or blood cells, to regain a pluripotent state similar to ESCs. This ability allows them to transform into any cell type in the body, making them valuable for regenerative medicine, disease modeling, and drug development. Scientists create iPSCs by introducing specific transcription factors (OCT4, SOX2, KLF4, and c-MYC) into adult cells, reverting them to an embryonic-like state [4, 9]. This technique enables the production of patient-specific stem cells without the ethical concerns associated with ESCs. Additionally, iPSCs have unlimited self-renewal capacity, meaning they can divide indefinitely under controlled conditions, making them highly useful for research and therapeutic purposes.

***Ethical advantage*:** Unlike ESCs, iPSCs do not require embryos, avoiding ethical controversies. iPSCs offer a promising method in treating dermatitis by promoting skin regeneration, decreasing inflammation, and restoring the skin barrier by generation of healthy skin cells.
***Generation of healthy skin cells*:** iPSCs can be differentiated into keratinocytes (skin cells) and fibroblasts, which are essential for maintaining healthy skin [6]. These new cells help replace damaged or inflamed skin, improving conditions such as AD (eczema), contact dermatitis, and psoriasis.
***Restoration of the skin barrier*:** Patients with dermatitis often have a weakened skin barrier, making them prone to irritation and infections. IPSC-derived keratinocytes can enhance the production of essential proteins such as filaggrin, which improves moisture retention and strengthens the skin barrier.
***Immune regulation and anti-inflammatory effects*:** iPSCs can be used to create immune-modulating cells, helping to reduce hyperactive immune responses seen in dermatitis. IPSC-derived therapies can regulate proinflammatory cytokines (such as IL-4, IL-13, and TNF-α), reducing chronic inflammation and flare-ups [4, 6]. Wound healing and tissue repair iPSC-derived skin grafts can be used to treat severe cases of dermatitis, where skin breakdown, ulcers, or infections occur. These stem cell–derived treatments promote collagen production and angiogenesis, accelerating healing and reducing scarring.
***Personalized therapy*:** iPSCs can be generated from a patient's own skin cells, creating personalized treatments that lower the risk of immune rejection. This makes iPSC-based therapies more compatible and effective for individuals with genetic forms of dermatitis or severe cases resistant to conventional treatments.

13.3 Role of Autoimmune System in Dermatological Skin Problem

Autoimmune diseases are main causes of morbidity and mortality worldwide, as evidenced by various studies. In recent years, the occurrence of these conditions has increased significantly. These conditions consist of chronic diseases in which the immune system incorrectly targets the body's healthy tissues [7]. A person's risk of developing these diseases, including skin disorders, depends on both genetic factors and environmental influences. Although the exact causes of most immune-related disorders remain uncertain, research suggests that their development is driven by a complex interplay of environmental, genetic, and epigenetic influences (Figure 13.1) [7].

The etiological variables that precipitate autoimmune disease are as follows:

***Genetics*:** Extensive research emphasizes the critical role of genetic predisposition in the complexity of autoimmune disorders. Many dermatological diseases exhibit higher concordance rates among identical twins compared to fraternal twins and are more frequently observed in individuals with a family history of such conditions [8]. The major histocompatibility complex plays a pivotal part in regulating immune responses and is the most

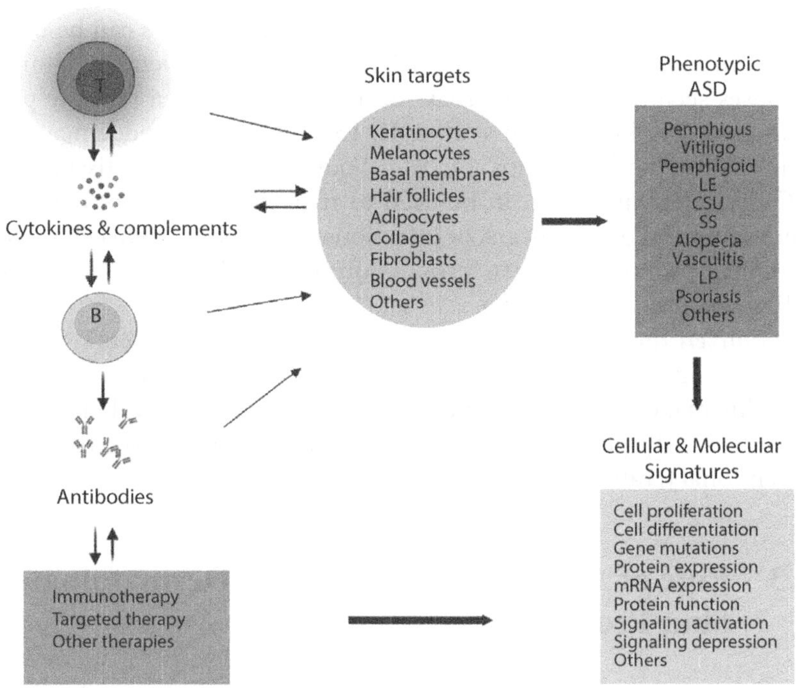

Figure 13.1 Immunomodulatory cytokine inflammatory mediators [7].

influential genetic component. Additionally, there are non-MHC genes associated with autoimmune susceptibility; however, identifying them remains challenging due to the intricate genetic landscape [7, 8]. The presence of a susceptible genetic locus, combined with specific autoantibodies, significantly surges the risk of developing autoimmune diseases in the future, despite the unchangeable nature of inherited genetic traits.

Environmental factors: Environmental influences such as UV radiation, vitamin D deficiency, infections, vaccinations, adjuvants, exposure to chemicals, dietary habits, stress, smoking, and certain medications can contribute to the onset or worsening of autoimmune conditions [8]. Once an individual's susceptibility to autoimmunity is identified, many of these environmental triggers can be modified or avoided. Sun exposure often exacerbates autoimmune skin conditions, as ultraviolet B radiation appears to have a proautoimmune effect [8, 9]. Additionally, various pathogens have been associated with autoimmune dermatological disorders, including the Epstein–Barr virus in systemic lupus erythematosus, herpesviruses in pemphigus, and infections caused by streptococcal, staphylococcal, and *Candida albicans* in psoriasis.

13.3.1 The Predictive Function of Autoantibodies in Dermatological Conditions

Autoantibody detection plays a pivotal role in identifying, classifying, and in assessing the prognosis of various skin-related autoimmune diseases. Beyond diagnosis, autoantibodies now serve a predictive function, as many can be identified before clinical symptoms appear, making them valuable for early detection, particularly in high-risk individuals. Mucocutaneous autoantibodies fall into two primary categories: type I, which involves self-antigens found in both skin and other tissues, and type II, which targets self-antigens exclusive to mucocutaneous cells [10]. Epidermal antigens produced by keratinocytes contribute to both blistering conditions, such as pemphigoid disorders, and nonblistering conditions such as vitiligo. In contrast, diseases such as dermatitis herpetiformis and cutaneous systemic lupus erythematosus primarily target dermal antigens. Although additional immune mechanisms play a role in disease progression, both types of autoantibodies serve as key diagnostic markers and may offer prognostic indicator [11].

13.3.2 Mechanism of ASD

To formulate precise predictions and targeted therapies, it is crucial to identify the unique cellular and molecular attributes. Immune cells, such as T and B cells, autoantibodies, and various cytokines, can target cellular components within the skin tissues [12]. Specific skin conditions can be identified by unique cellular and molecular features using techniques such as microarray analysis method, flow cytometry, immunofluorescence, immunohistochemistry, proteomics, metabolomics, and single-cell RNA sequencing, combined with advanced data analysis.

13.4 Therapeutic Value of Stem Cell in Different Allergic Skin Disease

13.4.1 Psoriasis

The disease is a prevalent immunologic chronic inflammatory dermatological condition characterized by erythematous, pruritic, scaly lesions, marked by a high prevalence, prolonged disease duration, and a propensity for relapse. The dysregulation of T_H1/T_H17 cells and the associated chemokines and cytokines, such as IL-17, IL-23, TNF-α, and interferon γ (IFN-γ), facilitates the development of psoriasis, which can be succinctly characterized as a malfunction of keratinocytes, inflammatory cells, and immune cells. Both of the subcutaneous (SC) and IV injections were utilized as methods of administration [12]. The IV injection dosage ranged from 0.5 to 3.0×10^6 cells/kg, whereas the SC injection dosage ranged from 0.1 to 2.0×10^8 cells. Investigation of a minimum 12 weeks is required for the efficient evaluation of IV injections and about 4 weeks for SC injections. MSC exosome has been formulated as an ointment for topical application and monitored over a 20-day period [13]. The PASI scoring is a predominant principle utilized as a key consequence in many research.

13.4.2 Systemic Sclerosis

Systemic sclerosis is an uncommon multisystem disorder that induces intensifying dermal fibrosis and internal organ system, characterized by microvascular disease, localized or diffused skin thickening and fibrosis, and aberrant activation of immunologic body system. Prolonged progression of the disease with consequences associated by various organ dysfunction markedly diminishes both the lifespan and quality of life of patients [14]. The disorder initiates with the initiation of the response of autoimmune system, resulting in endothelium impairment and the subsequent generation of autoantibodies and cytokines. Therapies such as suppression of the immune system, immunomodulators, and antifibrotic drugs have been utilized to alleviate pain; enhancement of organ function and the efficacy of IV immunoglobulins have been demonstrated. Most contemporary clinical investigations provide allogeneic or autologous MSCs *via* IV infusion. Numerous forms of MSCs from umbilical cord of adults, adipose tissue, Wharton jelly, and bone marrow are utilized. A minimum of 12 weeks is required for skin score assessment and monitoring of adverse occurrences [17]. The most often reported standard dosage of MSCs was 1.0×10^6 cells.

13.4.3 Dermatomyositis

Dermatomyositis is an uncommon, multisystem autoimmune inflammatory myopathy distinguished by the characteristic Gottron sign and papules, heliotrope rash, hemorrhages in the nail, and weakening of the symmetrical proximal muscle. Seventy-two patients may have one or more cutaneous signs, and there is no clinical correlation between the dermal manifestations and myopathy. Two clinical trials are presently underway utilizing human umbilical cord–derived MSC treatment for dermatomyositis by IV infusion. The dose-limiting toxicity is the principal key outcome in the investigation process. However, the optimal dosage and frequency require further evidence related to clinical data [15].

13.4.4 Atopic Dermatitis

AD is a persistent and recurrent dermal condition characterized by severe itching. Nearly about 10% of adults and 20% of children in high-income nations are plagued with AD. Intricate factors such as exposure to allergen and skin barrier impairment lead to adaptive immunological diseases and atypical inflammatory reaction. The therapeutic efficacy of MSC has been proven by numerous investigations [16]. For instance, MSC infusion may inhibit mast cell degranulation by elevating TGF-β1 and prostaglandin E_2; re-establishing the balance among T_H1, T_H2, and T_H17 cells; and reducing chemokines and cytokines. MSC-derived exosomes, extracts and conditioned medium, were observed to alleviate the disease. Both IV and SC infusions of MSCs were efficacious in mice models. In ongoing clinical trials, the IV injection of human umbilical cord–derived stem cells is the predominant therapy, with dosages ranging from 2.5×10^7 to 3.0×10^8 cells [17].

13.5 Role of Cutaneous Stem Cells in Regeneration of Different Tissues

Tissue regeneration and restoration are a meticulously regulated and interrelated sequence of procedure that results in the repairing of tissue functionality and integrity. Disruption in typical tissue regeneration process results in the formation of chronic wounds [17]. Numerous therapies and treatments have been formulated for chronic injuries and wounds, exhibiting differing levels of efficacy.

13.5.1 Clonal Conversion of Epidermal Cells

Significant development has been made in recognizing precise markers for isolating and enriching the stem cells. Researchers have leveraged advanced mouse models and cutting-edge methodologies such as lineage tracing, fate mapping, transplantation, label-retaining studies, and functional assays. Innovative techniques for cell culture reveal that human epidermal keratinocytes exist as a divergent population with distinct clonogenic properties. Based on morphological characteristics, three distinct colony types—holoclones, microclines, and paraclones, emerge from individual keratinocytes, each classified by its proliferative potential. Holoclones, recognized by their large, circular structure, consist of small, uniform cells with the highest propagating capacity [18, 19]. The colonies, widely regarded as stem cell derivatives, exhibit strong self-renewal abilities, express high levels of β1 integrin and K14, and generate both paraclones and meroclones. Meroclones, considered transient amplifying cells, consist of a diverse cell population with inconsistent growth potentials, leading to the formation of additional meroclones and paraclones upon reseeding. As these cells exit the stem cell niche, their p63 expression declines significantly. In contrast, paraclones appear as small, irregularly shaped colonies that represent cells belonging to postmitosis phase [19]. The cells display a limited lifespan replication and show elevated expression of involucrin, a key marker of terminal differentiation. The clonal conversion follows a trajectory under standard conditions, which is irreversible. Advances in single-cell analysis and live-cell imaging continue to refine our understanding of these transitions, offering deeper insights into epidermal regeneration and stem cell biology.

13.5.2 Epidermal Stem Cells Involved in Tissue Repair Postinjury

In response to injury, stem cells from both the interfollicular epidermis and hair follicle actively contribute to wound re-epithelialization. Experiments related to fate mapping demonstrate that K15-positive hair follicle bulge stem cells temporarily participate in this process following full-thickness wounds in mice [20]. However, these cells gradually disappear from the epidermis within a few weeks, suggesting that although stem cell hair follicles support initial healing of wounds, they are not essential for the ongoing maintenance of the interfollicular epidermis.

Interestingly, in mutant mice that are lacking hair follicles, the initial phases of re-epithelialization are delayed in linear incisional lacerations on the tail skin. Despite this delay, complete epidermal closure still occurs, likely driven by interfollicular epidermal stem cells, highlighting their regenerative potential [21]. Additionally, cells expressing Gli1, Lrig1, Lrg5, and MT24 play a very important and critical role in maintaining the homeostasis of the pilosebaceous unit. Upon skin injury, these cells become activated and contribute to interfollicular epidermis healing, illustrating the remarkable flexibility of epidermal stem cells in tissue repair.

13.5.3 MicroRNAs as Regulators of Epidermal Stem Cell Maintenance and Wound Healing

MicroRNAs are small, noncoding RNA molecules that modulate gene expression posttranscriptionally by either inhibiting messenger RNA translation or promoting its degeneration. A single miRNA can target multiple numbers of genes, whereas a single gene may be modulated by several miRNAs [22]. As key regulators of gene expression, miRNAs influence essential biological processes, including homeostasis, cell survival, and differentiation. However, their dysregulated expression is linked to the progression of the disorder. Studies identify several miRNAs that are either uniquely expressed or differentially regulated in the mouse epidermis in comparison to other skin lineages. Various cell types release miRNAs, which can be transported through exosomes into the bloodstream, enabling intercellular communication and modulating target cell functions [21, 22]. Exosomes that are derived from bone marrow MSCs are enriched with specific miRNAs that enhance the proliferation of both standard and diabetic wound fibroblasts. Similarly, keratinocyte-secreted exosomes regulate melanin production in melanocytes. Epidermal progenitor cells exhibit a high concentration of specific exosomal subunits, which play a major role in sustaining their proliferative capacity and prevention of premature differentiation. These findings underscore the significance of miRNAs, particularly through their exosomal transport, in preserving epidermal stem cell function [22].

13.6 Advantages of Stem Cell Therapy for Different Dermatological Diseases

13.6.1 Cutaneous Wound Healing

Typical wound healing process of human skin is not flawless, even under optimal conditions, as delayed healing and scar formation remain inevitable outcomes. In cases of prolonged wound healing, MSCs demonstrate significant potential. MSCs facilitate migration, angiogenesis, epithelialization, and granulation tissue development, accelerating wound healing and enhancing regenerative processes to ultimately reduce scarring [23]. In both acute and chronic wounds, MSCs are delivered straight to wounds *via* injection or spraying, enabling the direct administration of stem cells in both mice and humans. This approach accelerates wound healing, promotes angiogenesis, releases paracrine signaling molecules, and enhances re-epithelialization [26]. Research indicates that hair follicle stem cells play a role in wound healing. In response to injury, stem cells from hair follicles and the interfollicular epidermis migrate to the wound site. Fate-mapping studies reveal that K15-positive hair follicle stem cells aid in the re-epithelialization of full-thickness wounds. After epidermal damage, follicular epithelial cells rapidly move toward the wound center, forming temporary cells that support acute wound repair [23].

13.6.2 Vitiligo

Vitiligo, a condition resulting from the depletion of melanocytes, is the leading factor for the development of depigmented, patchy skin [10]. Repigmentation in vitiligo depends on melanocytes from the hair follicle unit, the lesion's border, and unaffected melanocytes within depigmented areas.

Treatments such as tacrolimus, phototherapy, and dermabrasion are often used and show effectiveness in some individuals. Advancements in stem cell therapy instill renewed optimism in vitiligo management [18]. The expression of PTEN in human vitiligo is enhanced, accompanied by the suppression of the AKT growth signaling pathway. This discovery indicates that PTEN plays a crucial role in melanocytic proliferation and viability. MSCs influence PTEN protein expression in melanocytes, reducing the harmful effects of PTEN overexpression and oxidative stress caused by H_2O_2. In vitiligo, dermal MSCs improve melanocyte transplantation by suppressing $CD8^+$ T-cell proliferation, promoting T-cell apoptosis, and modulating cytokine and chemokine production [24].

13.6.3 Alopecia

Alopecia is characterized by hair loss in one or multiple areas, leading to bald patches. Autologous hair transplantation remains the prime standard, but its use is constrained by the limited availability of donor material and decreased graft viability [23, 25]. Research indicates that injecting a combination of human cultured dermal papillary cells and $CD200^+$ and K15-marked hair epithelial cells into nude mice stimulates new hair growth, emphasizing the potential of stem cell therapy for alopecia treatment. A cutting-edge approach for androgenic alopecia involves using centrifuged hair follicle stem cells alongside dermal MSCs. Studies have shown that this mixture enhances hair growth and increases hair density at injection sites, making it a promising option for treating male pattern androgenetic alopecia [25]. Stem cell therapy continues to emerge as a promising treatment for alopecia by addressing the underlying irrational causes of hair loss, facilitating the regeneration of entire hair follicles, and promoting the development of new ones.

13.6.4 Melanoma

Melanoma was once thought to arise solely from cancerous melanocyte cells; however, emerging research reveals that it can also originate from cellular transformations in melanocytes, MSCs, or both. Studies demonstrated that MSCs actively migrate to both primary and metastatic tumor sites, guided by chemokines and cytokines secreted by the tumor microenvironment; when preloaded with *in vitro* anticancer agents such as paclitaxel, MSCs evolve into drug-releasing therapeutic cells that effectively suppress cancer cell proliferation [25]. In melanoma treatment, MSCs play a dual role by acting as carriers for targeted therapies such as adipose-derived MSCs expressing TRAIL, interferon α/γ, epithelium-derived pigment factor, and cytosine deaminase, simultaneously generating a diverse range of bioactive chemokines and tumor-inhibiting molecules. Furthermore, MSCs undergo genetic engineering to introduce tumor-suppressor genes that disrupt key pathways driving melanoma progression and metastasis [26]. This innovative approach harnesses MSCs not only as carriers of therapeutic agents but also as active participants in reshaping the tumor microenvironment, ultimately hindering melanoma's growth and spread.

13.6.5 Antiaging Therapies

Aging is a natural process that affects all organs, resulting in a gradual decline in physiological function. The visible changes in the skin are a clear manifestation of this progression. Although body cells deteriorate over time, stem cells aid in their replacement. The diminished regenerative and reparative capacity of stem cells significantly contributes to aging, especially when comparing the secretory functions of adipose stem cells and placental stem cells in facial rejuvenation [27].

Studies indicate that proteins released by adipose stem cells promote cell adhesion and migration, facilitating wound healing and tissue repair. Additionally, the melanin index rises more notably with the use of adipose stem cell–conditioned medium. Likewise, proteins from placental stem cells exhibit greater efficacy in supporting angiogenesis, cell proliferation and differentiation, cell survival, collagen degradation, and immune modulation [26, 27]. Both types of conditioned media contribute to enhancing facial skin quality. By boosting regenerative and reparative functions, stem cell therapy presents a promising strategy for slowing aging and addressing age-related conditions.

13.7 Different *In Vivo* and *In Vitro* Models for Preclinical SCT

13.7.1 Genetically Modified Mouse Models

Transgenic modified animals play a crucial role in researching human diseases, gene therapy, drug development, animal cloning, and organ transplantation. These models are created by isolating a specific transduced gene and incorporating it into the genome of a fertilized egg, which then develops to the blastocyst stage before being implanted into a surrogate mother [28]. Various techniques facilitate gene transfer, which includes physical ex-microinjection, mechanical ex- (gene gun, electroporation, laser irradiation), chemical ex- (calcium phosphate precipitation, DEAE-dextran), and biological methods (lipids, viruses, plasmids). Among these, microinjection remains widely used, along with ESC injection, retroviral infection of embryos, somatic cell nuclear transfer, and genetically engineered sperm. Development of transgenic animals started with the introduction of viral genes, such as simian virus-40 leukemia genes, into mouse blastocysts through microinjection and exposure to retroviruses. Later on, researchers successfully integrated a human growth hormone gene into mouse embryos, resulting in larger mice that transmitted the trait to their offspring [29]. Additionally, various transgenic animal models help in studying different forms of skin dermatitis and inflammation, providing valuable insights into disease mechanisms and potential treatments.

13.7.1.1 Flaky Tail Mice

Possess a genetic mutation that prevents them from producing functional filaggrin in the cornified layer of their epidermis. This deficiency leads to the absence of characteristic keratohyalin F granules and impairs the proteolytic processing of filaggrin. These mice begin to develop eczematous skin lesions resembling human AD around 28 weeks of age when housed under pathogen-free conditions [29]. This is accompanied by a gradual increase in blood immunoglobulin E (IgE) and IgG1 levels. Eight weeks later, these mice develop epidermal thickening, increased dermal $CD4^+$ cell infiltration, and elevated mRNA expression of IL-17, IL-6, and IL-23, whereas IL-4, IL-13, and IFN-γ remain unchanged [28, 33]. As the condition advances, 32-week-old filaggrin-deficient mice exhibit more severe skin alterations along with increased IL-4 mRNA levels, suggesting a shift toward T_H17-driven skin inflammation and age-related eczematous dermatitis.

13.7.1.2 C1 Apolipoprotein Transgenic Mice

Initially designed as a model for hyperlipidemia, human apolipoprotein C1 transgenic mice have also been fundamental key in researching inflammatory skin diseases. These mice naturally develop severe, age-related dermatitis, marked by mild epidermal thickening, hyperkeratosis, parakeratosis, scaling, lichenification, dermal inflammatory cell infiltration, and pruritus. Due to their impaired skin barrier, they serve as a valuable model for exploring the impact of barrier dysfunction in the development of AD [29].

13.7.1.3 NC/Tnd Mice

Derived from Japanese fancy mice, NC/Tnd mice originate from the NC/Nga strain and serve as a high-fidelity model for spontaneous Alzheimer disease. When kept in conventional housing without air regulation, these mice develop pruritic dermatitis between 6 and 8 weeks of age [22]. Their skin lesions exhibit epidermal hyperplasia, mast cell degranulation, and inflammatory cell infiltration, along with the accumulation of dendritic cells in the hyperplastic epithelium [22, 28]. Strong T_H2 immune response is evident, as excessive IL-4 and T_H2-specific chemokines are produced, paralleling early-stage AD in humans. Notably, their skin ceramide levels are significantly reduced even before dermatitis onset, leading to increased transepidermal water loss.

13.7.1.4 Transgenic IL-4 Mice

Mice with IL-4 overexpression in their skin spontaneously develop pruritus and chronic dermatitis by the age of 4 months. They serve as a viable model for studying Alzheimer disease due to similarities with human pathology [28].

13.7.1.5 Knockout RelB Mice

RelB knockout mice develop spontaneous dermatitis characterized by hyperkeratosis, acanthosis, infiltration of CD4+ T cells and eosinophils, and increased serum IgE levels, mirroring human AD. However, these mice do not exhibit pruritus [29, 30].

13.7.1.6 SC Chymotryptic Enzyme Transgenic Mice

Mice genetically modified to overexpress the human SC chymotryptic enzyme in suprabasal epidermal keratinocytes develop skin conditions resembling AD at or after 7 to 8 weeks of age [30].

13.7.1.7 IL-31 Transgenic Mice

By 2 months of age, IL-31 transgenic mice develop pruritus, mild to moderate hair loss, and significant ear skin thickening. Histological analysis shows hyperkeratosis, acanthosis, increased inflammatory cell infiltration, and elevated mast cells, closely mimicking human AD. Notably, serum IgE levels remain stable, and the T_H2 response has not yet been evaluated [28, 30].

13.7.1.8 NOA Mice

NOA mice develop ulcerative skin lesions, increased mast cell accumulation, elevated serum IgE levels, and frequent scratching behavior. Their characteristics make them a valuable model for investigating AD, especially the involvement of mast cells and IgE in disease progression [30].

13.7.1.9 Transgenic TSLP Mice

Following doxycycline induction, TSLP transgenic mice develop erythematous skin changes within 2 to 3 weeks, progressing to chronic AD-like symptoms, including persistent erythema, mild xerosis, crusting, and erosions by weeks 3 to 4. Histological analysis shows T_H2-dominated inflammation with increased IL-4, IL-5, and TNF-α expression [29]. Additionally, these mice display elevated serum IgE and IgG1 levels, accompanied by a decline in IgG2a levels.

13.7.1.10 Caspase 1 and IL-18 Transgenic Mice

Mice with epidermal keratinocyte-specific overexpression of the human caspase 1 precursor gene show elevated IgE and IgG1 levels by 8 weeks of age [29]. By 16 weeks, they develop mild pruritic dermatitis around the eyes and ears. Histological analysis highlights significant acanthosis, papillomatosis, parakeratosis, and intracellular edema, alongside extensive lymphocyte, neutrophil, and mast cell infiltration, although eosinophils are absent. This model replicates intrinsic Alzheimer disease features. Similarly, IL-1β–deficient CASP1 and IL-18 transgenic mice develop analogous dermatitis [25].

13.7.1.11 Cathepsin E Knockout Mice

Cathepsin E knockout mice on a C57BL/6 background spontaneously develop pruritic, erosive skin lesions, often associated with *Staphylococcus aureus* colonization. These mice provide a valuable model for studying the role of antigen-presenting cells in the pathology of Alzheimer disease [31].

13.8 Preclinical Test Method for Different Skin Disease

Following subject assessment, researchers are conducting preclinical investigations to determine the therapeutic efficacy of the concept for the targeted disease, while simultaneously evaluating the safety profile of the final formulation in an appropriate animal model. The primary aim of preclinical experimentation is to generate robust scientific evidence that supports the feasibility of transitioning to clinical trials [32]. Several critical criteria are necessitating consideration before advancing to human studies, including (i) establishing feasibility and scientific justification (e.g., validating biological activity, isolating active compounds *in vitro*, and elucidating mechanistic pathways *in vivo*), (ii) determining pharmacodynamics and pharmacokinetic properties (e.g., verifying safe initial dosage range), (iii) optimizing the administration paradigm (e.g., ensuring a biocompatible delivery method, assessing repeat dosing safety, and validating dosing intervals), (iv) characterizing pharmacological activity and potential toxicological responses (e.g., conducting acute and chronic toxicity profiling), (v) identifying potential adverse biological interactions (e.g., genotoxicity, carcinogenicity, immunogenicity, and neurotoxicity assessments), and (vi) making a data-driven decision regarding progression or discontinuation of therapeutic development [32, 36].

Researchers are continuously evaluating multiple parameters through diverse *in vitro* and *in vivo* methodologies, including characterizing cellular phenotypes, specifying optimal dosage concentrations, refining delivery strategies, and systematically assessing both therapeutic efficacy and biosafety [32].

13.8.1 Stem Cell Source Determination

With growing expectations for stem cell–based regenerative therapies, the scope of applications and stem cell sources is broadening, shifting from autologous treatments to allogeneic and iPSC approaches. The risks associated with these therapies vary depending on the cell manufacturing process and clinical experience, requiring a standardized evaluation of all stem cell treatments [33]. Therefore, identifying the advantages and limitations of each stem cell type remains essential for determining their optimal therapeutic efficacy across various disorders. This approach enables the development of disease-specific stem cell therapies by matching the appropriate stem cell type to the corresponding condition. We systematically examine the characteristics of different stem cell types.

Mesenchymal stem cells: MSCs are lineage-committed cells that primarily differentiate into mesenchymal lineages such as adipocytes, chondrocytes, and osteocytes. Under suitable *in vitro* conditions, they also have the potential to differentiate into adipocytes, neurons, glioma cells, and skeletal muscle cells. These stem cells are mainly sourced from myeloid and adipose tissues, but they are now being isolated from a variety of sources, including the retina, liver, gastric mucosa, tendon, cartilage, placenta, cord blood, and peripheral blood [34, 39]. The defining characteristic of MSCs lies in their immunosuppressive properties, as they actively inhibit the proliferation of activated T lymphocytes by discharging immunosuppressive cytokines and suppressing programmed cell death signaling. This function positions them as promising candidates for treating immune-related inflammatory disorders.

Hematopoietic stem cells (HSCs): HSCs continuously differentiate into all cellular components of the hematopoietic system, including myeloid cells and lymphocytes, within the bone marrow and thymus [35]. These stem cells are mainly obtained from adult bone marrow, the placenta, and umbilical cord blood. However, they carry immunological risks, including the potential for transplant rejection [36]. Despite these challenges, HSCs are proving to be effective in treating various conditions, including leukemia, malignant lymphoma, regenerative anemia, congenital metabolic disorders, congenital immune deficiencies, refractory autoimmune diseases, and solid tumors. Significantly, HSCs are the only type of stem cells currently approved by the Food and Drug Administration (FDA) for therapeutic use [37].

Embryonic stem cells (ESCs): ESCs are maintained as established cell lines through *in vitro* culture. These pluripotent cells have the ability to differentiate into almost any cell type in the human body and can be guided toward specific lineages using external factors in culture media or genetic modification methods [38]. However, their implantation into a suitable host may lead to teratoma formation, consisting of cell types originating from the endoderm, mesoderm, and ectoderm.

Induced pluripotent stem cells: IPSCs are theatrically reprogrammed stem cells generated from adult somatic cells such as fibroblasts [39]. These cells exhibit characteristics similar to ESCs, including self-renewal and pluripotent differentiation potential [40]. However, uncertainties persist regarding their cell-specific regulatory mechanisms, gene expression patterns, and epigenetic modifications. These factors contribute to potential risks such as tissue chimerism and cellular dysfunction.

13.8.2 Cell Dose Specification

Determining the ideal dosage for stem cells or stem cell–derived products in therapy is an ongoing process through *in vivo* and *in vitro* studies. Establishing a safe and effective treatment threshold is essential, with researchers aiming to identify the minimum effective dose whenever possible [41]. Administering stem cells to sensitive regions, such as the central nervous system and heart, presents challenges for conventional dosage assessments. Therefore, if nonclinical investigations successfully validate therapeutic potential, early-phase human clinical trials may commence with doses demonstrating potential efficacy [42]. Stem cells are continuously being explored across numerous disease models, based on their type and biological characteristics [43]. The route of administration plays a significant role and should not be overlooked in favor of cell quantity. Studies consistently indicate that the engraftment efficiency of transplanted cells remains lower than the total number of cells administered, often failing to sustain long-term viability within the target organ [37]. Stem cell delivery methods are classified into local and systemic administration. Local delivery involves precise injections using specialized techniques, such as intraperitoneal (IP), intramuscular, and intracardiac injections. Conversely, systemic administration relies on vascular routes, including IV and intra-arterial methods [44]. Conferring to research, IV administration remains the most commonly used technique, followed by intrasplenic and IP methods.

Preclinical testing for skin diseases involves a variety of *in vitro* and *in vivo* models to evaluate the efficacy and safety of potential treatments before clinical trials [45]. Psoriasis *in vitro* models such as keratinocyte cultures and 3D skin equivalents help study inflammation, whereas *in vivo* models such as the imiquimod-induced psoriasis model in mice mimic the disease's inflammatory response [46]. AD (eczema) is tested using keratinocyte and fibroblast cultures, along with murine models induced by allergens such as house dust mites or chemicals such as DNFB. In acne vulgaris, sebocyte cell cultures and *Cutibacterium acnes* infection models are widely used *in vitro*, whereas *in vivo* studies often involve *C. acnes*–induced inflammation in mice or rabbits [47]. In case of wound healing disorders, scratch wound assays in keratinocytes and fibroblasts, as well as 3D skin models, provide insights into cell migration and repair mechanisms, while *in vivo* excisional and diabetic wound models are used to assess healing rates [48]. Skin cancer melanomas are studied using melanoma cell lines (A375, B16F10) *in vitro*, while xenograft models and UV-induced carcinogenesis in mice serve as *in vivo* models. In vitiligo, melanocyte cultures help to investigate depigmentation processes, whereas mouse models induced with monobenzone or genetically modified for autoimmune susceptibility provide insights into disease progression. *In vitro* contact dermatitis studies utilize keratinocyte inflammation assays and allergen-induced hypersensitivity models in *in vivo* method. Fungal skin infections such as dermatophytosis and candidiasis are tested using fungal growth inhibition assays and 3D-reconstructed skin models, alongside murine infection models [49]. Lastly, alopecia research involves dermal papilla cell cultures for hair follicle studies, whereas androgenetic alopecia and chemotherapy-induced hair loss models in mice help in understanding hair regeneration. These preclinical methods are essential in advancing dermatological treatments before they reach human trials.

13.8.3 Optimizing Cell Transplantation through Strategic Modifications to Enhance Safety and Delivery Efficiency

Every medical intervention carries both advantages and potential risks, making the administration of unproven stem cell therapies unsafe. With the growing demand for regenerative treatments, the use of various delivery methods—including spinal cord, SC, and intramuscular injections—continues to evolve alongside therapeutic advancements [49]. The risks associated with stem cell treatments vary based on factors such as cell type, production techniques, and clinical expertise. Therefore, all stem cell therapies require standardized evaluation to ensure safety and efficacy. Furthermore, stem cells with self-renewing capabilities and pluripotent differentiation potential must be included in final products only under strictly regulated and justified conditions. At present, HSCs and MSCs are the only stem cells considered safe for regenerative medicine, as they are derived from natural sources. However, the clinical use of iPSCs and ESCs remains challenging due to risks such as immune rejection, teratoma formation, and genomic instability, which restrict their broader therapeutic application.

13.9 FDA-Approved Clinical Phase Trial Drugs for Dermatological Disease [6]

Each year, the FDA approves approximately 40 to 50 novel drugs for public use, making it challenging for physicians to stay updated on evolving treatment options [6, 50]. In the United States, investigational new drugs can be tested on humans only through clinical trials under an IND (Investigational New Drug) application. Since clinical trials are conducted under diverse conditions, adverse reaction rates observed in one trial cannot be directly compared to those of another and may not accurately reflect real-world patient outcomes. The FDA's primary criterion for approving drug marketing applications is a comprehensive benefit–risk assessment. Between 2011 and 2022, the FDA approved 46 new molecular entities for dermatologic indications, with 27 of these reviewed by the Division of Dermatology and Dentistry under the Centre for Drug Evaluation and Research, which oversees the regulation and approval of dermatologic drugs. Clinical trials are conducted in four phases to ensure drug safety and efficacy [50]. Phase I involves small-scale studies to assess safety, dosage range, and potential side effects. Phase II expands the participant pool to evaluate therapeutic efficacy. Phase III consists of large-scale, randomized, and controlled multicenter trials to confirm the drug's effectiveness and safety [51]. Phase IV occurs postmarketing to monitor long-term adverse effects across diverse populations. The FDA ensures subject safety at every stage of drug development, with phases II and III requiring rigorous scientific evaluation to determine effectiveness [52]. Since the early 2000s, targeted therapies for psoriasis have gained significant traction in clinical research, leading to continuous advancements in treatment options and improved patient outcomes. The success of clinical trials has resulted in the FDA approving various drugs for dermatological conditions. One of the most widely used treatment approaches for skin disorders is topical therapy, which offers distinct advantages over oral and injectable medications [51, 52]. By directly delivering the drug to the affected area, topical treatments minimize systemic exposure, reducing the risk of drug interactions and potential toxicity to other organs. However, certain patient and disease-specific factors can influence topical drug absorption, potentially increasing systemic exposure and the likelihood of off-target adverse effects.

Table 13.1 shows the list of various approved FDA drugs for dermatological skin disorders [6].

Table 13.1 Approved drugs for skin dermatological disorders.

Sl. no.	Disease	Drug	Mechanism of action	Year of approval
1	Atopic dermatitis	Dupilumab	IL4RA receptor blockade	2017
2	Acne vulgaris	Clascoterone	Androgen receptor inhibitor	2020
3	Plague psoriasis	Tapinarof	Aryl hydro-carbon receptor agonist	2022
4	Severe alopecia	Ritlecitinib	JAK3 inhibitor	2023
5	Nonsegmental vitiligo	Ruxolitinib	JAK1 ans 2 inhibitor	2022
6	Actinic keratosis	Tirbanibulin	Microtubule inhibitor	2020
7	Pustular psoriasis flares	Spesolimab-sbzo	IL36 receptor antagonist	2022
8	Psoriasis	Deucravacitinib	Tyrosine kinase 2 receptor	2022
9	Atopic dermatitis	Upadacitinib	JAK-1 inhibitor	2022
10	Molluscum contagiosum	Berdazimer	Nitric oxide release	2023

13.10 Conclusions and Future Directions

Numerous preclinical and clinical studies have highlighted the growing significance of stem cell therapy in managing dermatological diseases. Extensive research using animal models, particularly mice and rats, has provided crucial insights into the prospective of stem cells in treating various skin-related disorders. These studies reinforce the broad applicability of stem cell–based treatments for conditions such as dermatitis, wound healing, and skin regeneration. Regulatory advancements have also played a pivotal role in shaping therapeutic approaches. The FDA has introduced novel treatments targeting interleukin-derived factors, which act as key inflammatory cytokine mediators in skin conditions such as dermatitis and atopic diseases. These developments underscore the evolving landscape of dermatological treatments, shifting toward precision medicine and targeted therapies. Looking ahead, the future of clinical trials in stem cell therapy is poised to explore new biochemical pathways involving enzymes and proteins that contribute to skin disorders. By identifying and targeting these molecular mechanisms, researchers aim to develop innovative regenerative strategies that go beyond symptomatic relief to achieve long-term disease modification. This review delves into the latest advancements in preclinical studies, clinical trials, and real-world case studies, emphasizing the transformative potential of stem cell therapy in dermatology.

AI disclosure statement: The author acknowledges the use of AI-based tools for minor language editing or grammatical correction. However, all intellectual contributions, analyses, and conclusions presented in this chapter are entirely the author's own.

References

1. Robertson, S.A. and Agrawal, A., Masectomy Skin Flap Necrosis: Challenges and solutions. *Breast Cancer*, 13, 9, 2017.
2. Kolb, L. and Bruker, F., *Atopic Dermatitis*, National Library of Medicine, 8, 2023.
3. Ding, D.C. and Shyu, W.C., Mesenchymal Stem Cells. *Cell Transplant.*, 20, 1, 2011.
4. Crombie, D.E., Pera, M.F., Delatycki, M.B., Using Human Pluripotent Stem Cells to study Friedreich ataxia Cardiomyopathy. *Int. J. Cardiol.*, 212, 1, 2016.
5. Cerneckis, J., Cai, H., Shi, Y., Induced pluripotent stem cells (iPSCs): Molecular mechanisms of induction and applications. *Signal Transduct. Target. Ther.*, 9, 112, 2024.
6. Sikdar, B. and Sil, A., Fresh Baked: An Overview of Newly FDA-Approved Drugs for Dermatological Usage. *Indian J Dermatol.*, 68, 6, 707–720, 2024.
7. Mao, X., Yamagami, J., Pan, M., Editorial: Mechanism and therapy of autoimmune skin diseases. *Front. Med.*, 10, 1143454, 2023.

8. Miller, F.W., The increasing prevalence of Autoimmunity and autoimmune diseases: An urgent call to action for improved understanding, diagnosis, treatment and prevention. *Curr. Opin. Immunol.*, 80, 2022.
9. Yasmeen, F., Pirzada, R.H., Ahmad., B., Choi, B., Choi, S., Understanding autoimmunity: mechanisms, predisposing factors and cytokines therapies. *Int. J. Mol. Sci.*, 25, 14, 2024.
10. Ito, M., Liu, Y., Yang, Z., Nguyen, J., Liang, F., Stem cells in the hair follicle bulge contribute to wound repair but not to homeostasis of the epidermis. *Nat. Med.*, 11, 2005.
11. Zhou, M.N., Zhang, Z.Q., Wu, J.I., Lin, F.Q., Fu, L.F., Dermal Mesenchymal Stem Cells (DMSCs) Inhibit Skin-Homing CD8+ T Cell Activity, a Determining Factor of Vitiligo Patients' Autologous Melanocytes Transplantation Efficiency. *PLoS One*, 8, 2013.
12. Gentile, P., Scioli, M.G., Bielli, A., Orlandi, A., Cervelli, V., Stem cells from human hair follicles: first mechanical isolation for immediate autologous clinical use in androgenetic alopecia and hair loss. *Stem Cell Investig.*, 4, 2017.
13. Chen, H., Niu, J.W., Ning, H.M., Pan, X., Li, X.B., Li, Y., Wang, D.H., Hu, L.D., Sheng, H.X., Xu, M., *et al.*, Treatment of psoriasis with mesenchymal stem cells. *Am. J. Med.*, 129, 3, 2016.
14. Sierra-Sepúlveda, A., Esquinca-González, A., Benavides-Suárez, S.A., Sordo-Lima, D.E., Caballero-Islas, A.E., Cabral-Castañeda, A.R., Rodríguez-Reyna, T.S., Systemic sclerosis pathogenesis and emerging therapies, beyond the fibroblast. *Biomed Res. Int.*, 2019.
15. DeWane, M.E., Waldman, R., Lu, J., Dermatomyositis: clinical features and pathogenesis. *J. Am. Acad. Dermatol.*, 82, 2, 2020.
16. Na, K., Yoo, H.S., Zhang, Y.X., Choi, M.S., Lee, K., Yi, T.G., Song, S.U., Jeon, M.S., Bone marrow-derived clonal mesenchymal stem cells inhibit ovalbumin-induced atopic dermatitis. *Cell Death Dis.*, 5, 7, 2014.
17. Lee, B.C., Kim, J.J., Lee, J.Y., Kang, I., Shin, N., Lee, S.E., Choi, S.W., Cho, J.Y., Kim, H.S., Kang, K.S., Disease-specific primed human adult stem cells effectively ameliorate experimental atopic dermatitis in mice. *Theranostics*, 9, 12, 2019.
18. Yu, Y., Application of stem cell technology in antiaging and aging-related diseases. *Adv. Exp. Med. Biol.*, 1086, 2018.
19. Capecchi, M.R., The new mouse genetics: altering the genome by gene targeting. *Trends Genet.*, 5, 3, 1989.
20. Brosnan, M.J., Chen, L.H., Wheeler, C.E., *et al.*, Phosphocreatine protects ATP from a fructose load in transgenic mouse liver expressing creatine kinase. *Am. J. Physiol.*, 260, 1991.
21. Wight, D.C. and Wagner, T.E., Transgenic mice: a decade of progress in technology and research. *Mutat. Res.*, 307, 2, 1994.
22. Strachan, T. and Read, A.P., *Human Molecular Genetics*, Wiley-Liss Inc., 12, 7, 2010.

23. Kurita, K., Burgess, S.M., Sakai, N., Transgenic zebrafish produced by retroviral infection of *in vitro* cultured sperm. *Proc. Natl. Acad. Sci. U.S.A.*, 101, 5, 2004.
24. Jin, H., He, R., Oyoshi, M., Geha, R.S. Animal models of atopic dermatitis. *J. Invest Dermatol.*, 129, 1, 31–40, 2009.
25. Oyoshi, M.K., Murphy, G.F., Geha, R.S., Filaggrin-deficient mice exhibit TH17-dominated skin inflammation and permissiveness to epicutaneous sensitization with protein antigen. *J. Allergy Clin. Immunol.*, 124, 3, 2009.
26. Nagelkerken, L., Verzaal, P., Lagerweij, T., *et al.*, Development of atopic dermatitis in mice transgenic for human apolipoprotein C1. *J. Invest. Dermatol.*, 125, 5, 2008.
27. Matsuda, H., Watanabe, N., Geba, G.P., *et al.*, Development of atopic dermatitis-like skin lesion with IgE hyperproduction in NC/Nga mice. *Int. Immunol.*, 9, 3, 1997.
28. Aioi, A., Tonogaito, H., Suto, H., *et al.*, Impairment of skin barrier function in NC/Nga Tnd mice as a possible model for atopic dermatitis. *Br. J. Dermatol.*, 144, 1, 2001.
29. Chan, L.S., Robinson, N., Xu, L., Expression of interleukin-4 in the epidermis of transgenic mice results in a pruritic inflammatory skin disease: an experimental animal model to study atopic dermatitis. *J. Invest. Dermatol.*, 117, 4, 2001.
30. Barton, D., HogenEsch, H., Weih, F., Mice lacking the transcription factor RelB develop T celldependent skin lesions similar to human atopic dermatitis. *Eur. J. Immunol.*, 30, 8, 2000.
31. Hansson, L., Backman, A., Ny, A., *et al.*, Epidermal overexpression of stratum corneum chymotryptic enzyme in mice: a model for chronic itchy dermatitis. *J. Invest. Dermatol.*, 118, 3, 2002.
32. Soumelis, V., Reche, P.A., Kanzler, H., *et al.*, Human epithelial cells trigger dendritic cell mediated allergic inflammation by producing TSLP. *Nat. Immunol.*, 3, 7, 2002.
33. Konishi, H., Tsutsui, H., Murakami, T., *et al.*, IL-18 contributes to the spontaneous development of atopic dermatitis-like inflammatory skin lesion independently of IgE/stat6 under specific pathogen-free conditions. *Proc. Natl. Acad. Sci. U.S.A.*, 99, 17, 2002.
34. Tsukuba, T., Okamoto, K., Okamoto, Y., *et al.*, Association of cathepsin E deficiency with development of atopic dermatitis. *J. Biochem.*, 134, 6, 2003.
35. Steinmetz, K.L. and Spack, E.G., The basics of preclinical drug development for neurodegenerative disease indications. *BMC Neurol.*, 9, 2009.
36. Polson, A.G. and Fuji, R.N., The successes and limitations of preclinical studies in predicting the pharmacodynamics and safety of cell-surface-targeted biological agents in patients. *J. Cereb. Blood Flow Metab.*, 166, 2012.

37. Ioannidis, J.P.A., Kim, B.Y.S., Trounson, A., How to design preclinical studies in nanomedicine and cell therapy to maximize the prospects of clinical translation. *Nat. Biomed. Eng.*, 2, 2018.
38. Poulos, J., The limited application of stem cells in medicine: A review. *Stem Cell Res. Ther.*, 9, 1, 2018.
39. Lee, J.M., Jung, J., Lee, H.-J., Jeong, S.J., Cho, K.J., Hwang, S.-G., Kim, G.J., Comparison of immunomodulatory effects of placenta mesenchymal stem cells with bone marrow and adipose mesenchymal stem cells. *Int. Immunopharmacol.*, 13, 2012.
40. Zhu, X., Klomjit, N., Conley, S.M., Ostlie, M.M., Jordan, K.L., Lerman, A., Lerman, L.O., Impaired immunomodulatory capacity in adipose tissue-derived mesenchymal stem/stromal cells isolated from obese patients. *J. Cell. Mol. Med.*, 25, 2021.
41. Müller, A.M., Huppertz, S., Henschler, R., Hematopoietic stem cells in regenerative medicine: Astray or on the path? *Transfus. Med. Hemother.*, 43, 2016.
42. Mosaad, Y.M., Hematopoietic stem cells: An overview. *Transfus. Apher. Sci.*, 51, 2014.
43. Castro, I., Howe, L., Tompkins, D.M., Barraclough, R.K., Slaney, D., Presence and seasonal prevalence of Plasmodium spp. in a rare endemic New Zealand passerine (tieke or Saddleback, Philesturnus carunculatus). *J. Wildl. Dis.*, 47, 2011.
44. Sharkis, S.J., Jones, R.J., Civin, C., Jang, Y.-Y., Pluripotent stem cell–based cancer therapy: Promise and challenges. *Sci. Transl. Med.*, 4, 2012.
45. Liu, S., Zhou, J., Zhang, X., Liu, Y., Chen, J., Hu, B., Song, J., Zhang, Y., Strategies to optimize adult stem cell therapy for tissue regeneration. *Int. J. Mol. Sci.*, 17, 2016.
46. Ryu, C.-M., Yu, H.Y., Lee, H.-Y., Shin, J.-H., Lee, S., Ju, H., Paulson, B., Lee, S., Kim, S., Lim, J., *et al.*, Longitudinal intravital imaging of transplanted mesenchymal stem cells elucidates their functional integration and therapeutic potency in an animal model of interstitial cystitis/bladder pain syndrome. *Theranostics*, 8, 2018.
47. Byun, J.S., Kwak, B.K., Kim, J.K., Jung, J., Ha, B.C., Park, S., Engraftment of human mesenchymal stem cells in a rat photothrombotic cerebral infarction model: Comparison of intra-arterial and intravenous infusion using mri and histological analysis. *J. Korean Neurosurg. Soc.*, 54, 2013.
48. Nacif, L., Ferreira, A.O., Maria, D.A., Kubrusly, M.S., Molan, N., Chaib, E., D'Albuquerque, L.C., Andraus, W., Which is the best route of administration for cell therapy in experimental model of small-for size syndrome in rats? *Acta Cir. Bras.*, 30, 2015.
49. Zhou, Q., Jin, J.F., Zhu, L.L., Chen, M., Xu, H.M., Wang, H.F., Feng, X.Q., Zhu, X.P., The optimal choice of medication administration route regarding intravenous, intramuscular, and subcutaneous injection. *Patient Prefer. Adherence*, 9, 2015.

50. Aleckovic, M. and Simon, C., Is teratoma formation in stem cell research a characterization tool or a window to developmental biology? *Reprod. Biomed. Online*, 17, 2008.
51. Lewis, F.S. and Marcus, K.A., Food and Drug administration role in Dermatologic Drug Development. *Dermatol. Clin.*, 40, 2022.
52. Torre, K. and Shahriari, M., Clinical trials in dermatology. *Int. J. Womens Dermatol.*, 3, 3, 2017.

Bibliography

1. Collier, R., Legumes, lemons and streptomycin: A short history of the clinical trial. *CMAJ*, 180, 2009.
2. Sedgwick, P., What are the four phases of clinical research trials? *BMJ*, 348, 2014.

Index

Printed and bound by CPI Group (UK) Ltd, Croydon, CR0 4YY
17/09/2026
14964283-0004